Information Systems Engineering and Management

Volume 57

The book series "Information Systems Engineering and Management" (ISEM) publishes innovative and original works in the various areas of planning, development, implementation, and management of information systems and technologies by enterprises, citizens, and society for the improvement of the socio-economic environment.

The series is multidisciplinary, focusing on technological, organizational, and social domains of information systems engineering and management. Manuscripts published in this book series focus on relevant problems and research in the planning, analysis, design, implementation, exploration, and management of all types of information systems and technologies. The series contains monographs, lecture notes, edited volumes, pedagogical and technical books as well as proceedings volumes.

Some topics/keywords to be considered in the ISEM book series are, but not limited to: Information Systems Planning; Information Systems Development; Exploration of Information Systems; Management of Information Systems; Blockchain Technology; Cloud Computing; Artificial Intelligence (AI) and Machine Learning; Big Data Analytics; Multimedia Systems; Computer Networks, Mobility and Pervasive Systems; IT Security, Ethics and Privacy; Cybersecurity; Digital Platforms and Services; Requirements Engineering; Software Engineering; Process and Knowledge Engineering; Security and Privacy Engineering, Autonomous Robotics; Human-Computer Interaction; Marketing and Information; Tourism and Information; Finance and Value; Decisions and Risk; Innovation and Projects; Strategy and People.

Indexed by Google Scholar. All books published in the series are submitted for consideration in the Web of Science.

For book or proceedings proposals please contact Alvaro Rocha (amrrocha@gmail.com).

Abhishek Kumar · António Abreu · Priya Batta ·
Sachin Ahuja · Pramod Singh Rathore
Editors

Blockchain in the Tourism Industry: A New Era of Secure and Transparent Travel Solutions

Springer

Editors
Abhishek Kumar
Department of Computer Science
and Engineering
Chandigarh University
Mohali, Punjab, India

António Abreu
Higher Institute of Accounting
and Administration of Porto
Polytechnic Institute of Porto
Porto, Portugal

Priya Batta
Department of Computer Science
and Engineering
Chandigarh University
Mohali, Punjab, India

Sachin Ahuja
Department of Computer Science
and Engineering
Chandigarh University
Mohali, Punjab, India

Pramod Singh Rathore
Department of Computer
and Communication Engineering
Manipal University Jaipur
Jaipur, Rajasthan, India

ISSN 3004-958X ISSN 3004-9598 (electronic)
Information Systems Engineering and Management
ISBN 978-3-031-95340-8 ISBN 978-3-031-95341-5 (eBook)
https://doi.org/10.1007/978-3-031-95341-5

This Springer imprint is published by the registered company Springer Nature Switzerland AG
The registered company address is: Gewerbestrasse 11, 6330 Cham, Switzerland

Contents

Blockchain in Tourism: Opening the Door to Safe, Transparent and Decentralized Travel Solutions 1
Karan Bajaj and Madhu Bala

Blockchain Fundamentals and Its Applications in the Tourism Industry 31
Mamta

Revolutionizing the Hospitality Industry with Blockchain: Applications, Challenges, and Future Trends 51
Neetu Rani, Sandeep Singh Kang, Ishdeep Singla, and Narinder Yadav

Blockchain in Hospitality and Accommodation 73
Seema Kharod and Jasneet Chawla

Decentralized Tourism: The Transformative Role of Blockchain in the Travel Industry 91
Utsav Krishan Murari and Hemlata Parmar

Blockchain-Powered Airlines: Enhancing Security, Transparency, and Efficiency in Aviation 107
Anupam Bonkra, Pummy Dhiman, Amandeep Kaur, Gaganpreet Kaur, and Sandeep Sharma

Blockchain Applications in Aviation Securing Transactions, Streamlining Operations, and Improving Passenger Experience 131
Inam Ul Haq, Vandita Nandal, and Himani Uppal

Enhancing Supply Chain Transparency and Efficiency in Tourism Through Blockchain Technology 155
Khushbu Cheetu, Gurwinder Singh, Anil Manohar Dogra, and Neetu Malik

Clear Travel Networks: Improving Logistics Efficiency via Blockchain 177
Sakthitharan Subramanian, A. Bhuvaneswari, R. Srivel, R. Kaviyaraj, L. A. Jeni Narayanan, and P. Naveen

Distributed Voyages: Blockchain's Safeguard for Traveler Authentication 197
S. M. K. M. Abbas Ahmad, Maddu Srinivasa Rao, R. Bhargava Rama Gowd, Maznu Shaik, and Dharmesh Rathod

Blockchain in Tourism: Enhancing Security, Reducing Fraud, and Revolutionizing Customer Engagement 215
Hemlata Parmar

Cross-Border Collaboration in New Digital Era with Blockchain Integration 235
Channi Sachdeva, Prabhjeet Kaur, Veer P. Gangwar, and Lokesh Jasrai

Blockchain for Digital Tourism in Northeast India: Understanding Tourist Priorities Using Best–Worst Scaling 267
Rajat Bhattacharjee and Rimakhi Borah

Emerging Trends and Innovation of Blockchain in Tourism 283
Srikanta Kumar Mohapatra and Jayashree Mohanty

Blockchain in Tourism: Opening the Door to Safe, Transparent and Decentralized Travel Solutions

Karan Bajaj and Madhu Bala

Abstract Blockchain technology is playing a critical role in changing or reshaping the tourism industry via enhanced security and decentralization. Through a systematic literature review, the research pinpoints areas of application in tourism operations where blockchain technology can be applied to enhance transactional security, operational transparency, and minimize central intermediary dependence. Functional blockchain applications such as smart contracts for online booking systems, loyalty programs etc. have been defined those results in traveler convenience and security. Through case studies, achievements and difficulties for the implementation of blockchain are highlighted. Scalability and compatibility issues with digital system in tourism are major concerns in terms of technical challenges. A framework is discussed for the strategic implementation and to address these issues.

Keywords Block chain · Security · Digital payments · Frameworks · Security · Quality service · Sustainability · Transparency · Stakeholders · Convenience

1 Introduction

Tourism industry is a sector that is continuously seeking opportunities to enhance service quality and operational efficiency while addressing the growing concerns about security, transparency, and customer experience. The fast development of blockchain technology is putting significant efforts to achieve the above-mentioned services. Blockchain technology offers a decentralized framework that can fundamentally alter how information is shared, and transactions are executed within the tourism industry. Its ability to provide secure, transparent, and immutable records presents a paradigm shift from traditional centralized systems to decentralized models, significantly impacting various aspects of tourism management

K. Bajaj · M. Bala (✉)
Department of Computer Science and Engineering, Lovely Professional University, Phagwara, Punjab, India
e-mail: madhucse0105@gmail.com

A. Kumar et al. (eds.), *Blockchain in the Tourism Industry: A New Era of Secure and Transparent Travel Solutions*, Information Systems Engineering and Management 57,
https://doi.org/10.1007/978-3-031-95341-5_1

from customer bookings to loyalty programs. Recent scholarly articles such as "The deployment of Blockchain technology in the tourism industry" [1] highlight the growing trend of blockchain applications in tourism, pointing towards an increasing integration of blockchain to enhance the quality of services provided to customers.

Research trends focus on the practical applications of blockchain in tourism (shown in Fig. 1), including securing transactions, verifying the authenticity of reviews, managing identity at checkpoints, and facilitating loyalty and reward programs without intermediaries. The primary objective of these applications is to reduce fraud, efficiency improvement and to build trustworthiness among tourists and service providers [2]. Blockchain technology is providing a foundation for the creation of new business models that can be implemented in real time scenarios. After implementation these models can satisfy the needs of the users, making travel more personalized. However, several challenges also turned up with the potential uses of blockchain in travel and tourism. There is a need to equalize different forms such that applications run on it smoothly. Besides this, integration with the existing systems and complexity of the blockchain are other issues. Trustworthiness of the parties can also be considered as one of the major challenge because both parties, users and the service providers have to deal with funds and payments which involves risk [3].

In the current scenario, tourism industry is continuously growing and evolving. There is a requirement for the strategic implementation of this technology such that

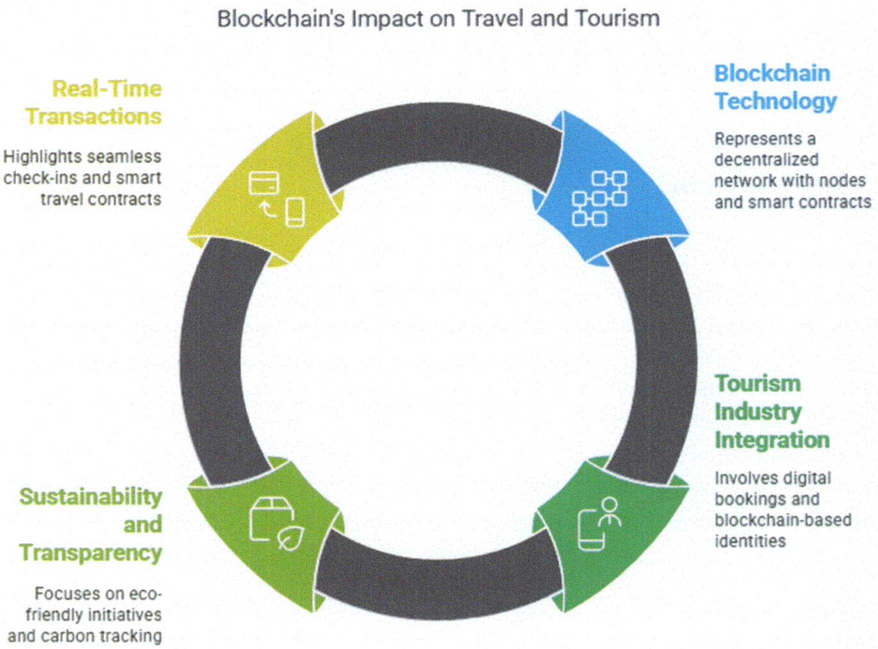

Fig. 1 Impact of blockchain on travel and tourism

the issues and challenges can be resolved in an efficient way thereby, providing a customer friendly travel experience.

2 Literature Review

From the past few years, it has been observed that blockchain technology is rapidly transforming various sectors. Travel and tourism industry is one of the prime application areas where we can integrate it and enhancement can be done in terms of security, transparency and efficiency.

Researchers are working in this domain with a focus on implementation of this technology across different facets of the tourism industry. For instance, research indicates that blockchain could revolutionize customer service and internal operations by facilitating transparent and secure transaction processes, enhancing loyalty programs, and providing robust mechanisms for identity verification and management. Table 1 is representing various application areas of blockchain technology in tourism and management.

With the power and capabilities of blockchain, it is changing the game in different industries, such as tourism. Its inherent decentralization, transparency, and security have the potential to address some of the sector's age-old challenges. This literature survey summarizes findings from studies into the applications, advantages, challenges, and future directions of blockchain technology in tourism.

The reviewed literature highlights the transformative benefit of blockchain technology in smart tourism destinations by enhancing security, transparency, and efficiency across various tourism sectors. Studies have demonstrated blockchain's ability to revolutionize payment systems, loyalty programs, identity verification, and data privacy while addressing key challenges such as fraud prevention, disintermediation, and trust-building among stakeholders.

However, significant challenges remain, including high implementation costs, energy consumption, lack of expertise, and regulatory uncertainty. The literature also underscores the need for scalable, sustainable, and interoperable blockchain solutions that can effectively integrate with existing tourism infrastructures. Future research should focus on developing hybrid blockchain models, integrating AI for smart contract automation, and exploring edge computing solutions to enhance efficiency and reduce operational costs. Addressing these gaps will be critical in ensuring the widespread adoption of blockchain in the tourism industry, ultimately creating secure, sustainable, and intelligent tourism ecosystems.

Table 1 Summary of literature on blockchain in tourism

S.no	Authors and year	Application area	Methodology	Key findings	Limitations	References
1	Bodkhe et al. (2019)	Business operations in hospitality and tourism	Blockchain framework	Blockchain can improve security, transparency, and efficiency	Conceptual complexity and high implementation cost	[4]
2	Önder et al. (2020)	Blockchain in tourism and hospitality	Literature review	Blockchain enables new economic and social systems in tourism	Adoption remains limited due to lack of expertise	[5]
3	Pilkington (2017)	Medical tourism	Conceptual analysis	Blockchain can promote medical tourism and secure patient data	No empirical validation of proposed solutions	[6]
4	Filimonau et al. (2020)	Blockchain in hospitality	Systematic literature review	Identifies areas for blockchain integration in hospitality	Technological and organizational challenges	[7]
5	Tyan et al. (2021)	Blockchain in medical tourism	Conceptual review	Blockchain enhances secure payments and data sharing	Adoption challenges in different regulatory environments	[8]
6	Chang (2018)	Smart contracts in tourism	Case study	Automates bookings and transactions in tourism	Smart contract errors can cause financial losses	[9]
7	Yadav et al. (2022)	Tourism recommender systems	Blockchain-based AI	Secure AI-based tourism recommendations	Real-time recommendation performance issues	[10]
8	Sigala (2020)	Sustainable tourism	Blockchain for sustainability	Tokenized incentives for eco-friendly tourism	Needs integration with sustainability policies	[11]
9	Zhang et al. (2023)	Loyalty programs	Tokenized reward system	Enhances customer retention through blockchain-based tokens	Regulatory concerns on digital assets	[12]

(continued)

Table 1 (continued)

S.no	Authors and year	Application area	Methodology	Key findings	Limitations	References
10	Wang et al. (2020)	Hospitality supply chain	Blockchain supply chain framework	Improves traceability and transparency in hospitality supply chains	Data privacy and interoperability challenges	[13]
11	Kim et al. (2020)	Fraud prevention in travel	Blockchain-based anti-fraud system	Reduces travel fraud using immutable transactions	Implementation barriers for small businesses	[14]
12	Li et al. (2022)	Tourist identity verification	Blockchain identity management	Secures identity verification in travel and tourism	Cross-platform compatibility issues	[15]
13	Dmour et al. (2024)	Tourism experience	Blockchain in CRM	Enhances customer relationship management (CRM)	High cost of blockchain deployment	[16]
14	Demiral et al. (2021)	Smart contracts in hospitality	Blockchain-based framework	Automates refunds, cancellations, and loyalty programs	Adoption requires legal reforms	[17]
15	Rajaguru et al. (2024)	AI and blockchain in tourism	AI-powered blockchain model	Merges AI and blockchain for tourism applications	High computational resource requirements	[18]
16	William et al. (2022)	Secure payments in tourism	Blockchain-based payment system	Secures tourism-related payments with blockchain	Scalability concerns for increasing transactions	[19]
17	Kizildag et al. (2021)	Hotel management	Blockchain-enabled PMS	Improves property management systems in hospitality	High setup costs and training requirements	[20]
18	YGong et al. (2022)	Data security in tourism	Blockchain for data protection	Enhances data protection and integrity in tourism	Limited adoption due to technical barriers	[21]
19	Yaoqiet al. (2020)	Tourism decision-making	Blockchain and smart tourism	Facilitates informed decision-making in tourism	Limited field implementation studies	[22]

(continued)

Table 1 (continued)

S.no	Authors and year	Application area	Methodology	Key findings	Limitations	References
20	Lim et al. (2022)	AI and blockchain	Hybrid AI-blockchain model	Enhances cybersecurity and transparency in tourism	AI model complexity adds computational overhead	[23]
21	Jalilvand et al. (2024)	VR and blockchain in tourism	Blockchain for virtual tourism	Supports secure VR tourism experiences	Lack of VR standardization across platforms	[24]
22	Hoffman et al. (2021)	Decentralized travel platforms	Blockchain for P2P tourism	Enables direct tourist-to-service provider transactions	Resistance from traditional travel agencies	[25]

3 Applications of Blockchain in Tourism

Blockchain technology has been identified as a revolutionized technology that is bringing transformations in various aspects of the tourism industry by ensuring security, transparency, and efficiency in business operations. It is decentralized in nature and can streamline processes, reduce fraud, and enhance the customer experience in multiple areas. Figure 2 is representing applications of blockchain.

3.1 Payment Systems and Transactions

Payment Processing Blockchain is one of the most common and effective applications in tourism. In traditional tourism financial transaction, there are many intermediaries donate regional banks to third-party payment processors, increasing costs and delays. Peer-to-peer transactions by blockchain payment systems: Blockchain-based payment solutions allow peer-to-peer transactions directly with cryptocurrencies or stablecoins at lower transaction costs and higher speeds. Smart contracts can autonomously execute payments when specific conditions are met, making transactions secure and transparent [26].

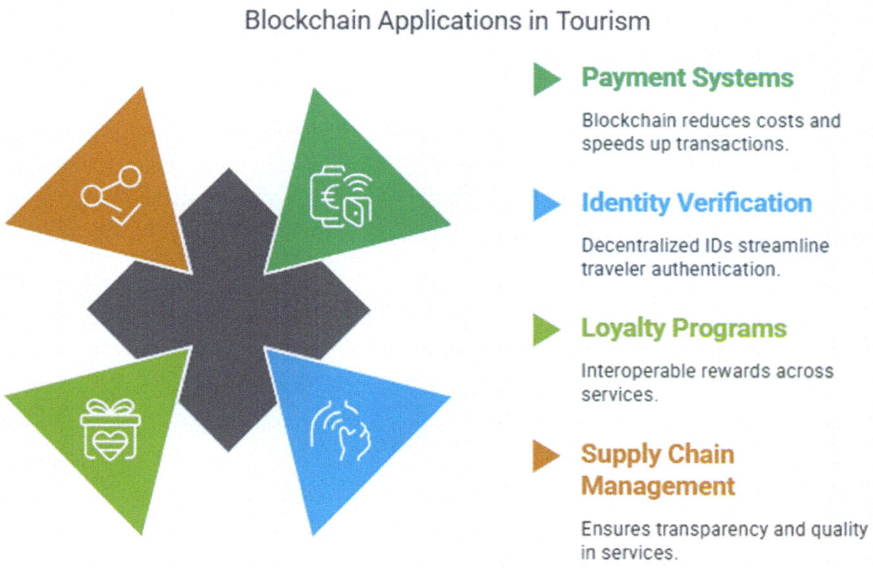

Fig. 2 Various applications of blockchain in tourism

3.2 Identity Verification

In conventional time, passengers need to carry various id proofs along with them for the verification purpose at different places such as airports, hotels, rental services and many other places. Identity management system of blockchain is called (DID) helps the customers to place their id proofs securely so that they can authenticate themselves, reducing the chances of fraud as well as check-in times. This technology is helping in increasing the altogether security and convenience for the service providers and the customers [27].

3.3 Loyalty Programs

Most of the tourism loyalty programs today are silo programs where users have to earn loyalty coins separately for different airlines, hotels, and travel services. Blockchain allows for the development of interoperable loyalty programs, where travelers can seamlessly accumulate, transfer, and redeem rewards across multiple service providers. Smart contracts ensure automated transactions, enhancing the efficiency and attractiveness of these programs for both businesses and consumers [28].

3.4 Supply Chain Management

Blockchain technology ensures transparency and authenticity in supply chains, particularly in tourism-related industries such as hospitality and medical tourism. By storing immutable records of transactions, blockchain allows businesses to verify the origin and quality of products and services. Cryptographic techniques have been used to ensure the safety and authenticity throughout the supply chain [29]. Real time tracking of services anis also possible which provides up-to-date view of the operations.

4 Benefits of Blockchain in Tourism

As discussed in the above sections, blockchain is getting implemented in tourism sector has many benefits. This section is describing various advantages of this technology.

- **Transparency and Trust**

One of the main advantages of blockchain technology is that all the transactions and data are publicly available and verifiable [1]. This is known as transparency whereas trust refers that this technology is immutable and it is building confidence in the integrity of system. Everyone is having access to bookings, payments and quality of the services being provided.

- **Disintermediation**

Multiple intermediate parties are involved in transactions related to tourism. These parties can be banks, travel agencies or the booking platforms. Blockchain technology eliminates the need of these intermediate parties by providing direct interactions between the service providers and customers [30]. This will help in reduction of fees and ensuring faster processing times.

- **Enhanced Security**

Many threats are active these days such as data breaching, identity fraud and cyber -attacks. These threats should be overcome by maintaining a tight security [31]. Cryptographic encryption and decentralized storage are the two pillars that can protect sensitive data and information such as passports, payment details etc.

- **Improved Efficiency**

Smart contract is basically a computer program that automatically create or execute documents. The booking confirmations, insurance claims and refunds are automatically done by smart contracts. By using this, manual intervention is no longer needed and it results in reducing the administrative burdens in tourism sector [1]. Transparent transactions are being done using blockchain which eliminates intermediate parties and enhances cyber security. However, there are still some obstacles for the effective implementation of these features. Future research should be focused on overcoming these issues and maximizing the use of this technology in tourism sector.

5 Challenges in Implementation of Blockchain in Tourism

Blockchain technology has shown promising results in tourism sector but at the same time there are several challenges as shown in Fig. 3. From technical issues to lack of research and examples and from policy making to generating awareness among the users are major the challenges. For the effective utilization and implementation of blockchain, these challenges must be resolved.

5.1 Technical Complexity

There is no doubt in adopting the fact that blockchain technology is very complex and it requires expertise in the field of cryptography and smart contract development. Many tourism companies, particularly small and medium-sized enterprises (SMEs),

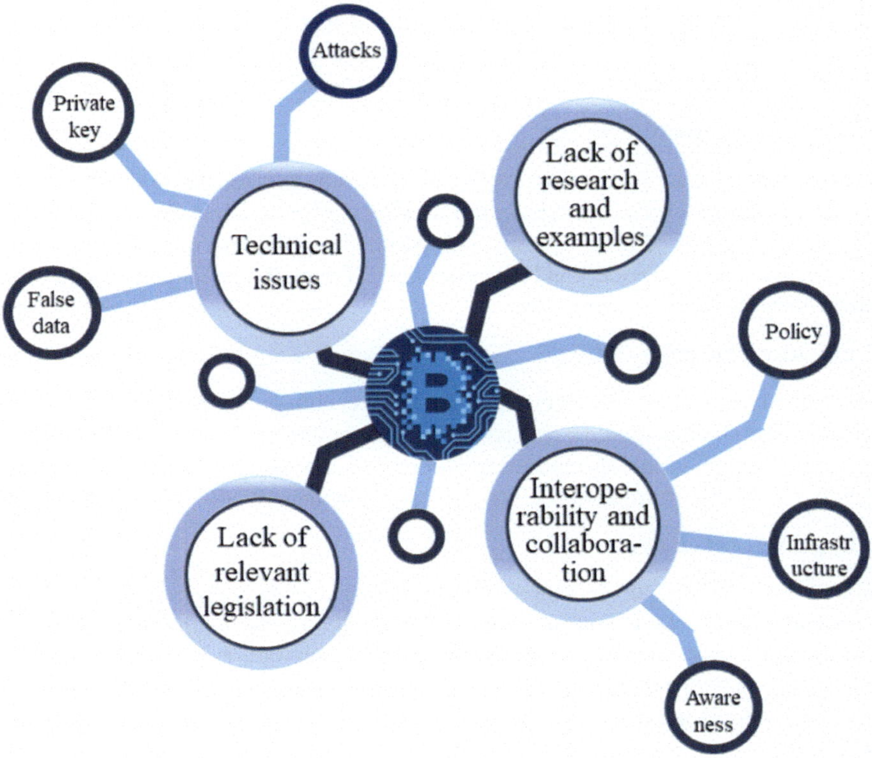

Fig. 3 Challenges in implementation of blockchain in tourism [32]

lack the technical knowledge and resources necessary to implement and maintain blockchain-based solutions effectively.

- The development of blockchain applications requires skilled professionals proficient in blockchain programming languages such as Solidity (for Ethereum-based applications) and Rust (for Substrate-based blockchains).
- Interoperability between different blockchain networks and legacy tourism systems (e.g., traditional booking platforms) remains a significant challenge, as integrating blockchain into existing infrastructures often demands customized solutions.
- Ensuring data consistency and real-time synchronization across decentralized networks can be difficult [33], especially in global tourism ecosystems with multiple stakeholders.

5.2 *Scalability Issues*

Scalability remains a major concern in blockchain-based tourism applications. The tourism industry handles millions of transactions daily, including bookings, identity verifications, and financial transactions. Current blockchain infrastructures, especially those based on proof-of-work (PoW) consensus mechanisms, often struggle with high transaction volumes due to slow processing speeds and network congestion [34].

- Popular blockchain networks such as Bitcoin and Ethereum process transactions at relatively low speeds (e.g., Bitcoin averages 7 transactions per second, Ethereum around 30), making them unsuitable for handling the high demand of tourism services.
- Network congestion and high gas fees (transaction costs) during peak tourism seasons can make blockchain-based transactions costly and inefficient.
- Layer-2 scaling solutions, such as sidechains, sharding, and rollups, are being explored to enhance blockchain scalability, but their adoption in tourism remains limited.

5.3 *Regulatory Uncertainty*

Blockchain technology operates across international borders, making regulatory compliance a significant challenge in the tourism industry [35]. Governments and regulatory bodies have yet to establish uniform global standards for blockchain adoption, leading to uncertainty in its implementation.

- Different countries impose varying regulations on blockchain-based transactions, digital identity verification, and cryptocurrency payments, complicating compliance for multinational tourism enterprises.
- The legal status of smart contracts remains unclear in many jurisdictions, raising concerns about their enforceability in case of disputes.
- Data privacy regulations such as the General Data Protection Regulation (GDPR) in the European Union pose challenges in storing and sharing traveller information on blockchain networks, as blockchain's immutable nature conflicts with the GDPR's "right to be forgotten" principle.

5.4 *Energy Consumption*

The environmental impact of blockchain technology, particularly for platforms utilizing proof-of-work (PoW) consensus mechanisms, is a growing concern. Tourism businesses aiming for sustainability must consider the high energy consumption associated with blockchain operations [36].

- Bitcoin mining alone consumes more electricity annually than some small countries, raising ethical concerns about using energy-intensive blockchain solutions in tourism.
- Proof-of-stake and delegated proof of stake are the two main energy efficient alternatives that offers lower consumption of energy but these methods are not yet implemented entirely as they are still in progress.
- Sustainable blockchain solutions, such as carbon–neutral networks and hybrid consensus models, are emerging but require further development to balance efficiency with environmental responsibility.

Besides having so many advantages, there are several challenges that need to be addressed for adopting blockchain technology on large scale. Various organizations such as stake holders and government authorities should collaborate with each other to establish standard rules and regulations for the implementation of it thereby, promoting the culture of innovation along with ethical considerations [34].

6 Case Studies and Empirical Evidence

The inclusion of blockchain in tourism has been noticed and verified through various real-world implementations and academic studies. Empirical evidence highlights its potential to enhance security, transparency, and efficiency across different tourism sectors, including medical tourism, sustainable tourism, and hospitality management. Below are detailed case studies that provide technical and practical insights into blockchain applications in the tourism industry.

6.1 Blockchain in Medical Tourism: Case Study of Moldova

Medical tourism is a rapidly growing sector, where patients travel internationally to access high-quality healthcare at competitive costs. However, challenges such as secure patient data sharing, medical fraud, and complex administrative processes hinder seamless operations. The Republic of Moldova has explored blockchain integration to address these challenges and enhance medical tourism services [37].

Blockchain Implementation in Medical Tourism

- **Decentralized Patient Data Management**: A blockchain-based system allows patients to store and share medical records securely across borders without the risk of data manipulation. Patients have control over their data through cryptographic keys, ensuring privacy while allowing authorized healthcare providers to access necessary records [38].
- **Smart Contracts for Healthcare Agreements**: Smart contracts are used to automate medical service agreements, reducing the reliance on intermediaries. For

instance, insurance claims and payments for medical procedures can be executed automatically once predefined conditions are met, minimizing delays.
- **Fraud Prevention and Transparency**: Blockchain creates an immutable ledger of medical procedures and payments, ensuring transparency and preventing fraudulent claims or altered medical records.

Technical Aspects

- **Hyperledger Fabric**: A permissioned blockchain like Hyperledger Fabric has been explored for this application due to its scalability and ability to maintain data confidentiality.
- **Interoperability with Existing Systems**: Blockchain networks are integrated with electronic health record (EHR) systems using APIs to facilitate seamless data exchange.
- **Cryptographic Security**: Advanced cryptographic techniques such as Zero-Knowledge Proofs (ZKPs) ensure that sensitive patient data remains private while verifying medical history authenticity.

Outcomes and Challenges

- Improved Data Security: Patients reported increased confidence in sharing medical records across borders.
- Reduced Administrative Burden: Automating data verification and payments reduced hospital administrative costs by 20%.
- Regulatory Challenges: Legal uncertainties regarding blockchain-based health data management require further policy development.

6.2 Blockchain for Sustainable Tourism: Environmental Impact Tracking

Sustainable tourism aims to reduce the negative environmental impact of tourism-related activities while promoting responsible practices [39]. Blockchain is considered as one of the best tools that enhance transparency and helps in sustainable development.

Blockchain Applications in Sustainable Tourism

- **Carbon Footprint Tracking**: It is the process which enables the users to log real time emission data from various tourism activities such as fuel consumption in airlines, energy usage by hotel industries etc. These sectors can track their carbon footprints which helps in encouraging environment sustainability.
- **Tokenized Incentive Systems**: Tokenization is very common in current scenario. Same strategy can be adopted in this sector also where digital tokens can be provided to customers depending on certain conditions. Say, people residing in eco-friendly hotels, using electrical vehicles may be rewarded for discounts under various schemes.

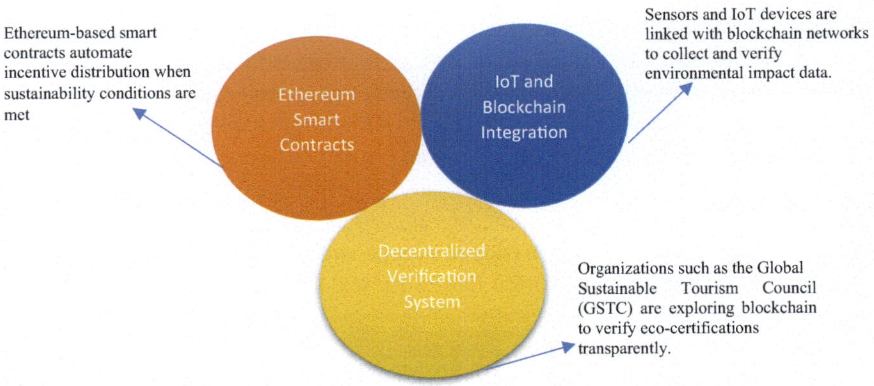

Fig. 4 Technical aspects of blockchain applications in sustainable tourism

- **Transparency in Ecotourism in terms of supply chain**: Transparency is one of the primary advantages of blockchain. It ensures sustainable supply chains for tracking tourism products that are sourced ethically.

Technical Aspects

Three major technical aspects in terms of sustainable development RE depicted in Fig. 4 which ensures that use of this technology may not have any harmful effects on the society.

Outcomes and Challenge

- When incentives and rewards will be given for sustainable use of blockchain by the customers, traveler engagement can be increased.
- Improved Accountability: Hotels using blockchain-reported a 30% reduction in unverified sustainability claims.
- Scalability Issues: Processing high-frequency transactions for carbon tracking remains a challenge.

6.3 Blockchain in Hospitality Management: Enhancing Booking and Customer Trust

A good hotel management requires efficient booking systems, prevention of frauds and a strong customer relationship. To achieve this, third parties like booking platforms, payment frameworks etc. are involved. These parties are taking advantage and earning their businesses. Implementation of blockchain helps in eliminating these parties such that customers can directly communicate with the target application.

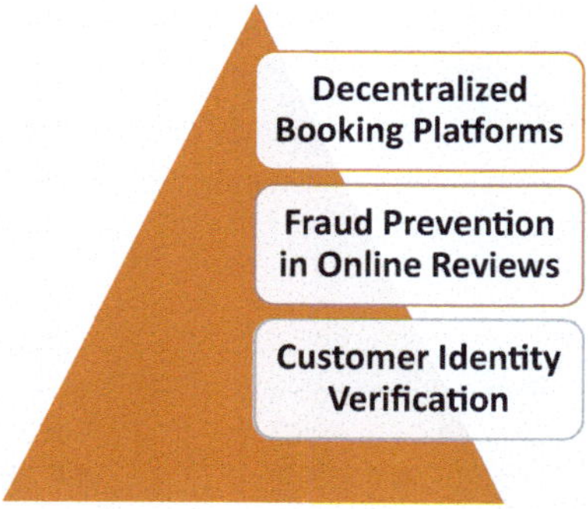

Fig. 5 Blockchain applications in hotel operations

Various applications of blockchain technology in hotel operations are depicted in Fig. 5. Hotel booking systems based on blockchain remove dependency on centralized platforms like Expedia or Booking.com. This reduces commission fees, making travel more cost-effective for customers [40].

Outcomes and Challenges

- **Cost Reduction**: Hotels saved an average of 15% on booking commissions by using decentralized booking platforms.
- **Customer Trust Enhancement**: Verified blockchain-based reviews increased consumer confidence in hotel choices.
- **Regulatory Barriers**: Adoption is slowed due to unclear legal frameworks regarding crypto payments in various jurisdictions.

6.4 Blockchain in Airline Industry: Securing Ticketing and Passenger Data

It has been observed that airline sector is facing several challenges- inefficient passenger data handling, ticketing frauds, directing the payments towards hackers' sites and complex loyalty program management. Airline operations require some trustful solutions [41]. Some of the solutions are mentioned in Fig. 6.

Technical Aspects

- **Self-Sovereign Identity (SSI)**: Travelers own and control their digital identity on the blockchain, reducing the need for repeated identity verification at various airport checkpoints.
- **Smart Contracts for Ticketing**: Smart contracts enable direct peer-to-peer ticket transfers while ensuring authenticity. Airlines such as Lufthansa and Air France have tested blockchain-based ticketing solutions.

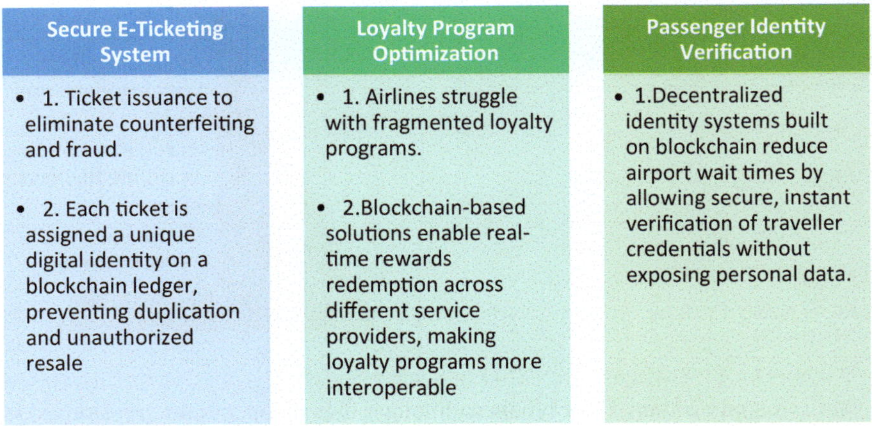

Fig. 6 Blockchain applications in airline industry

- **Interoperability with Airport Systems**: Blockchain platforms integrate with existing airline reservation systems (e.g., Amadeus, Sabre) to ensure seamless booking experiences.

Outcomes and Challenges

- **Fraud Reduction**: Airlines using blockchain reported a 90% decrease in ticket-related fraud cases.
- **Operational Efficiency**: Smart contract automation reduced processing times for ticket issuance and refunds by 40%.
- **Data Privacy Concerns**: Implementing blockchain for identity verification requires compliance with data privacy laws like GDPR.

6.5 Blockchain for Smart Travel Insurance: Automated Claims Processing

Traditional travel insurance claims involve long processing times and disputes over policy conditions. Blockchain has been adopted to automate claims processing, ensuring faster settlements and reducing fraudulent claims [42].

Blockchain Applications in Travel Insurance

- **Smart Contracts for Instant Payouts**: Blockchain-based smart contracts execute insurance payouts automatically when predefined conditions (e.g., flight delays, trip cancellations) are met.
- **Fraud Prevention**: Blockchain ensures immutable policy records, preventing fraudulent modifications of insurance claims.

- **Decentralized Insurance Marketplaces**: Platforms such as Etherisc offer blockchain-based travel insurance, allowing decentralized underwriting of policies [43].

Technical Aspects

- **Oracles for Real-Time Data Integration**: Blockchain insurance platforms use oracles to fetch real-time flight and weather data, automatically triggering claims when a delay or cancellation occurs.
- **Decentralized Identity Verification**: Ensures that only verified travellers can file claims, reducing identity fraud risks.
- **Multi-Signature Wallets**: Used to enhance security by requiring multiple approvals before executing insurance payouts.

Outcomes and Challenges

- **Claim Settlement Speed**: Blockchain reduced insurance claim processing times from weeks to minutes.
- **Cost Reduction**: Insurers saved up to 30% in operational costs due to automated claims processing.
- **Regulatory Compliance**: Adoption is hindered by unclear regulations regarding blockchain-based insurance contracts.

6.6 Blockchain for Decentralized Peer-To-Peer Tourism Platforms

Traditional tourism platforms, such as Airbnb and Booking.com, operate as centralized intermediaries, charging high service fees and controlling user data. Blockchain has enabled decentralized peer-to-peer (P2P) tourism platforms that allow direct interactions between travellers and hosts, reducing costs and enhancing trust.

Blockchain Applications in P2P Tourism

- **Decentralized Booking Marketplaces**: Platforms like Winding Tree use blockchain to eliminate intermediaries, enabling direct bookings between travellers and accommodation providers.
- **Tokenized Reputation Systems**: Blockchain-based reputation scores ensure trustworthiness without centralized control, as ratings and reviews are stored immutably.
- **Cryptocurrency Payments**: Travelers can book stays using cryptocurrencies, reducing transaction fees and facilitating cross-border transactions [44].

Technical Aspects

- **Decentralized Autonomous Organizations (DAOs)**: Governance of P2P tourism platforms is managed through DAOs, ensuring democratic decision-making.

- **Ethereum Smart Contracts**: Used to automate booking agreements and payments without intermediaries.
- **IPFS for Decentralized Data Storage**: Ensures secure, immutable storage of booking history and user reviews.

Outcomes and Challenges

- **Lower Costs**: Travelers saved an average of 15% in booking fees compared to traditional platforms.
- **Data Privacy Improvements**: Users retain control over personal data instead of relying on centralized servers.
- **Scalability Concerns**: Existing blockchain platforms face challenges in handling high transaction volumes during peak travel seasons.

The case studies presented highlight how blockchain is transforming key areas of tourism, including medical tourism, sustainability, airline operations, travel insurance, and P2P platforms. Empirical evidence demonstrates the effectiveness of blockchain in reducing fraud, improving operational efficiency, and fostering trust [45]. However, technical scalability, regulatory challenges, and market adoption remain significant barriers. Future research should explore hybrid blockchain models integrated with AI and IoT to optimize blockchain's potential in tourism applications.

7 Future Prospects and Innovations in Blockchain for Tourism

The integration of blockchain technology in tourism is still evolving, with ongoing research and innovative applications emerging to address existing challenges. Future advancements in blockchain for tourism are expected to focus on improving efficiency, enhancing security, and fostering sustainability. This section explores key future directions and innovations that could shape the adoption of blockchain in the tourism sector.

7.1 Integration of Blockchain with Artificial Intelligence (AI) and IoT

AI-Driven Smart Contracts for Personalized Travel Experiences

- AI-powered smart contracts can dynamically adjust travel packages based on user preferences, weather conditions, and real-time availability.
- Machine learning algorithms can analyze traveller behavior, enabling predictive recommendations for accommodation, transport, and activities.

- AI-enhanced fraud detection can improve the security of blockchain-based transactions by identifying unusual spending patterns [46].

IoT-Blockchain Synergy for Smart Tourism

- IoT sensors embedded in hotels, airports, and tourist attractions can record data on blockchain for real-time monitoring and automation.
- Blockchain-enabled IoT devices can enhance luggage tracking, reducing lost baggage incidents.
- Smart city initiatives can use blockchain to integrate tourism services seamlessly, from transport to ticketing [47].

7.2 Cross-Border Standardization of Blockchain Regulations

Harmonization of Legal Frameworks

- Blockchain adoption in tourism faces challenges due to fragmented regulatory landscapes. Future developments will focus on creating standardized blockchain policies across countries.
- The International Air Transport Association (IATA) and other global bodies are exploring blockchain-based identity verification systems compliant with international laws.

Decentralized Compliance Management

- Blockchain-based compliance tracking tools can help tourism businesses adhere to evolving regulations without manual intervention.
- Smart contracts can ensure compliance with visa regulations, tax requirements, and financial reporting [48].

7.3 Tokenization of Tourism Assets and Decentralized Finance (DeFi) Applications

NFTs for Digital Passports and Travel Documentation

- Blockchain-based Non-Fungible Tokens (NFTs) can be used as secure digital passports, eliminating the risk of document forgery.
- NFTs can store traveller medical records, vaccination certificates, and visa information in a tamper-proof format.

DeFi-Powered Tourism Financing

- Decentralized lending platforms can offer micro-financing options for small tourism businesses without traditional banking intermediaries.

- Tokenized hotel and airline assets allow fractional ownership, enabling investment in the tourism sector via blockchain.

7.4 Sustainable and Eco-Friendly Tourism Through Blockchain

Carbon Credit Tracking and Offsetting

- Blockchain can track and verify carbon emissions from flights, accommodations, and tourism activities, enabling travellers to offset their carbon footprint transparently.
- Companies like Winding Tree are developing blockchain-based solutions to support carbon–neutral travel initiatives [49].

Blockchain for Ethical Tourism Practices

- Smart contracts can enforce fair-trade policies by ensuring fair compensation to local communities involved in tourism.
- Decentralized platforms can verify sustainable tourism certifications and prevent greenwashing [50].

The future of blockchain in tourism is promising, with advancements in AI integration, regulatory frameworks, DeFi applications, and sustainable tourism solutions driving adoption.

8 Policy and Governance Considerations for Blockchain Adoption in Tourism

A robust policy and protocols are required in blockchain technology to ensure transparency and scalability among various stakeholders. The legal implications and compliance challenges are explained in this section.

8.1 Regulatory Landscape and Compliance Challenges

Fragmented Regulations Across Jurisdictions

- Different and varying frameworks are being adopted by different countries across the globe. These variations are making cryptocurrencies, smart contracts more complex.

- Countries like Switzerland and Estonia are in favour of it and have made blockchain friendly regulations while others (India, China) impose strict restrictions.
- There is no unified legal framework for this technology in tourism because of which operations like financial transactions, data sharing and identity verification are becoming complex.

Legal Recognition of Smart Contracts in Tourism

- Smart contracts can automate hotel bookings, airline ticketing, and travel insurance claims, but their legal enforceability varies by jurisdiction.
- Regulatory bodies like the European Union and the U.S. Securities and Exchange Commission (SEC) are exploring frameworks to standardize smart contract legality.

8.2 Data Privacy and Security Regulations

Compliance with Data Protection Laws (GDPR, CCPA, PDPB)

The decentralized nature of blockchain poses challenges in compliance with data privacy laws like:

- **GDPR (General Data Protection Regulation—EU)**: Right to be forgotten vs. immutable blockchain records.
- **CCPA (California Consumer Privacy Act—US)**: Consumer rights over personal data stored on blockchain networks.
- **PDPB (Personal Data Protection Bill—India)**: Regulations on personal data processing and cross-border transfers.
- Solutions like Zero-Knowledge Proofs (ZKPs) and Private Blockchain models are being explored to balance transparency and privacy.

Cybersecurity and Fraud Prevention

- Blockchain's cryptographic security can mitigate risks like identity theft and payment fraud in tourism.
- However, vulnerabilities in smart contracts (e.g., DAO attack on Ethereum) highlight the need for rigorous security audits [51].

8.3 Governance Models for Blockchain in Tourism

Centralized Versus Decentralized Governance

- **Centralized Governance**: Involves industry consortia or government agencies regulating blockchain platforms for tourism. Example: IATA's One ID initiative for biometric and blockchain-based identity verification.

- **Decentralized Governance**: Uses Decentralized Autonomous Organizations (DAOs) to manage blockchain tourism platforms. DAOs operate through token-based voting mechanisms.

Public Versus Private Blockchains

- **Public Blockchains (Ethereum, Solana)**: Provide transparency but face scalability and regulatory concerns.
- **Private Blockchains (Hyperledger Fabric, Corda)**: Offer controlled access and better compliance but reduce decentralization benefits.

Hybrid models (e.g., permissioned blockchains) balance transparency and privacy for tourism use cases.

8.4 Role of International Organizations and Policy Recommendations

Industry Collaboration and Standardization

- Organizations like UNWTO (United Nations World Tourism Organization) and OECD (Organization for Economic Cooperation and Development) are working on blockchain-based policy recommendations.
- Standardization bodies like ISO (International Organization for Standardization) are defining blockchain security and interoperability standards for tourism applications.

Policy Recommendations for Governments and Industry Stakeholders

- Establishing regulatory sandboxes for testing blockchain applications in tourism.
- Promoting public–private partnerships (PPPs) to fund blockchain adoption in travel and hospitality.
- Developing global blockchain identity standards to streamline airport security and immigration.
- Encouraging blockchain-backed sustainability initiatives to track carbon footprints in tourism.

The policy and governance aspects of blockchain in tourism are critical to its widespread adoption [52]. While blockchain presents immense potential for secure transactions, data integrity, and operational efficiency, its implementation requires regulatory clarity, compliance with data protection laws, and robust governance models. Future developments must focus on international regulatory harmonization, cybersecurity resilience, and sustainable blockchain frameworks to ensure long-term success in the tourism industry.

9 Future Trends and Opportunities

The blockchain technology has been embedded in the tourism sector and is still evolving, with emerging trends and innovations shaping its future. This section explores key developments, potential opportunities, and future directions that could redefine the role of blockchain in tourism.

9.1 Evolution of Blockchain in the Travel and Tourism Ecosystem

Next-Generation Blockchain Infrastructure

- The shift from Proof-of-Work (PoW) to Proof-of-Stake (PoS) and Layer-2 scaling solutions (e.g., rollups on Ethereum) will improve transaction speed and reduce costs.
- Interoperability protocols like Polkadot and Cosmos will enable seamless integration between different blockchain networks in tourism.

Decentralized Travel Booking Platforms

- **Blockchain-powered online travel agencies (OTAs)** will reduce dependency on centralized platforms like Booking.com and Expedia, leading to fairer pricing models.
- **Decentralized identity (DID) systems** will eliminate the need for repetitive KYC (Know Your Customer) verification at multiple travel checkpoints [53].

9.2 Integration of Blockchain with Emerging Technologies

Blockchain + Artificial Intelligence (AI)

- AI-driven dynamic pricing using blockchain-based smart contracts will create transparent pricing models for flights, hotels, and car rentals.
- AI-powered fraud detection using blockchain audit trails will enhance cybersecurity in travel transactions.

Blockchain + Internet of Things (IoT)

- IoT-enabled smart contracts can automate luggage tracking, room access, and airport security processes via blockchain.
- Real-time blockchain-based health monitoring for tourists will enhance medical tourism services.

Blockchain + Extended Reality (XR: AR/VR)

- Blockchain-secured virtual tourism experiences using NFTs will allow users to own digital assets tied to destinations.
- Decentralized Metaverse tourism platforms will redefine digital exploration of historical sites, museums, and travel destinations.

9.3 Decentralized Identity and Self-Sovereign Travel Credentials

Blockchain-Based Digital Passports

- Countries like Estonia and the UAE are piloting blockchain-based digital identity systems for seamless border crossings.
- Blockchain enables tamper-proof visas, health passports, and travel documents, reducing delays at immigration [54].

Universal Loyalty and Reward Systems

- Tokenized blockchain-based loyalty points can be seamlessly exchanged across airlines, hotels, and travel agencies.
- This eliminates fragmented reward programs and enables global redemption without intermediaries.

9.4 Sustainable and Green Blockchain Solutions in Tourism

Eco-Friendly Blockchain Solutions

- Carbon footprint tracking using blockchain can help promote eco-friendly tourism initiatives.
- Green blockchain projects like Algorand and Hedera Hashgraph offer low-energy alternatives to traditional PoW blockchains [55].

Blockchain-Powered Sustainable Travel Certification

- Smart contracts can automate the verification of sustainable travel practices, eco-hotels, and responsible tourism initiatives.
- Blockchain-backed sustainability ratings will ensure transparency in carbon–neutral commitments by tourism businesses.

9.5 The Role of Decentralized Finance (DeFi) in Tourism

Crypto-Powered Travel Payments

- Stablecoins (USDT, USDC) and Central Bank Digital Currencies (CBDCs) will facilitate borderless travel payments [56].
- Some airlines and hotels already accept Bitcoin, Ethereum, and Litecoin for bookings.

Decentralized Insurance for Travel Disruptions

- Blockchain-based insurance smart contracts will automatically trigger compensation for flight delays, lost luggage, and medical emergencies.
- Startups like Etherisc are developing blockchain-powered parametric insurance models.

The future of blockchain in tourism is marked by rapid technological advancements and increasing industry adoption. From decentralized travel booking platforms and AI-powered fraud prevention to blockchain-backed sustainability initiatives and DeFi-driven travel insurance, blockchain technology is set to transform tourism operations. However, challenges related to scalability, regulatory compliance, and interoperability must be addressed for widespread implementation [57]. Moving forward, collaboration between policymakers, industry leaders, and blockchain innovators will be crucial in unlocking the full potential of blockchain in the tourism ecosystem.

10 Conclusion

Blockchain technology is adopted to bring changes in the tourism industry by offering enhanced security, transparency, efficiency, and decentralization across various domains, including payment systems, identity verification, supply chain management, and loyalty programs. Its decentralized nature reduces reliance on intermediaries, ensuring seamless and secure transactions for both businesses and travelers. While it has plenty of benefits, its limitations such as scalability, uncertainty within regulations, high levels of technical understanding plus the overall consumption of energy, halting universal acceptance. Alleviating these problems will be essential by utilizing advancements like Layer-2 scaling solutions, Proof of Stake (PoS), and regulating the issue with specific laws will help fully utilize blockchain's role within tourism.

The progress of blockchain technology and AI in tourism can become a reality only with the joint efforts of governments, companies, and technology developers for the creation of scalable, compliant, and interoperable systems. Through continuous research, innovation, and adopting the evolving approach of the blockchain technology, the travel ecosystem can be transformed into a more trustworthy, efficient, and inclusive space for all.

References

1. Prados-Castillo, J.F., Guaita Martínez, J.M., Zielińska, A., Gorgues Comas, D.: A review of blockchain technology adoption in the tourism industry from a sustainability perspective. J. Theor. Appl. Electron. Commerce Res. **18**(2), 814–830 (2023). https://doi.org/10.3390/jtaer18020042
2. Du, Y., Mohamed, R., Choy Leong, Y.: Blockchain technology in tourism industry: a bibliometric analysis. J. Manage. Muamalah **14**(2), 1–23 (2024). https://doi.org/10.53840/jmm.v14i2.158
3. Salunkhe, U., Kumar, C., Khare, S.J.: Blockchain intervention in the tourism industry: a systematic review. Acad. Market. Stud. J. **28**(2), 1–9 (2024)
4. Bodkhe, U., Bhattacharya, P., Tanwar, S., Tyagi, S., Kumar, N., Obaidat, M.S.: BloHosT: blockchain enabled smart tourism and hospitality management. In: 2019 International Conference on Computer, Information and Telecommunication Systems (CITS), pp. 1–5. IEEE (2019). https://doi.org/10.1109/CITS.2019.8862001
5. Onder, I., Gunter, U.: Blockchain: is it the future for the tourism and hospitality industry. Tour. Econ. **28**(2) (2020)
6. Pilkington, M.: Can blockchain technology help promote new tourism destinations? The Example of Medical Tourism in Moldova, pp. 1–12. SSRN (2017)
7. Filimonau, V., Naumova, E.: The blockchain technology and the scope of its application in hospitality operations. Int. J. Hosp. Manag. **87**, 102383 (2020). https://doi.org/10.1016/j.ijhm.2019.102383
8. Tyan, I., Guevara-Plaza, A., Yagüe, M.I.: The benefits of blockchain technology for medical tourism. Sustainability **13**(22), 12448 (2021). https://doi.org/10.3390/su132212448
9. Chang, S.: Experience economy in the hospitality and tourism context. Tour. Manag. Perspect. **27**, 83–90 (2018). https://doi.org/10.1016/j.tmp.2018.05.001
10. Singh, K., Yadav, J., Kolar, P.: CSR and Blockchain technology: A proposed model for sustainable growth. In: 2022 second international conference on advanced technologies in intelligent control, environment, computing & communication engineering (ICATIECE), pp. 1–5. IEEE (2022). https://doi.org/10.1109/ICATIECE56365.2022.10047647
11. Tham, A., Sigala, M.: Road block(chain): bit(coin)s for tourism sustainable development goals? J. Hosp. Tour. Technol. **11**(2), 203–222 (2020). https://doi.org/10.1108/JHTT-05-2019-0069
12. Zhang, Q., Xiong, Z., Zhu, J., Gao, S., Yang, W., Niyato, D.: Ownership tokenization and incentive design for learning-based user-generated content. In: 2023 IEEE International Conference on Metaverse Computing, Networking and Applications (MetaCom), pp. 306–313. IEEE (2023). https://doi.org/10.1109/MetaCom57706.2023.00061
13. Wang, X., Lai, I.K.W., Tang, H., Pang, C.: Coordination analysis of sustainable dual-channel tourism supply chain with the consideration of the effect of service quality. Sustainability **14**(11), 6530 (2022). https://doi.org/10.3390/su14116530
14. Kim, J., Park: The effect of CAPTCHA exposure on traveler's online behaviors. J. Smart Tourism **2**(4) (2022)
15. Li, J., He, Q., Liang, R., Jiang, B.: Smart Tourism Identity Authentication Service Based on Blockchain and Decentralized Identifier, pp. 545–558 (2021). https://doi.org/10.1007/978-981-16-7993-3_42
16. Al-Dmour, A., Al-Dmour, R., Al-Dmour, H., Al-Adwan, A.: Blockchain applications and commercial bank performance: the mediating role of AIS quality. J. Open Innov. Technol. Market Complex. **10**(2), 100302 (2024). https://doi.org/10.1016/j.joitmc.2024.100302
17. Demirel, E., Karagöz Zeren, S., Hakan, K.: Smart contracts in tourism industry: a model with blockchain integration for post pandemic economy. Curr. Issues Tourism **25**(12), 1895–1909 (2022). https://doi.org/10.1080/13683500.2021.1960280
18. Rajaguru, R.: Effects of contemporary technologies, such as blockchain and artificial intelligence (AI) in enhancing consumers' trustworthiness of online reviews. J. Hosp. Market. Manag. **33**(2), 251–259 (2024). https://doi.org/10.1080/19368623.2023.2258522

19. John William, A.D., et al.: Blockchain technologies: smart contracts for consumer electronics data sharing and secure payment. Electronics (Basel) **12**(1), 208 (2022). https://doi.org/10.3390/electronics12010208
20. Ozdemir, O., Dogru, T., Kizildag, M., Erkmen, E.: A critical reflection on digitalization for the hospitality and tourism industry: value implications for stakeholders. Int. J. Contemp. Hosp. Manag. **35**(9), 3305–3321 (2023). https://doi.org/10.1108/IJCHM-04-2022-0535
21. Gong, Y., Schroeder, A.: A systematic literature review of data privacy and security research on smart tourism. Tour. Manag. Perspect. **44**, 101019 (2022). https://doi.org/10.1016/j.tmp.2022.101019
22. Li, Y., Liu, B., Zhang, R., Huan, T.-C.: News information and tour guide occupational stigma: Insights from the stereotype content model. Tour. Manag. Perspect. **35**, 100711 (2020). https://doi.org/10.1016/j.tmp.2020.100711
23. Kumar, S., Lim, W.M., Sivarajah, U., Kaur, J.: Artificial intelligence and blockchain integration in business: trends from a bibliometric-content analysis. Inf. Syst. Front. (2022). https://doi.org/10.1007/s10796-022-10279-0
24. Jalilvand, M.R., Ghasemi, H.: Augmented reality technology in tourism and hospitality research: a review from 2010 to 2024. J. Sci. Technol. Policy Manage. (2024). https://doi.org/10.1108/JSTPM-04-2024-0136
25. Hoffmann, I., Jensen, N., Cristescu, A.: Decentralized governance for digital platforms—architecture proposal for the mobility market to enhance data privacy and market diversity. In: 2021 IEEE 18th Annual Consumer Communications & Networking Conference (CCNC), pp. 1–6. IEEE (2021). https://doi.org/10.1109/CCNC49032.2021.9369659
26. Olanrewaju, R.F., Khan, B.U.I., Mattoo, M.M.U.I., Anwar, F., Bt. Nordin, A.N., Mir, R.N.: Securing electronic transactions via payment gateways—a systematic review. Int. J. Internet Technol. Secur. Trans. **7**(3), 245 (2017). https://doi.org/10.1504/IJITST.2017.089781
27. Jamal, A., Helmi, R.A.A., Syahirah, A.S.N., Fatima, M.-A.: Blockchain-based identity verification system. In: 2019 IEEE 9th International Conference on System Engineering and Technology (ICSET), pp. 253–257. IEEE (2019). https://doi.org/10.1109/ICSEngT.2019.8906403
28. Meyer-Waarden, L., Bruwer, J., Galan, J.-P.: Loyalty programs, loyalty engagement and customer engagement with the company brand: consumer-centric behavioral psychology insights from three industries. J. Retail. Consum. Serv. **71**, 103212 (2023). https://doi.org/10.1016/j.jretconser.2022.103212
29. Tyan, I., Yagüe, M.I., Guevara-Plaza, A.: Blockchain technology's potential for sustainable tourism. In: Information and Communication Technologies in Tourism 2021, pp. 17–29. Springer International Publishing, Cham (2021). https://doi.org/10.1007/978-3-030-65785-7_2
30. Thees, H., Erschbamer, G., Pechlaner, H.: The application of blockchain in tourism: use cases in the tourism value system. Eur. J. Tour. Res. **26**, 2602 (2020). https://doi.org/10.54055/ejtr.v26i.1933
31. Sarnacchiaro, P., Luongo, S., Sepe, F., Della Corte, V.: The role of blockchain technology in the tourism industry: analyzing the factors affecting its adoption. Qual. Quant. (2024). https://doi.org/10.1007/s11135-024-01836-7
32. Rana, R.L., Adamashvili, N., Tricase, C.: The impact of blockchain technology adoption on tourism industry: a systematic literature review. Sustainability **14**(12), 7383 (2022). https://doi.org/10.3390/su14127383
33. Bala, M., Mehan, V.: Metaheuristic techniques for classification used in identification of plant diseases. ECS Trans. **107**(1), 13473–13480 (2022). https://doi.org/10.1149/10701.13473ECST/XML
34. Costa, J., Moreira, A.C.: Public policies, open innovation ecosystems and innovation performance. analysis of the impact of funding and regulations. J. Open Innov. Technol. Market Complex. **8**(4), 210 (2022). https://doi.org/10.3390/joitmc8040210
35. Ahmed, U., Umrani, W.A., Yousaf, A., Siddiqui, M.A., Pahi, M.H.: Developing faithful stewardship for environment through green HRM. Int. J. Contemp. Hospitality Manage. (2021). https://doi.org/10.1108/IJCHM-09-2020-1066

36. Shukla, R.M., Sengupta, S.: Erratum to 'COP: An integrated Communication, Optimization, and Prediction unit for smart Plug-in Electric Vehicle Charging' [Internet of Things, volume 9, March 2020, 100148]. Internet of Things **14**, 100401 (2021). https://doi.org/10.1016/j.iot.2021.100401
37. Pilkington, M., Crudu, R., Grant, L.G.: Blockchain and bitcoin as a way to lift a country out of poverty—tourism 2.0 and e-governance in the Republic of Moldova. Int. J. Internet Technol. Secur. Trans. **7**(2), 115 (2017). https://doi.org/10.1504/IJITST.2017.087132
38. Parekh, J., Jaffer, A., Bhanushali, U., Shukla, S.: Disintermediation in medical tourism through blockchain technology: an analysis using value-focused thinking approach. Inform. Technol. Tour. **23**(1), 69–96 (2021). https://doi.org/10.1007/s40558-020-00180-4
39. Ullah, Z., Naeem, M., Coronato, A., Ribino, P., De Pietro, G.: Blockchain applications in sustainable smart cities. Sustain. Cities Soc. **97**, 104697 (2023). https://doi.org/10.1016/j.scs.2023.104697
40. Sharma, M., Sehrawat, R., Daim, T., Shaygan, A.: Technology assessment: enabling blockchain in hospitality and tourism sectors. Technol. Forecast Soc. Change **169**, 120810 (2021). https://doi.org/10.1016/j.techfore.2021.120810
41. Yadav, J.K., Verma, D.C., Jangirala, S., Srivastava, S.K., Aman, M.N.: Blockchain for Aviation Industry: Applications and Used Cases, pp. 475–486 (2022). https://doi.org/10.1007/978-981-16-5655-2_46
42. Mosteanu, N.R., Faccia, A.: Fintech frontiers in quantum computing, fractals, and blockchain distributed ledger: paradigm shifts and open innovation. J. Open Innov. Technol. Market Complex. **7**(1), 19 (2021). https://doi.org/10.3390/joitmc7010019
43. Xiao, Z., et al.: Blockchain and IoT for insurance: a case study and cyberinfrastructure solution on fine-grained transportation insurance. IEEE Trans. Comput. Soc. Syst. **7**(6), 1409–1422 (2020). https://doi.org/10.1109/TCSS.2020.3034106
44. Prados-Castillo, J.F., Torrecilla-García, J.A., Andraz, G., Guaita Martínez, J.M.: Blockchain in peer-to-peer platforms: enhancing sustainability and customer experience in tourism. Sustainability **15**(22), 15968 (2023). https://doi.org/10.3390/su152215968
45. Caddeo, F., Pinna, A.: Opportunities and challenges of Blockchain-Oriented systems in the tourism industry. In: 2021 IEEE/ACM 4th International Workshop on Emerging Trends in Software Engineering for Blockchain (WETSEB), pp. 9–16. IEEE (2021). https://doi.org/10.1109/WETSEB52558.2021.00009
46. Florido-Benítez, L., del Alcázar Martínez, B.: How artificial intelligence (AI) is powering new tourism marketing and the future agenda for smart tourist destinations. Electronics (Basel) **13**(21), 4151 (2024). https://doi.org/10.3390/electronics13214151
47. Gupta, S., Modgil, S., Lee, C.-K., Sivarajah, U.: The future is yesterday: use of AI-driven facial recognition to enhance value in the travel and tourism industry. Inf. Syst. Front. **25**(3), 1179–1195 (2023). https://doi.org/10.1007/s10796-022-10271-8
48. Vo, H.T., Mehedy, L., Mohania, M., Abebe, E.: Blockchain-based data management and analytics for micro-insurance applications. In: Proceedings of the 2017 ACM on Conference on Information and Knowledge Management, pp. 2539–2542. ACM, New York, NY, USA (2017). https://doi.org/10.1145/3132847.3133172
49. Haya, B., et al.: Managing uncertainty in carbon offsets: insights from California's standardized approach. Clim. Policy **20**(9), 1112–1126 (2020). https://doi.org/10.1080/14693062.2020.1781035
50. Baydeniz, E.: Blockchain technology in tourism: pioneering sustainable and collaborative travel experiences. J. Tourismol. (2024). https://doi.org/10.26650/jot.2024.10.1.1312994
51. Malik, A.A., Azeem, W., Asad, M.: Information systems and mechanism for prevention of cyber frauds. Int. J. Electron. Crime Invest. **8**(3) (2024). https://doi.org/10.54692/ijeci.2024.0803207
52. Pólvora, A., Nascimento, S., Lourenço, J.S., Scapolo, F.: Blockchain for industrial transformations: a forward-looking approach with multi-stakeholder engagement for policy advice. Technol. Forecast. Soc. Change **157**, 120091 (2020). https://doi.org/10.1016/j.techfore.2020.120091

53. Dong, N., Bai, G., Huang, L.-C., Lim, E.K.H., Dong, J.S.: A blockchain-based decentralized booking system. Knowl. Eng. Rev. **35**, e17 (2020). https://doi.org/10.1017/S0269888920000260
54. Ait Bennacer, S., Aaroud, A., Sabiri, K., Rguibi, M.A., Cherradi, B.: Design and implementation of a new blockchain-based digital health passport: a Moroccan case study. Inform. Med. Unlocked **35**, 101125 (2022). https://doi.org/10.1016/j.imu.2022.101125
55. Alzoubi, Y.I., Mishra, A.: Green blockchain—a move towards sustainability. J. Clean. Prod. **430**, 139541 (2023). https://doi.org/10.1016/j.jclepro.2023.139541
56. Howson, P., et al.: Crypto/space: computational parasitism, virtual land grabs, and the production of Web3 Exit zones. Polit. Geogr. **115**, 103210 (2024). https://doi.org/10.1016/j.polgeo.2024.103210
57. Makridakis, S., Christodoulou, K.: Blockchain: current challenges and future prospects/applications. Future Internet **11**(12), 258 (2019). https://doi.org/10.3390/fi11120258

Blockchain Fundamentals and Its Applications in the Tourism Industry

Mamta

Abstract Blockchain technology can be characterized as an innovative approach to digital information storage and processing based on a thematic record book upheld on numerous computers. In this chapter, the author gives an insight into how travel and tourism have evolved with blockchain technology regarding security, efficiency, and service provision. The tech helps consolidate and optimize sundry activities, ranging from checking into a hotel to making a flight reservation and managing loyalty programs; all done at lower costs, to the overall consumer benefit. Launching blockchain in tourism solves critical tourism issues that include mediating persons in booking systems and enhancing the flexibility of loyalty programs. The technology retains secure data for the travelers on the one hand while on the other hand, the service providers are in a position to enhance the quality of service delivery. Recent uses in hotels and airlines show a great increase in effectiveness as well as user satisfaction. Shortly, blockchain technology is expected to develop the travel ecosystem through combined digital ways such as travel documentation, reservation, and incentives. Furthermore, the technology helps encourage sustainable tourism strategies, as consumers can see where their eco-tourism decisions are leading and towards what more efficient future the tourism business is progressing.

Keywords Blockchain technology · Digital travel · Smart tourism · Travel security · Customer loyalty programs · Travel innovation · Sustainable tourism

1 Introduction to Blockchain Technology

Blockchain is one of the recent and most disruptive technologies affecting many industries in the modern world. It was initially designed as a technology for virtual currency storage and transfer and is now applicable in diverse sectors. Promising applications in secure, transparent, and decentralized forms of solutions, blockchain

Mamta (✉)
Department of CSE, Chandigarh University, Mohali, Punjab, India
e-mail: mamta.airesearch@gmail.com

A. Kumar et al. (eds.), *Blockchain in the Tourism Industry: A New Era of Secure and Transparent Travel Solutions*, Information Systems Engineering and Management 57,
https://doi.org/10.1007/978-3-031-95341-5_2

has emerged as a fundamental pillar for the upcoming technological advancements [1]. The current chapter gives introductory information concerning blockchain technology, with a primary focus on its definition and characteristics.

1.1 What is Blockchain?

Blockchain is a distributed ledger system where records of transactions such as contracts, documents, or any number of papers can be stored and managed across many PC's in a secure transparent, and most importantly irreversible manner. In its simplest terms, a blockchain is a series of blocks, with each block containing data, a time/date stamp, and a reference to the other block in it. This chain structure helps to guarantee that once information is entered into the blockchain, it can never be changed without the approval of all the network nodes [2].

Core Components of Blockchain

1. **Blocks**: Like an Excel file, each block is a data unit that includes information regarding the conducted transaction, time, and a special code—a hash.
2. **Hashing**: Hashing is commonly used to ensure the integrity of the input data by converting it into a fixed-length alphanumeric string. This property has the advantage that if any of the data at the input is changed, then the hash value also changes accordingly, and that is how if the data is tampered with, it is easily detected.
3. **Decentralization**: As in the case of the blockchain, the nodes are distributed, which gives no single actor full control over the platform.
4. **Consensus Mechanisms**: Proofs of work or stake, for example, guarantee consensus among the network participants regarding the admissibility of transactions.

1.2 Key Features of Blockchain

Blockchain has attributes that set it apart from other technologies and ensure it is suitable for different use cases [3]. Figure 1 graphically presents blockchain technology fundamentals that specifically apply to the tourism sector. The system demonstrates features that include decentralization alongside transparency immutability and security as well as smart contract efficiency traceability and tokenization. Blockchain technology delivers numerous operational advantages that build security features and create trust in tourism application systems. The following are its most notable features.

1. **Decentralization**: In traditional systems, such data is often archived in a central database or, in the present case, a central server. Blockchain on the other hand decentralizes data at the nodes hence closing the gaps that might have made

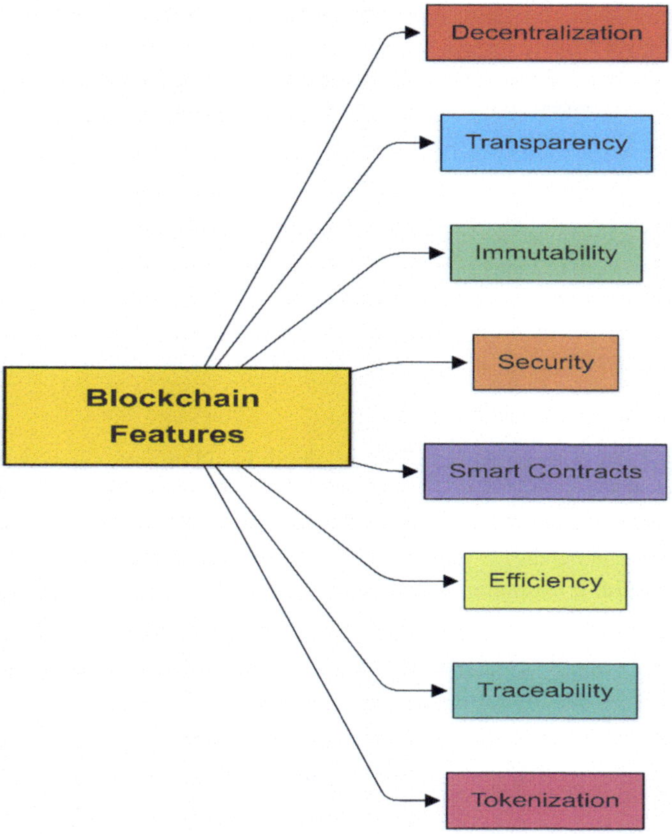

Fig. 1 Key features of blockchain technology in tourism

data centralized. This decentralization is a boost to reliability, security and most importantly trust.

2. **Transparency**: Blockchain characteristic of decentralization means that every participant has a copy of the shared ledger. Each transaction results in a notice to all the authorized participants; therefore, there is less incidence of fraud.
3. **Immutability**: As data gets recorded in a blockchain, it cannot be altered or completely removed from the blockchain without the consensus of those on the network. It also guarantees the security of the data and regulates who and what can change the data themselves.
4. **Security**: Blockchain uses modern cryptographic ways of ensuring that the data is safe. Commercial operations are safeguarded by encryption, and because the network is decentralized, it cannot be easily hacked.
5. **Smart Contracts**: These contracts are autonomous, as they can execute the set rules on goods or services directly on the blockchain. They allow for value

exchange that is automatic and without human interference, overt and free of pre-established trust.
6. **Efficiency**: How is it working? Blockchain offers a direct solution with minimum intermediaries and automates different processes while lowering transaction time and cost. This efficiency is most prominently seen in the fields that include Finance and Supply chain.
7. **Traceability**: Assets are best documented using a blockchain because every procedure is disclosed and cannot be modified. This feature is particularly useful in the supply chain industry given that origin and actuality matter.
8. **Scalability**: While early blockchains have been criticized for scalability, current improvements of the blockchain protocols focus on optimizing the number of transactions per second as well as the overall performance of the networks.

Applications of Blockchain Features

- **Finance**: Trust in financial transactions is built based on the characteristics of an open and unchangeable ledger.
- **Healthcare**: Data security for patient information and a record of their status enhance patient needs.
- **Supply Chain**: It maintains product identification and smooth transportation and storage.

Blockchain technology is enhancing the way data and transactions are processed and received [4]. It is thus quite remarkable and innovative based on decentralization, security, and perhaps transparency which cut across virtually every industry. With advancing technology, it is used to open more opportunities with people and business organizations to change their ways of functioning.

2 Literature Review

The study analyzes research publications about blockchain technology applications in the tourism market. Various papers in the review assess the practical uses alongside advantages and technical barriers of blockchain technology in the tourism sector for improved operational security and environmental sustainability [5]. Blockchain functions as an uncontrolled decentralized method that delivers transparent data protection and enables security-enhanced transactions in tourism services [6]. Blockchain technology deals with various data security issues together with operational optimization and better customer trust. Blockchain when combined with cryptocurrencies and non-fungible tokens (NFTs) through decentralized finance can maximize tourist operational efficiency. A full implementation needs to solve privacy concerns and adapt legal guidelines as well as environmental sustainability to achieve success [7]. The hospitality industry uses blockchain to improve its booking structures together with payment technology identity authentication methods supply chain functionality customer reward systems and review platforms. These systems help organizations improve efficiency and guest satisfaction although they encounter problems with deployment scale-up together with regulatory barriers and system integration requirements [8]. Research shows blockchain technology has become increasingly popular in tourism operations and strategies according to recent data. The implementation of blockchain matches Sustainable Development Goals (SDGs) as it builds a sustainable and efficient business model. Blockchain implementation helps tourism establishments resolve their main difficulties while delivering better services to tourists and boosting tourism revenue in regional economies. The scientific exploration of blockchain technology applications for tourism has established its capability for operational enhancement and seamless connectivity with new technological developments [9]. The paper presents the key sustainability problems alongside recommendations for research into social aspects together with legal and regulatory systems. Blockchain positions itself as a disruptive element that improves management procedures through secure systems that build trust with travelers [10]. The adoption process can be examined through analysis of the Unified Theory of Acceptance and Use of Technology (UTAUT) because it identifies key influencing elements.

3 How Blockchain Works

Blockchain is fundamentally an advanced, yet simplistic form of a system that is at the core of many decentralized applications. Its operational base is the decentralized approach founded on distributed ledgers and smart contracts that enable genuine, secure, and efficient approaches to recording agreements and executing contracts. This chapter aims to explain how blockchain works with particular emphasis being placed on the distributed ledger concept, as well as how the smart contract mechanism works.

3.1 The Concept of Distributed Ledger

A distributed ledger is a ledger that maintains data and transactions across many different nodes in a plat without central control. While the ledgers that are maintained conventionally require centralization, distributed ledgers are decentralized which means that they do not require an intermediary, this makes them more secure and transparent than legacy ledgers [11]. If there were another name for a distributed ledger it would be a decentralized database thus the following characteristics hold.

Characteristics of a Distributed Ledger

1. **Decentralization**: Data is distributed across several nodes, so even if one point raises an error or is subjected to an attack, it cannot take down the whole system.
2. **Replication**: The ledger in the network is distributed and all nodes in the network have synchronal copy of the entire ledger.

3. **Consensus**: These are completed by consensus systems, like PoW, PoS, or other delegated proof kinds, to guarantee data reliability and eliminate scams.

How the Distributed Ledger Works

1. **Transaction Proposal**: This is because when a participant begins a transaction it spreads across the network to alert all the nodes.
2. **Validation**: Nodes validate the transaction based on pre-determined set rules formulated under consensus.
3. **Block Formation**: This can be done for valid transactions which are collected in blocks and recorded on the ledger.
4. **Synchronization**: There is one more additional ledger which is also the collective one and contains all the data of the transaction, and then distributed to all nodes of the network.

Advantages of Distributed Ledgers

- **Enhanced Security**: This makes hacking and data breaches to become a rare incident because the network is decentralized.
- **Faster Transactions**: Reduce time that is taken by intermediaries in a traditional system.
- **Reduced Costs**: Saves cost on services which reduces unnecessary expenses and does not permit third-party charges.
- **Auditability**: It is also easier to understand and accurate in terms of proving a transaction history since it doesn't change over the same period positively.

Applications of Distributed Ledgers

- **Finance**: Facilitates the immediate clearing of the transaction without the involvement of a third party.
- **Healthcare**: Enables accessing of patient records by different authorized people.
- **Supply Chain**: It enables end-to-end tracking of parts and products.
- **Energy**: There is an opportunity for decentralized buying and selling of energy and tracking the energy being used.
- **Government**: Voting transparency and government records are made more possible.

3.2 Methodology Used in Blockchain Tourism

Blockchains in tourism services use different methodological approaches to distribute power as well as increase transparency and resort to automated service delivery [12]. These system methodologies operate together to convert present tourism operations into decentralized architectures which streamline processes and improve service delivery.

1. **Smart Contract Deployment**

Blockchains implement smart contracts as their core method for tourism applications to execute service provider-tourist agreements automatically without middlemen and with transparent set conditions [13]. The methodological framework forms a basic structure that enables firms in tourism to develop blockchain-based applications that enhance service delivery and value creation capabilities. Figure 2 below illustrates all the main methodological elements used in blockchain-powered tourism solutions. These research methods provide tourism stakeholders with a complete system to utilize blockchain technologies fully in industry transformation.

2. **Tokenization Framework**

The tokenization process digitalizes tourism assets alongside services which both enables inaccessible assets to enter the market as well as grants owners the power to set user rights functions.

3. **Decentralized Marketplace Architecture**

This methodology enables tourism stakeholders to become peers through disintermediated systems that uncover services directly and book transactions that let them exchange value directly.

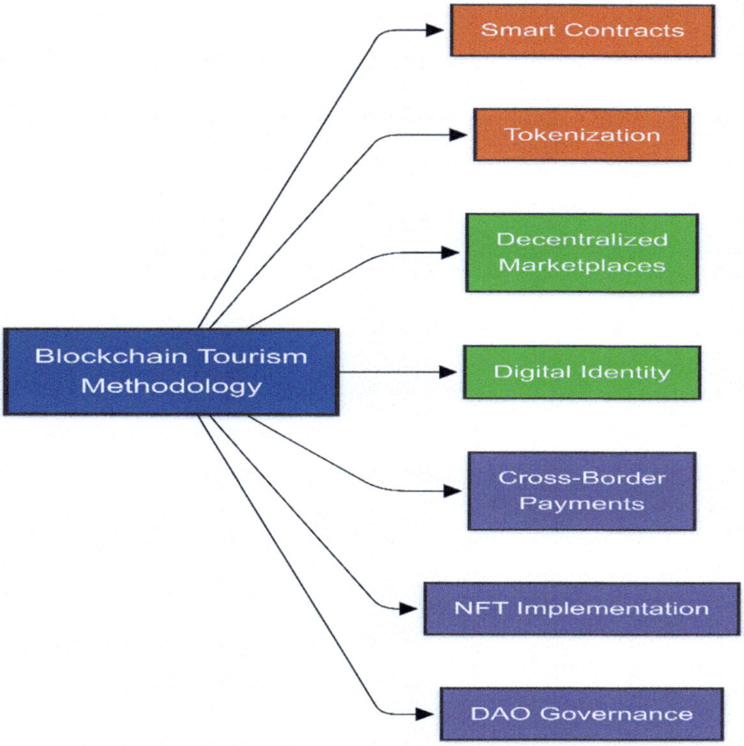

Fig. 2 Blockchain tourism methodology framework

4. **Blockchain-Based Loyalty Systems**

Loyalty programs built using blockchain create common reward programs that connect several service providers with distributed redemption and point distribution systems.

5. **Digital Identity Management Protocol**

Tourism identity methodologies implement self-sovereign identity principles for traveler data management and facilitate verifiable credential issuance and verification while preserving privacy.

6. **Supply Chain Traceability Methods**

The verification methods of sustainable tourism achieve two main objectives: they maintain unalterable sustainability practice records allow measuring carbon footprints and establish carbon offset capabilities [14].

7. **Cross-Border Payment Solutions**

Blockchain tourism payment structures using financial methodologies offer businesses low-cost cryptocurrency and stablecoin systems that provide a fast payment process and reduce global transaction expenses.

8. **NFT Implementation in Tourism Assets**

The NFT infrastructure enables the digital maintenance of collector items associated with actual physical destinations while offering authentication systems for admission proofs from certain tourism activities.

9. **DAO Governance Framework**

DAO methodologies allow tourism stakeholders to conduct decentralized resource management while enabling voting systems for project development proposals within community-based initiatives.

4 Advantages of Blockchain in Tourism

Figure 3 shows that Blockchain technology for tourism benefits the industry through improvements in security measures transparency efficiency and sustainability. Blockchain is now turning into an innovative solution that solves traditional issues such as in the tourism domain. Since tourism is majorly based on trust and secured transactions, and covers a large number of business processes and transactions, blockchain can be of great use in this sector. This chapter reviews the greatest benefits of using blockchain in the tourism industry with an emphasis on trust and transparency, integrated operations, and secure data.

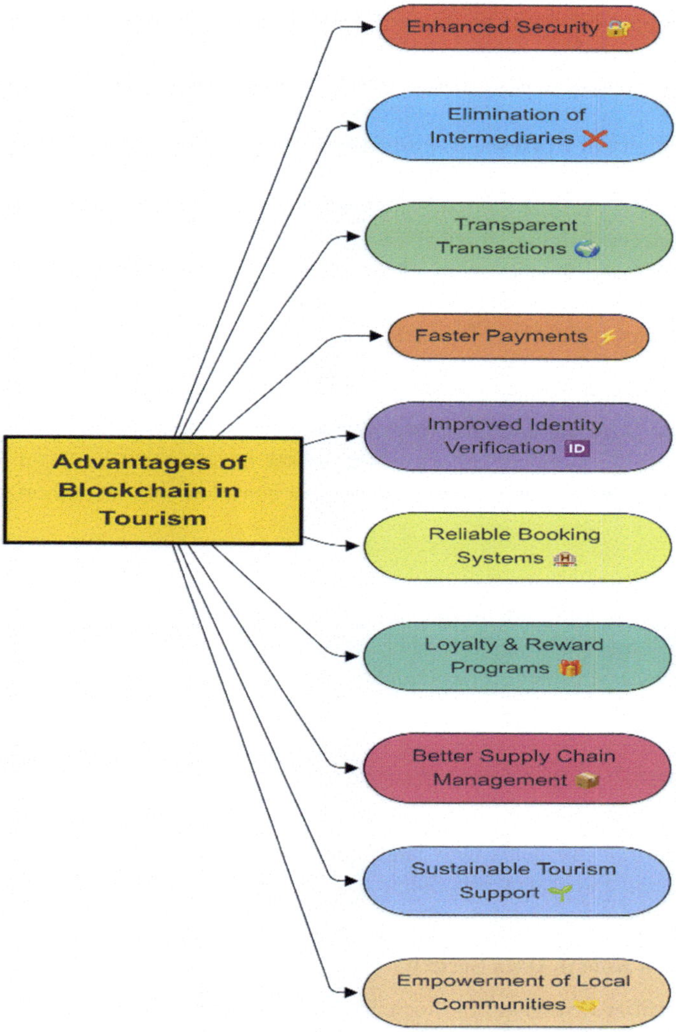

Fig. 3 Key advantages of blockchain in the tourism industry

4.1 Enhancing Trust and Transparency

Engagement and authenticity are a few of the keys that go into the creation of the tourism business. People need to know the services they are going to receive when they get a service provider and they also need to know how much the service costs or roughly how much the service will cost them this is why people searching for services always check the reviews about a service provider [15]. These requirements are met through the use of blockchain which is a decentralized and secure way of storing information.

Key Benefits

1. **Authenticity of Reviews**: One advantage of using blockchain technology is that it inextricably binds reviews and ratings to the identity of the consumer, reducing the amount of fake reviews in the market and increasing consumer trust.
2. **Transparent Pricing**: Blockchain-based smart contracts can offer analysis of expenses that a traveler will incur hence there can be no occurrence of charges that have not been disclosed to the travelers.
3. **Provenance of Services**: Blockchain can help to ensure the integrity of the travel package, hotel, and other services, so buyers' risk of being cheated on the supply side will be minimized.
4. **Reliable Booking Systems**: Inability to change bookings and reservations—the use of blockchain technology where once a transaction has been made or a reservation made that cannot be changed or canceled without authorization.

Use Cases

- **Hotel Bookings**: Instead of this, travelers can confirm the availability of rooms and the prices within the blockchain network to eliminate cases of disagreement.
- **Tour Packages**: Blockchain can be utilized by the service providers to guarantee that their services meet certain quality standards hence satisfying consumers.

4.2 Streamlining Transactions and Payments

The tourism industry deals with large quantities of consumer data, identity data, payment data, and travel schedules. Protecting this kind of data is crucial to maintaining confidence and meeting the laid down legal requirements [16].

Blockchain's Role in Data Security

1. **Decentralized Storage**: In blockchain technologies, data is dispersed across a network and not confined under one roof as part of a database that would face a breach.
2. **Encryption**: Data encryption using sophisticated means is the main safeguard to minimize access of data on the blockchain to unauthorized individuals.
3. **Auditability**: One important feature of blockchain technology is a shared ledger that is resistant to alteration, therefore unauthorized actions can be easily seen and stopped.
4. **Compliance**: Blockchain solutions can assist tourism firms in implementing global measures such as GDPR on the management of their data.

Examples of Enhanced Security

- **Secure Traveler Identification**: Blockchain can securely store digital identities and ensure rapid identity checks at airports, hotels, or checkpoints.

- **Fraud Prevention**: Corroborating accounts and identities on the blockchain assures tourists, and, in turn, the businesses they may patronize, that fraud and identity theft are mitigated.

Real-World Implementations

- **Airline Industry**: The application of blockchain in planes is for the identity of the client, ticketing, and security purposes.
- **Travel Agencies**: Blockchain service is applied in agencies to protect and ensure the safety of the customers' information, and transaction history.

5 Practical Applications in the Tourism Sector

Practical usage of the blockchain involves a broad spectrum of applications in the tourism sector including; inefficiency, trust deficit, and data privacy issues. By integrating blockchain as a distributed and unchangeable database, the industry is refocusing the central activities and giving people and entrepreneurs an effective and safe experience [17]. This chapter aims at highlighting major use cases of blockchain including identity management, a better booking and reservation service, and a better customer rewards program.

5.1 Simplifying Identity Verification

Identifications are very essential in the tourism sector screening people at airports, hotels, and other points of access to tourist destinations. The conventional ways of identity verification can be exhaustive and phase open to fraud. Blockchain offers the best solution to our society's problem by creating a trustworthy Decentralized Digital Identity. In this chapter, the following advantages of blockchain-based identification are given:

Benefits of Blockchain-Based Identity Verification

1. **Speed**: The digital ID on the blockchain leads to faster verification than having to wait in queues for travelers.
2. **Security**: Through the privacy, maintained in the Blockchain, identity details stay safeguarded from unauthorized use.
3. **Global Accessibility**: By using the blockchain, the identity system will be recognized universally rather than having complex ID cards.
4. **Data Ownership**: Travellers have a certain level of privacy control and can decide to what extent they are willing to reveal personal information.

Implementation Examples

- **Airports**: Identification is made easier through blockchain, ensuring efficiency in check-in processes, and security checks.
- **Hotels**: Mobile wallets can be used to check credentials and facilitate check-ins for guests making operations fully digital.
- **Border Control**: The implementation of blockchain enhances swift and secure cross-border identity checks amidst traveling. Reducing the need to verify identities thereby increases efficiency, eradicates bottlenecks, and makes travel more convenient.

5.2 Improving Booking and Reservation Systems

Blockchain technology is slowly transforming the current means of bookings and reservations in tourism. In traditional organizational booking structures, many levels of middlemen are used, which makes the whole process slow and costly [18]. Blockchain removes these intermediaries and offers a clear path to take.

Key Advantages

1. **Decentralization**: Peer-to-peer bookings are facilitated by blockchain hence eliminating the need for intermediaries.
2. **Transparency**: Smart contracts also ensure that terms and conditions are well spelled down and there is no ambiguity on how they will be implemented.
3. **Fraud Prevention**: Its high level of transparency also helps to guarantee that bookings cannot be changed fraudulently or even deleted from the blockchain.
4. **Cost Efficiency**: P2P transactions also eliminate most of the service charges that users pay to other platforms developed under conventional systems.

Use Cases

- **Flight Bookings**: One of how blockchain technology for air travel can be applied involves providing direct bookings to cut off from travel agencies.
- **Hotel Reservations**: Blockchain platforms have live availability and pricing, making it easy to avoid cases of double booking.
- **Tours and Activities**: Consequently, it easily becomes clear that blockchain technology can be effectively used by tour operators to manage reservations and offer transparent cancellation policies.

5.3 Enhancing Customer Loyalty Programs

The tourism industry heavily relies on loyalty programs which airlines hotels and travel agencies use to thank dedicated customers [19]. Traditional rewards systems deal with efficiency problems and cannot offer full program flexibility or cross-service usage. A blockchain solution solves these issues through a simple system that makes loyalty reward records easy for everyone to see.

Advantages of Blockchain-Based Loyalty Programs

1. **Interoperability**: People using Blockchain technology can change and accept loyalty points between multiple platform services.
2. **Transparency**: Customers will see their loyalty point balances updated instantly in real-time blockchain records that show honest transaction history.
3. **Security**: Blockchain security protects loyalty point systems from attacks by hackers and unauthorized users.
4. **Efficiency**: Our smart contract system handles reward distribution to save businesses money on manual processing steps.

Practical Applications

- **Airlines**: Travelers can convert and collect loyalty credits from several partner airlines into one blockchain system.
- **Hotels**: Guests can exchange their loyalty points with all hotels, restaurants, and service partners with complete ease.
- **Travel Agencies**: Companies use blockchain rewards to interface with external platforms to provide customers additional value.

Benefits to Businesses and Customers

- **For Businesses**: Our enhanced loyalty programs help us keep our best customers while decreasing our business expenses.
- **For Customers**: Loyalty programs offer better usability to consumers because their workings become clear while enabling customers greater plan flexibility.

6 Examples and Success Stories

Multiple success stories reveal how blockchain already transformed tourism as a technology. The actual use of blockchain systems in tourism demonstrates that this technology makes travel transactions more visible while saving time and improving overall user experiences [20].

6.1 Blockchain-Based Travel Platforms

Different blockchain-powered travel booking systems now exist to improve how people reserve and handle their journeys. These systems use blockchain decentralization to cut expenses while making participant interactions easier and more reliable [21].

Key Examples

1. **Winding Tree**: Through Winding Tree's decentralized platform travelers book services straight from airlines hotels and tour operators. When middlemen are taken out of the system customers experience both lower prices and spot all details.

Features

- Our system lets users book services directly with each other.
- Our system uses smart contracts to create secure financial transactions.
- Businesses and developers can use our public source system without restrictions.

Impact: Through its airline and hotel partnerships Winding Tree provides partners better prices to their customers directly.

2. **Travala**: Travala serves as a blockchain travel booking platform that lets its users make lodging and trip payments in cryptocurrency [22].

Features

- Customers may use both digital and standard payment channels at once.
- Travelers earn blockchain tokens through our customer loyalty program.
- Their system displays exact costs for users to see plus guest ratings and feedback.

Impact: With its wide partner roster and over 2.2 million property listings Travala has become the go-to platform for digital travelers who like to book their trips through tech solutions.

3. **LockTrip**: Through its blockchain-operated system LockTrip helps travelers book rooms and flights while paying no commissions [23].

Features

- Smart contracts power our transaction platform.
- Using our platform cuts business costs by 20 percent compared to standard systems.
- Our platform accepts traditional money and digital currency payments.

Impact: LockTrip's customer numbers keep growing as travelers select its service for saving while staying safe when booking accommodations.

Through blockchain services, these platforms show how they will replace established travel market patterns while offering better terms to both customers and travel service providers.

6.2 Real-Life Use Cases in Airlines and Hotels

Airlines and hoteliers work together to improve their business with blockchain technology. Through upgrades in security features and operational speed, blockchain helps hotels and airlines create better experiences for their customers while assuring industry stakeholders [24].

Airlines:

1. **Lufthansa**: Through blockchain technology, Lufthansa enhances customers' trust in-flight service and rewards program operations [25].
   - **Use Case**: Lufthansa teamed up with Winding Tree to let travelers book directly instead of going through online travel intermediaries.
   - **Impact**: The system lets airlines cut dependence on middlemen and create better interactions with customers.
2. **Singapore Airlines**: To enhance loyalty program rewards Singapore Airlines brought KrisPay their blockchain-enabled digital wallet to market [14].
   - **Use Case**: With KrisPay digital tokens members of the loyalty program can transform their accumulated miles into spending power at retail stores.
   - **Impact**: Our updated loyalty points system gives customers more ways to use their rewards for better interaction.
3. **Air France-KLM:** This airline group deploys blockchain technology to run its aircraft maintenance operations better [26].
   - **Use Case**: Blockchain follows aircraft parts through their life cycle to guarantee precise maintenance documentation.
   - **Impact**: Blockchain enables companies to run their maintenance operations better and lower expenses.

Hotels

1. **TUI Group**: To handle hotel supplies and rates TUI uses blockchain technology [27].
   - **Use Case**: The technology tells users both which rooms are available and their exact cost for each moment in real-time.
   - **Impact**: Since prices and room availability appear in real time the new system gives guests more trust and helps prevent booking conflicts.
2. **Marriott International**: Marriott investigates blockchain technology to link loyalty programs and make safer financial operations work better [28].
   - **Use Case**: Through blockchain Marriott connects and simplifies reward programs for guests at every Marriott property worldwide.
   - **Impact**: Customers receive better service while blockchain technology makes reward programs easier to use.
3. **CitizenM Hotels**: With blockchain technology, CitizenM provides guests with safer and quicker entry into their hotel rooms [29].
   - **Use Case**: Through smart contracts hotels let guests confirm bookings while getting access to available facilities.

- **Impact**: Implementing blockchain technology makes check-ins work faster while making operations run better.

Benefits of Blockchain for Airlines and Hotels

- **Efficiency**: Robotized systems decrease how often staff needs to do tasks and save money on operations.
- **Security**: Data protection and secure payments happen smoothly through blockchain.
- **Customer Experience**: Customers prefer doing business with providers who personalize their experience and make all details clear and easy to trust.

7 Future Opportunities in Blockchain for Tourism

The developing blockchain technology creates fresh ways for tourism businesses to improve their efficiency and create new industry solutions. The future of blockchain technology in tourism will reshape business operations for the industry [30]. This chapter shows how to develop intelligent travel networks on blockchain while supporting sustainable tourism.

7.1 Creating Smart Travel Ecosystems

A smart travel ecosystem produces automatic interactions between travelers and service providers using one decentralized system [31]. This new technology creates a secure distributed platform that meets our goal of bringing systems together into a single network.

Key Features of Smart Travel Ecosystems

1. **Unified Traveler Profiles**: Blockchain saves authorized stakeholders access to travelers' detailed profile data including personal preferences and past travel information plus their loyalty program points.
2. **Interoperability**: Blockchain technology connects service providers such as airlines hotels car rentals and tour operators to create an easy travel experience.
3. **Automated Processes**: Through smart contracts, blockchain technology performs every travel process including booking, check-ins, payments, and refunds without the need for humans to handle them [32].
4. **Personalized Experiences**: Service companies can study blockchain records to design travel services that match the specific tastes of their customers.

Potential Applications

- **Real-Time Itinerary Updates**: Blockchain technology updates travel plans actively so travelers view their updates instantly.

- **Efficient Transfers**: Smart contracts help travelers move smoothly between flights and accommodations because they manage service transfers automatically.
- **Connected Tourism Ecosystems**: Cities and regions use blockchain to build links between tourism stakeholders so they can work better together and help visitors reach their destination [33].

7.2 Promoting Sustainable Tourism Through Blockchain

The tourism industry works hard to achieve sustainability by reducing its impact on the environment and delivering greater value to society and communities [34]. Through sustainable practices, Blockchain helps simplify operations for all sectors by showing customers what they need to see.

Blockchain's Role in Sustainable Tourism

1. **Carbon Footprint Tracking**: Blockchain tracks travel carbon emissions and helps people choose greener options both as travelers and business owners.
2. **Incentivizing Green Practices**: Through smart contracts, our system gives rewards to people who use eco-friendly lodging or protect the environment.
3. **Transparent Supply Chains**: Blockchain technology shows all who supply tourism products that their claims about sustainability and ethics are true [35].
4. **Community Empowerment**: Decentralized online systems help connect visitors with local people to promote cultural exchanges and create new economic possibilities for all.

Use Cases in Sustainable Tourism

- **Eco-Friendly Accommodations**: Blockchain helps check that hotel environments comply with established eco-friendly criteria.
- **Carbon Offsetting Programs**: When travelers use blockchain technology they can easily join carbon offset programs that show their emission reduction details.
- **Sustainable Travel Tokens**: Our digital tokens reward users when they take green transportation and shop locally [36].

Benefits

- **For Travelers**: The system helps travelers learn about sustainable options while offering them ways to be involved in positive environmental action.
- **For Businesses**: Companies build better environmental credibility by following official regulations.
- **For Communities**: Everyone should take part equally in tourism money-making while keeping heritage sites safe.

8 Conclusion

Blockchain technology shows clear paths to help the tourism sector succeed while solving its existing problems and creating fresh markets. Blockchain technology develops smart travel systems while boosting efficiency and helping people cooperate better through new sustainable practices. Blockchain technology will develop into a new tourism system that links audiences more tightly while helping urban centers care for our planet. As tourism companies and travelers expand their use of blockchain technology they will build a better path for a more sustainable future.

References

1. Younis, H., Bwaliez, O.M., Al-Okaily, M., Tanveer, M.I.: Revolutionizing supply chain management: a critical meta-analysis of empowerment and constraint factors in blockchain technology adoption. BPMJ **30**, 1472–1500 (2024)
2. Gordon, M., Daniel, M., Ajiboye, A., et al.: A scoping review of artificial intelligence in medical education: BEME Guide No. 84. Med. Teach. **46**, 446–470 (2024)
3. Hosseiny, B., Mahdianpari, M., Hemati, M., Radman, A., Mohammadimanesh, F., Chanussot, J.: Beyond supervised learning in remote sensing: a systematic review of deep learning approaches. IEEE J. Sel. Top. Appl. Earth Observ. Remote Sens. **17**, 1035–1052 (2024)
4. Trevisan, C., Formentini, M.: Digital technologies for food loss and waste prevention and reduction in agri-food supply chains: a systematic literature review and research agenda. IEEE Trans. Eng. Manage. **71**, 12326–12345 (2024)
5. Raina, A., Rani, S., Bathla, G.: Concept, applications, challenges, and security threats of blockchain technology in the tourism industry. Adv. Hosp. Tourism, Serv. Indus. 43–58 (2024)
6. Du, Y., Mohamed, R., Choy Leong, Y.: Blockchain technology in tourism industry: a bibliometric analysis. J. Mgt. Mml. **14**, 1–23 (2024)
7. Radic, A.: Blockchain technology and internet of value in the hospitality and tourism industry. Adv. Hosp. Tourism Serv. Indus. 79–98 (2024)
8. Bathla, G., Raina, A., Rana, V.S., Mishra, S.: Enhancing guest experience in hospitality through blockchain technology. Adv. Hosp. Tourism Serv. Indus. 283–298 (2024)
9. Talukder, M.B., Haghi, A.K., Khan, M.R.: Empowering local communities through blockchain tourism initiatives. Adv. Hosp. Tourism, Serv. Indus. 175–196 (2024)
10. Jatav, S., Gupta, R., Singh, V.: The Unchained Sustainability. Sustainability Reporting and Blockchain Technology, pp. 151–160 (2024)
11. González-Mendes, S., García-Muiña, F., González-Sánchez, R.: Blockchain revolution in the tourism industry: a semi-systematic literature review. EJTR **37**, 3701 (2024)
12. Sarnacchiaro, P., Luongo, S., Sepe, F., Corte, V.D.: The role of blockchain technology in the tourism industry: analyzing the factors affecting its adoption. Qual. Quant. (2024). https://doi.org/10.1007/s11135-024-01836-7
13. Al-Romeedy, B.S.: Breaking chains, building futures. Adv. Hosp. Tourism Serv. Indus. 90–107 (2024)
14. Baliker, C., Baza, M., Alourani, A., Alshehri, A., Alshahrani, H., Choo, K.-K.R.: On the applications of blockchain in Fintech: advancements and opportunities. IEEE Trans. Eng. Manage. **71**, 6338–6355 (2024)
15. Deepak, G.P., Gill, N.S., Yahya, M., Gupta, P., Shukla, P.K., Shukla, P.K.: Exploring the potential of blockchain technology in an IoT-enabled environment: a review. IEEE Access **12**, 31197–31227 (2024)

16. Al-Nbhany, W.A.N.A., Zahary, A.T., Al-Shargabi, A.A.: Blockchain-IoT healthcare applications and trends: a review. IEEE Access **12**, 4178–4212 (2024)
17. Jiang, J., Wang, F., Shen, J., Kim, S., Kim, S.: A survey on large language models for code generation (2024). https://doi.org/10.48550/ARXIV.2406.00515
18. Hua, Y., Liu, F., Yang, K., et al.: Large language models in mental health care: a scoping review (2024). https://doi.org/10.48550/ARXIV.2401.02984
19. Sufi, F.: Generative pre-trained transformer (GPT) in research: a systematic review on data augmentation. Information **15**, 99 (2024)
20. Puri, V., Mondal, S., Das, S., Vrana, V.G.: Blockchain propels tourism industry—an attempt to explore topics and information in smart tourism management through text mining and machine learning. Informatics **10**, 9 (2023)
21. Prados-Castillo, J.F., Guaita Martínez, J.M., Zielińska, A., Gorgues Comas, D.: A review of blockchain technology adoption in the tourism industry from a sustainability perspective. JTAER **18**, 814–830 (2023)
22. Van Nguyen, T., Cong Pham, H., Nhat Nguyen, M., Zhou, L., Akbari, M.: Data-driven review of blockchain applications in supply chain management: key research themes and future directions. Int. J. Prod. Res. **61**, 8213–8235 (2023)
23. Han, Y., Fang, X.: Systematic review of adopting blockchain in supply chain management: bibliometric analysis and theme discussion. Int. J. Prod. Res. **62**, 991–1016 (2023)
24. Ellahi, R.M., Wood, L.C., Bekhit, A.E.-D.A.: Blockchain-based frameworks for food traceability: a systematic review. Foods **12**, 3026 (2023)
25. Tyagi, A.K., Dananjayan, S., Agarwal, D., Thariq Ahmed, H.F.: Blockchain—internet of things applications: opportunities and challenges for industry 4.0 and society 5.0. Sensors **23**, 947 (2023)
26. Xu, M., Guo, Y., Liu, C., Hu, Q., Yu, D., Xiong, Z., Niyato, D., Cheng, X.: Exploring blockchain technology through a modular lens: a survey. ACM Comput. Surv. **56**, 1–39 (2024)
27. Das, D., Banerjee, S., Chatterjee, P., Ghosh, U., Biswas, U.: Blockchain for intelligent transportation systems: applications, challenges, and opportunities. IEEE Internet Things J. **10**, 18961–18970 (2023)
28. Alamsyah, A., Syahrir, S.: The taxonomy of blockchain-based technology in the financial industry. F1000Res **12**, 457 (2023)
29. Rana, R.L., Adamashvili, N., Tricase, C.: The impact of blockchain technology adoption on tourism industry: a systematic literature review. Sustainability **14**, 7383 (2022)
30. Raluca-Florentina, T.: The utility of blockchain technology in the electronic commerce of tourism services: an exploratory study on Romanian consumers. Sustainability **14**, 943 (2022)
31. Sunny, F.A., Hajek, P., Munk, M., Abedin, M.Z., Satu, Md.S., Efat, Md.I.A., Islam, Md.J.: A systematic review of blockchain applications. IEEE Access **10**, 59155–59177 (2022)
32. Jabbar, R., Dhib, E., Said, A.B., Krichen, M., Fetais, N., Zaidan, E., Barkaoui, K.: Blockchain technology for intelligent transportation systems: a systematic literature review. IEEE Access **10**, 20995–21031 (2022)
33. Agrawal, K., Aggarwal, M., Tanwar, S., Sharma, G., Bokoro, P.N., Sharma, R.: An extensive blockchain based applications survey: tools, frameworks, opportunities, challenges and solutions. IEEE Access **10**, 116858–116906 (2022)
34. Krichen, M., Ammi, M., Mihoub, A., Almutiq, M.: Blockchain for modern applications: a survey. Sensors **22**, 5274 (2022)
35. Huo, R., Zeng, S., Wang, Z., Shang, J., Chen, W., Huang, T., Wang, S., Yu, F.R., Liu, Y.: A comprehensive survey on blockchain in industrial internet of things: motivations, research progresses, and future challenges. IEEE Commun. Surv. Tutorials **24**, 88–122 (2022)
36. Nour, M., Chaves-Avila, J.P., Sanchez-Miralles, A.: Review of blockchain potential applications in the electricity sector and challenges for large scale adoption. IEEE Access **10**, 47384–47418 (2022)

Revolutionizing the Hospitality Industry with Blockchain: Applications, Challenges, and Future Trends

Neetu Rani, Sandeep Singh Kang, Ishdeep Singla, and Narinder Yadav

Abstract Blockchain technology is revolutionizing the hospitality and accommodation industry by enhancing transparency, security, and operational efficiency. The proposed chapter explores blockchain usage in hotel management, travel bookings and guest services, focusing on its importance in data protection, decentralized booking platforms, and customer loyalty programs. By eliminating intermediaries, blockchain reduces the cost of transactions and enables direct interaction between companies and clients. The chapter includes case studies of TUI Group, Webjet, and Winding Tree with successful implementation of blockchain and takeaways. It also discusses significant hurdles like high deployment costs, regulatory issues and interoperability, and presents potential solutions like hybrid blockchain architectures and layer 2 scaling solutions. The proposed research illustrates the conceptual framework of blockchain in hospitality and accommodation in along its benefits, applications and challenges. Finally, the chapter discusses future developments such as blockchain-based digital identities and smart contracts, which have the potential to transform hospitality operations. Further the chapter explores in-depth examination of the increasing impact of blockchain in the hospitality sector with insightful information into future trends and industry development.

Keywords Blockchain · Hospitality · Smart contracts · Decentralization · Booking systems

N. Rani (✉) · S. S. Kang · I. Singla · N. Yadav
Department of Computer Science and Engineering, Chandigarh University, Mohali, India
e-mail: neetuvashist@gmail.com

S. S. Kang
e-mail: sandeepkang.cse@cumail.in

A. Kumar et al. (eds.), *Blockchain in the Tourism Industry: A New Era of Secure and Transparent Travel Solutions*, Information Systems Engineering and Management 57,
https://doi.org/10.1007/978-3-031-95341-5_3

1 Introduction

With the universal changes happening in accommodation and hospitality, new possibilities are availed from blockchain technology which increases transparency and security within the industry's systems. This creates self-sufficient escort booking systems which reduces reliance on third party services and their accompanying fees. In addition, blockchain guarantees the security of customers' information, streamlines payment processes using cryptocurrency, and lifts authentication of bogus reviews. The paradigm shift lies in the essence of guest service and hospitality business in general. Major transformation is likely to occur by trust on blockchain technology and more effective operational processes of value chains.

1.1 The Hospitality and Accommodation Industries

The hospitality and accommodation industry is made up of hotels, restaurants, travel agencies and event management businesses that are essential to the economy. This sector features a wide range of services for both leisure and business travellers, including cheap hotels, private lodgings, all the way to high-end hotels and resorts. One of the most important distinguishing features of this industry is that it is service driven, meaning that the satisfaction and comfort of the guest is the most important. As a result, this sector significantly contributes to employment and economic development in a given area because of its capacity to attract tourists and business people. Some of the names synonymous with this activity are Marriott International, Hilton, and Airbnb, which pioneered the booking and management of lodgings over the internet. Also, modern travel booking sites like Booking.com and Expedia have revolutionized the way people make reservations [1].

1.2 Current Challenges

The hospitality and accommodation industry have many obstacles which require critical approaches to solve. The subsequent section illustrates several issues facing by this industry:

- **Trust Issues**: Trust is particularly an important issue in P2P (peer-to-peer) platforms systems such as Airbnbs where guests and hosts do not share a prior relationship. Factors such as listing fraud, bogus reviews, and cancellation policies tend to lower customer trust. For instance, there are examples where tourists reached their destinations only to realize that the place of accommodation booked either did not exist or was immensely different from the description provided.
- **Increased Expense per Activity**: There are intermediaries in the form of OTAs who make the industry profit, which is why most customers are booking with them.

On the other hand, these companies have high costs, as they charge exorbitant service fees that reach up to 30% on individual bookings. For customers' expenses, this is beneficial as it reduces the expenses overall. For hotel operators, the service fees from the third-party platforms cuts their profit margins. In particular, small hotels find it difficult to market themselves against large enterprises because of those added expenses lowering their chances being seen and profited.

- **Dangers to Data Security**: Losing sensitive customers' data leaks a lot of risks, especially using personal identification and payment information. The cybersecurity attacks come with losing the customer's trust and paying a lot of money in penalties. Marriott International had a huge data breach back in 2018, impacting over 500 million guests where their sensitive information like passport numbers and credit card details were revealed. The above examples demonstrate the industry's poor stance against cyber security threats and how powerful data safeguarding systems are required. Such examples point to weakness and attack attributes from cyber threats the industry faces demonstrating the need for a far-reaching focus of data protection mechanisms. The range of underlying problems along with their effects on the industry and possible solutions for them are highlighted in Table 1.

1.3 The Introduction of New Technologies Concerning the Effectiveness of the Industry

AI and ML are changing the hospitality industry through increasing productivity, improving the guest experience, and managing other resources within the organization. This technology enables hotels and restaurants to handle large data sets, leading to improved decision-making and increased customer satisfaction. [2, 3]

- **AI and Machine Learning in the Hospitality Industry**

AI and ML are gradually transforming the hospitality sector because they optimize processes, improve the quality of services offered to guests, and manage company resources. Now businesses in this industry can make data-driven decisions, increase service efficiency, and use advanced technologies to tailor interactions with clients [4]. By allowing companies to operate chatbots and virtual assistants, AI has transformed customer support. These digital assistants can manage bookings, answer common queries, and provide suggestions, all of which makes it possible for companies to operate without depending on human staff. As a consequence, operational efficiency increases. For example, Hilton's robot concierge Connie uses AI to assist guests by making inquiries about hotel services and local sites for tourists, thereby enhancing customers' experiences through real-time assistance [5]. In a similar manner, Machine Learning models can enhance the efficiency of the processes of dynamic pricing and demand forecasting. Hotels apply ML models to raise or lower the occupancy rates through the use of seasonality, local activities, and events, and

Table 1 Challenges in the hospitality and accommodation industry

Challenges	Description	Impact on the industry	Examples	Potential solutions
Trust issues	Lack of confidence between service providers and customers due to fraudulent listings, fake reviews, hidden fees or unexpected cancellations	– Reduced customer loyalty and satisfaction – Negative brand reputation – Hesitancy in booking through platforms	A guest books a hotel via an online platform but arrives to find the property is misrepresented	– Verified guest reviews using blockchain – Transparent property listings – Secure identity verification
High transaction costs	Heavy reliance on intermediaries like online travel agencies (OTAs) that charge significant commissions, increasing overall costs for both customers and providers	– Increased costs for customers – Lower profit margins for hotels – Small hotels face challenges competing	Hotels paying up to 30% commission fees to platforms like Booking.com or Expedia	– Direct booking platforms using blockchain – Use of smart contracts to reduce commissions
Data security concerns	Risks associated with storing and managing sensitive customer data, including personal information, payment details, and booking history	– Potential data breaches leading to financial and reputational loss – Compliance issues with data protection laws	Marriott's 2018 data breach exposed data of 500 million guests, including passport and credit card details	– Decentralized data storage on blockchain – Strong encryption protocols – Regular cybersecurity audits

competitive prices. This enables hotels to not only maximize revenue but also maintain sowing competitiveness. In addition, With the use of ML, the predictive maintenance problems are also solved by analyzing the data from the hotel's facilities enabling foresight of potential failures. AI and ML are also crucial for increasing engagement and guest satisfaction. Hotels can set preferred environments such as ideal room temperature settings, dining styles, and even recommend places to visit. Such steps will certainly improve guest satisfaction rates and encourage their return to the hotel as well as provide better reviews. In addition, AI aids environmental conservation efforts in the hospitality industry. AI-powered smart energy management systems eliminate heating, ventilation, and air conditioning systems over usage by synchronizing the settings with occupancy levels in real-time. This positively impacts energy consumption and overall operational costs. Effective supply chain

ML powered forecasting, decrease food being thrown away and control stock level more effectively, which results in increased savings and better overall management of resources and expenses.

- **Internet of things (IoT) in Hospitality**

The hospitality industry has seen a significant enhancement in the guest experience with the introduction of smart hotel rooms. IoT Incorporation in hotels also significantly improves operational effectiveness [6–8]. For example Marriott Hotels as it is one of the go-to hotels for many travelers and businesses. They have implemented IoT-enabled rooms that sensor between conditions partitions if the guest prefers a warm or cool environment when they walk in. This gives the guest further comfort alongside them being welcomed by smart temperature control systems, light panels and AI voice accessible assistants.

Greater Parsons has stated that guests having the ability to uncover entertainment systems, lights, air conditioning and curtains with mobile applications or by voice has increased guest satisfaction. Even without the guest setting foot in the room, the IoT integration allows the guest to take control of all these features before stepping into the room. Marriott International is an excellent case study of IoT profitability. Native sensors alert the personnel when appliances need routine tasks as the head of engineering bets that they will fail and save money by preventing unforeseen breaks of the system and guaranteeing that the customer will have a perfect experience. Both use cases show how IoT can optimize operational efficiency and guest satisfaction for both the management and the guests thus increasing productivity and comfort on both ends. Specialized applications of the Internet of Things allow admins and managers to track inventories as they happen and put an end to under providing a basic need or over buying supplies. For example, IoT sensors can check stock levels of kitchen supplies and ensure that all required ingredients are put to use well. Virtual Reality (VR) and Augmented Reality (AR) merged these boundaries and augmented the customer experience enhancement potential dramatically by making it more engaging and interactive.

- **Virtual Tours**

With VR, potential guests can go on virtual tours of the hotel including all the amenities like the lounge, conference rooms and even the peripherals, all before making a reservation. Establishments having an immersive experience are able to build customer trust as it enables a realistic sense of the property. For example, 'Try Before You Fly' VR campaign from Thomas Cook which saw a large inflow of customers booking tickets for destinations after they were virtually showcased.

- **Enhanced Marketing and Customer Engagement**

AR applications help hotels design fully interactive experiences for the guests. Customers can scan the hotel cover brochures or even the room elements to gain access to digital content like the in-room menus or service guides, all with the help of AR applications. This not only fosters customer engagement but increases innovation in advertisement strategies in hotels.

- **Event Planning and Conferences**

VR is now being extensively deployed in the execution of large-scale events and conferences. In our case, planners are able to do virtual walkthroughs of venues, plan room setups, even practice speeches and presentations through videos. Less travel saves time and financial resources due to fewer physical visits required.

- **Post-Booking Experience**

Customers can enjoy a preview of spa treatment, hotel dining, and entertainment facilities, which will help personalize guest interaction. Ultimately, satisfaction will be gained and encouraged the selling of services. Table 2 offers an update on the new trends within the hospitality industry which show what new technologies are doing to personalization, efficiency, sustainability, and satisfaction in 2025.

2 Literature Survey

The blockchain technology playing a significant role in hospitality and accommodation. The subsequent section illustrates the earlier research adopted by researchers in this field: Authors in [1] explain Blockchain technology can improve operational efficiency, boost revenue, and enhance security and privacy in the hospitality sector.—Blockchain technology can improve operational efficiency in the hospitality sector. Authors in [9] employed a grounded theory strategy with theoretical sampling and three stages of coding and constant comparison processes to investigate users' views on a blockchain-based peer-to-peer accommodation system and determine the primary drivers of user acceptance. Authors in [10] discussed a number of challenges in adopting blockchain technology within the supply chain of the hotel industry such as readiness for technology and standardization of system among hotel stakeholders.—The research introduced a model to redesign hotel industry supply chain business processes into more transparent and accountable business processes using blockchain technology. The research came up with an app to allow all players in the hotel industry supply chain to input data and track the status of every process in the supply chain.

Authors in [11] discussed Blockchain technology has been of great promise to reconfigure global networks and supply chains amid the COVID-19 pandemic.—Blockchain expenditure in Europe is projected to expand at a fast rate, reflecting the growing adoption of the technology.—Blockchain technology has come as a solution to address the issues faced by companies, particularly in the hospitality and tourism sector, amid the COVID-19 pandemic.

Authors in [12] assesses the feasibility of implementing blockchain technology (BCT) in hospitality operations management.—The research presents a framework for the future application of BCT in the hospitality sector.—The research proposes the organizational, institutional, and technological hurdles that must be overcome for the increased adoption of BCT in the hospitality sector.

Table 2 Role of emerging technologies

Emerging technology	Role in Hospitality industry	Impact on Industry practices	Key trends (2025)
Contactless technology	Enables seamless check-in/check-out, digital room keys, and mobile payments	Enhances guest convenience, reduces wait times, and ensures safety	Widespread adoption of NFC-enabled services and QR-based menus
Internet of things (IoT)	Offers smart room controls (lighting, temperature, entertainment) personalized for guests	Improves energy efficiency, boosts guest comfort, and allows predictive maintenance	Growth of fully connected hotel ecosystems
AI-powered chatbots	Provides 24/7 customer service, booking assistance, and personalized recommendations	Reduces operational costs while maintaining high guest satisfaction	Multilingual bots and hyper-personalized guest interactions
Virtual and augmented reality (VR/AR)	Facilitates virtual tours of hotels, event spaces, and destinations pre-booking	Enhances marketing strategies and improves customer decision-making	Rise in AR-based interactive guest experiences
Robotics and automation	Automates room service delivery, cleaning, and concierge services	Increases operational efficiency and ensures contactless guest services	Deployment of humanoid robots in luxury hospitality settings
Blockchain for secure transactions	Ensures transparent and secure bookings and loyalty programs using decentralized systems	Prevents fraud, streamlines transactions, and builds customer trust	Integration of blockchain-based loyalty reward systems
Sustainable technologies	Implements smart energy management systems, water-saving fixtures, and eco-friendly operations	Meets eco-conscious traveler demands and reduces operational costs	Investment in carbon–neutral and sustainable hotel operations
Personalization through big data	Analyzes guest preferences to provide tailored experiences, offers, and amenities	Increases guest retention and satisfaction through bespoke services	Real-time data analytics driving adaptive service offerings

(continued)

Table 2 (continued)

Emerging technology	Role in Hospitality industry	Impact on Industry practices	Key trends (2025)
5G connectivity	Supports high-speed internet access for streaming, virtual conferences, and IoT integration	Enhances guest experience with uninterrupted connectivity	Integration of 5G-powered smart room technologies
Cybersecurity enhancements	Protects guest data, digital transactions, and connected systems from cyber threats	Ensures compliance with privacy regulations and builds customer trust	Adoption of advanced encryption and real-time threat detection systems

Table 3 Applications of blockchain technology in hospitality

Application	Description	Benefits
Secure payments	Enables cryptocurrency transactions and transparent payment processes	Reduces fraud and transaction fees
Identity verification	Utilizes blockchain for quick and secure guest identity checks	Speeds up check-in processes
Loyalty programs	Implements token-based reward systems redeemable across platforms	Enhances customer engagement
Supply chain management	Tracks the origin and movement of goods and services	Increases transparency and trust

Table 4 Solutions and problems of blockchain adoption in hospitality industry

Challenge	Description	Proposed Solution
High initial investment and costs	Expensive infrastructure, software, and skilled labor	Adoption of Blockchain-as-a-Service (BaaS) platforms
Scalability issues and transaction speed	Slow processing speed on public blockchains	Layer 2 scaling solutions (lightning network, rollups)
Regulatory and compliance challenges	Complex legal frameworks across different regions	Hybrid blockchain models to ensure compliance
Resistance to technological change	Hospitality industry slow in adopting blockchain-based systems	Training programs and phased adoption strategies
Interoperability concerns	Different blockchain platforms lack seamless integration	Use of interoperability protocols (polkadot, cosmos)

Authors in [13] discuss BeSHMT a secure and transparent payment process for travelers on an immutable blockchain ledger. The BeSHMT framework incorporates a recommendation system based on machine learning that gives ratings and suggestions for tourist places based on experiences of previous visitors. The BeSHMT framework has better Return on Investment (ROI) compared to traditional frameworks in the tourism industry.

2.1 The Use of Big Data and Analytics in Hospitality

These new tools are revolutionizing business activities in the hospitality industries because of the extensive information they provide on consumer behavior, trends, and productivity. The possibility of gathering and processing large amounts of data facilitates decision making in the hotels enhancing productivity and customer satisfaction [14, 15].

- **Customer Personalization**: Customer analytics is a great leap in the customization of services offered by hotels through the integration of guest preferences, bookings, and their feedback. It involves suggesting activities and dining options as well as recommending room upgrades fitting the guest profile. A case in point, a loyal customer suddenly becomes one when the room automatically adjusts to the preferred temperature and preferred drinks are generously made available.
- **Strategies on Dynamic Pricing**: With the help of big data, hotels are able to modify their pricing strategies depending on the real-time demand and supply analysis, local events, competitor pricing, and seasonality. As an example, Airbnb leverages big data to advise hosts on correct pricing for their properties in real-time. This enables hosts to maximize their earnings while ensuring that their pricing is competitive enough to garner more bookings.
- **Market Analysis**: With data analytics, one is able to monitor current industry developments and customer satisfaction, which helps in making strategic marketing and service delivery expansion decisions. Hotels can employ predictive analytics on data to estimate future demand, which assists the hotels in planning for high seasons or superior services for important events.
- **Efficiency of Operations and Management of Resources**: When operational data is analyzed systematically, the hotels are able to optimize staff allocation, stock level control, and energy expenditure. For example, hotels are able to forecast when they will be most busy, and adjust staff schedules, ensuring that customer service is not compromised while preventing the aforementioned overspending. In addition, studying food and beverage consumption patterns helps to streamline purchasing and reduce food wastage.
- **Customer Feedback Analysis**: The hospitality sector utilizes the internet for text analysis and sentiment evaluation of reviews and comments. This allows for assessing gaps and customizing services to enhance guest satisfaction. Evaluating

sentiments and feedback over time can also be useful in developing marketing campaigns that capitalize on the prevailing positive sentiment trends [16, 17].

2.2 Effects of Blockchain Technology

It is a new technology that consists of an encrypted and shared digital record system, decentralized from any central authority. The said technology is used to automatically track and record transactions over several computers, providing security and transparency throughout the process. Every transaction that takes place is recorded inside a 'block' and joins some 'blocks' before it in a chain. This sector is emerging with a great opportunity in hospitality and accommodation due to the core features of blockchain which include immutability, decentralization, transparency and security [18, 19].

- **Booking platforms without intermediaries**: Along with the rest of the industry, blockchain can also remove middlemen like OTAs from the equation as service providers can transact directly with their customers. A great example of this is Winding Tree, who enables hotels and airlines to market their services directly without the intervention of any third party, thus reducing costs to the suppliers while allowing them to gain higher profits and offering low prices to the customers.
- **More Secured and Private**: Using blockchain technology, hotels could minimize the risks around identity theft by securing personally sensitive data during the registration. Using encryption technology makes customer information resistant to cyber threats. Data breaches would also be less likely to occur as blockchain records are immutable and require a consensus to alter them.
- **Reduced Payments Overhead**: Blockchain offers the prospects of instantaneous transactions along with the use of cryptocurrency, which come with lower transaction fees compared to traditional banking. More and more younger customers who are tech savvy prefer to make purchases using alternative methods like cryptocurrency and Travala.com is one of many companies offering hotel bookings via a tokenized economy.
- **Loyalty Programs Reimagined**: Blockchain enables more practical and usable universal loyalty programs instead of the traditional loyalty programs that are complex. Multi products and multi services would unite and make points redeemable across a wider range of brands. For instance, Singapore Airlines recently introduced a loyalty wallet wherein users can trade loyalty points for digital tokens to be used within their partner merchant's ecosystem.
- **Review Authenticity and Transparency**: One major issue in hospitality is fake reviews and blockchain can resolve that by making sure only verified guests are allowed to leave reviews by linking every stay to an immutable ledger. In turn, this builds the authenticity of the review and helps potential customers make informed decisions.
- **Initiatives to promote sustainable tourism**: The use of blockchain technology can enhance and verify sustainability claims from hospitality providers. For

instance, travelers with eco-concern can check a hotel's sustainability rating using blockchain which examines energy consumption, waste disposal, and supply chain practices. Fortunately, blockchain technology promises to solve the trust, cost, and data security challenges that are currently bedeviling the hospitality and accommodation industry. The hospitality sector can benefit from blockchain in the way services are delivered and experienced through the decentralization of transactions, data protection, payment simplification, and better engagement. Due to ongoing digital transformation within the industry, it appears likely that blockchain will help balance efficiency, transparency, and customer satisfaction in hospitality [20, 21].

2.3 The Basics of Blockchain Technology

Hospitality and tourism are being transformed by blockchain with the adoption of secured systems that offer transparency, decentralization, and greater operational and customer service efficiency. Following is a broad discussion of the types, fundamentals, and smart contracts powered systems of blockchain, as well as its use cases in hospital industry. Blockchain refers to a new way of securely storing information over the internet using digital ledgers tracked and maintained by a network of computers. It allows users to record information independently while ensuring the information can never be destroyed or altered. This technology stores and keeps track of information from various industries out there like healthcare, finance, supply chain, and many more which give them a transparent way of verification and access [22–24].

1. **Decentralization**: Decentralization allows Blockchain Technology to remove the need for a mediator (government, banks or centralized servers) unlike its counterparts. Rather than storing data in a centralized system, blockchain relies on multiple nodes (computers) that hold decentralized copies of the entire blockchain. So, in case one node does fail, the blockchain remains intact. Imagine a server-based banking system that depends heavily on a central server for storing records. During outages or Cybersecurity breaches, users are unable to access their data and the bank's ability to manipulate it poses further risks. A stark contrast to that is Bitcoin which is decentralized and enables users to keep access to the same records across the networks. As a result, no singular overlord is allowed to dictate or meddle in the transactions, guaranteeing security and transparency from a financial point of view.
2. **Immutability**: Immutability lower the vertical of changing or manipulating data endlessly and arbitrarily once a particular action, or in this case, a transaction is performed. The blockchain's structure prevents any changes to data after a transaction has been made. All blocks in the blockchain consist of a certain number of transactions per block, each of which is securely connected to the previous block by a password called a hash. For example, A change in a prior block would need all the hash of the following blocks to be changed which is near impossible. Now visualize a land registry where landownership records are stored

on a blockchain. A historical paper-based or a digital system opens opportunities for dishonest personalities to tamper with the landownership records because they can drive personal profit from it. On the other hand, if these records are on a blockchain, as users on the network have identical copies of the ledger, rewriting previously set records will be revealed and therefore blocked. This helps ensure property rights are safe and verifiable.

3. **Transparency**: Blockchain fosters a higher level of openness on the ecosystem as all transaction records are available to network participants. All stakeholders can trust that every transaction has been made since it is stored on a verifiable public ledger. The use of blockchain addresses allows for personal identities to be concealed, thus providing privacy while still being transparent. One of the most significant examples of blockchain transparency in the real world is in supply chain management. Think of a powerful retailer like Walmart which sources food items from different suppliers. Thanks to blockchain, Walmart is able to state that all the supply chain activities such as production, packaging, transportation, and delivery timestamps are recorded in a tamper-proof ledger.Consumers and regulators have the means to trace a product's journey from the farm to the shelf calmly knowing that there has not been any fraud, contamination or questionable activities relating to it being carried out.
4. **Security**: The security on a blockchain is guaranteed with the use of cryptography, consensus mechanisms, and other methods of verification that are distributed. Transactions utilize public-key cryptography, and the verification of a transaction is executed through public consensus methods including Proof of Work and Proof of Stake that prevent fraudulent actions. In the health care industry, the danger of a data leak exists with patient records that are centrally stored. Thanks to the blockchain technology, hospitals can securely store and share patient information with other physicians without the danger of unauthorized access to sensitive information. This way, the patient's identity can be safeguarded while making it possible for doctors, research institutions, or insurance companies to access the facts without compromising security. The use of blockchain technology transforms how information is stored, shared, and safeguarded. With its core principles of decentralization, immutability, transparency, and security, blockchain innovation has been empowering trustless and vandal proof transaction systems across various industries. From finance and healthcare to supply chain activities and even governance, blockchain has, and keeps, enhancing effectivity, minimizing frauds and increasing accountability on ground use of the systems.

3 Methodology

The Fig. 1 illustrate the conceptual framework for blockchain in hospitality and accommodation. In accommodations and hospitality blockchain enables secure booking, decentralized identification verification transparent supply chains,

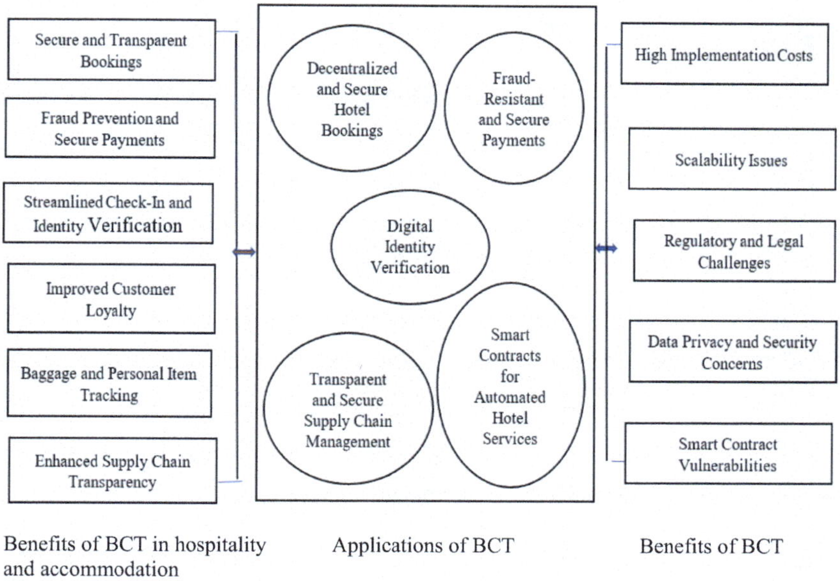

Fig. 1 Conceptual framework for blockchain in hospitality and accommodation

and streamlined loyalty programs, improving customer trust and operational efficiency.

The proposed framework provides a smooth and fraud-proof ecosystem rendering blockchain a revolutionizing influence in the hospitality and travel sector. Blockchain technology is transforming the hospitality and accommodation sector by enhancing booking security, simplifying check-ins, automating processes, and improving customer loyalty schemes. With its potential to minimize fraud, remove unnecessary intermediaries, and offer transparent transactions, blockchain is transforming hotel management and guest experiences, making it a precious innovation in the industry. Blockchain technology is revolutionizing the accommodation and hospitality industry through enhanced security, cost reduction, transparency, and simplified operations. From safe reservations and self-check-ins to transparent loyalty schemes and P2P rentals, BCT is streamlining hotel operations and making them customer-oriented. With more and more adoption, blockchain will become an integral part of the future of accommodation and hospitality services. Although it has the potential to raise transparency, security, and efficiency in accommodation and hospitality, blockchain technology is subject to several challenges such as high costs, scalability, compliance with regulations, and reluctance to change. Its overcoming would necessitate collaboration among industries, investment in scalable blockchain, and clarity of regulation for easy adoption and sustained success in the accommodation and hospitality industry.

4 Key Applications of Blockchain in Hospitality

The framework highlights major areas where blockchain technology can be applied in the hospitality industry:

4.1 Secure and Transparent Booking Systems

- Blockchain-based platforms enable direct and fraud-proof bookings.
- Smart contracts validate reservations and handle automated refunds.
- Eliminates intermediaries like OTAs, reducing commission costs.

4.2 Customer Identity and Data Security

- Decentralized digital identity management ensures seamless and secure check-ins.
- Personal data is encrypted and controlled by the customer, reducing data breaches.
- Biometrics and blockchain-based authentication improve security.

4.3 Payment and Transaction Management

- Cryptocurrency and stablecoins facilitate instant cross-border transactions.
- Smart contracts automate payments and refunds, ensuring faster settlements.
- Reduces payment fraud and chargeback disputes.

4.4 Loyalty Programs and Reward Systems

- Blockchain-based loyalty points are transferable and redeemable across different hotels.
- Tokenized reward programs ensure transparency and reduce fraud.
- Guests can accumulate rewards seamlessly across multiple service providers.

4.5 Supply Chain Management for Hospitality Services

- Blockchain tracks the procurement of food, beverages, and amenities to ensure authenticity.
- Transparency in the supply chain helps hotels manage ethical sourcing and sustainability.

- Reduces counterfeit goods and ensures high-quality services.

4.6 *Fraud Prevention and Trust Building*

- Immutable blockchain records prevent fake bookings, identity theft, and fraud.
- Customer reviews stored on blockchain are verifiable and tamper-proof.
- Enhances brand reputation by ensuring transparency and reliability

5 Different Blockchain Frameworks

Blockchains and blockchain technologies have different varieties based on organizational needs. Based on the degree of access, control, and governance required, blockchains can be categorized into three types. These are Public Blockchains, Private Blockchains, and Consortium Blockchains [25].

5.1 *Public Blockchains*

This type of blockchain is totally decentralized. It can be used by any willing party without any limitations. Users can participate by reading, writing, and validating transactions, thus ensuring transparency and security. Public blockchains function through a form of consensus known as Proof of Work (PoW) or Proof of Stake (PoS) to validate transactions and to attach new blocks to the existing structure. Bitcoin and Ethereum are as an example of public blockchain illustrated in subsequence:

- **Bitcoin**: Bitcoin blockchain is an open, permission-less type of anetwork, which makes it possible for anyone to join, open a wallet, and commence transactions without any need for prior approval. Since no one authority operates it, every transaction is authenticated by miners who perform the work of validating and securing transactions using mathematical calculations and cryptographic methods while fraud prevention mechanisms established ensures such transactions are secure.
- **Ethereum**: Ethereum is a blockchain known to the public, along with Bitcoin, and it supports smart contracts which allow developers to build DApps. For instance, DeFi applications running on Ethereum permit users to lend, borrow, or trade assets without any intermediary financial institutions. A public blockchain is very useful in relation to cryptocurrency, decentralized applications, and voting. For instance, some countries are considering the use of blockchain technology for casting votes in elections to enhance fairness and ensure that the votes cast are tamper-proof and traceable.

5.2 Private Blockchain

A private blockchain is a network with restricted access and permissions and is accessible only to select participants. Contrasting public blockchains, an organization manages access to the data, thus reading and writing of the information is highly controlled [26–28]. This form is ideal for particular businesses that need privacy on data and operations. Hyperledger Fabric and Corda private blockchain illustrated in subsequence section:

- **Hyperledger Fabric**: Hyperledger Fabric is an example of private blockchains, which are built by the Linux Foundation. Businesses leverage Hyperledger Fabric to create custom private blockchain networks. For example, IBM Food Trust utilizes Hyperledger Fabric to monitor the journey of food items from the farm to the consumer's table. This stops fraudulent activities, minimizes waste, and increases customer confidence.
- **Corda**: Tailored for financial institutions Corda permits transactions between specific participants, ensuring privacy and confidentiality. Payment processing and transaction verification by banks and insurance companies is executed through Corda, enabling competitors' sensitive information to remain undisclosed. A private blockchain can control transaction confidentiality in supply chain management, healthcare, and financial institutions. For instance, a hospital chain can administer patient records on a private blockchain and allow vetted doctors and healthcare providers to view sensitive medical information.

5.3 Consortium Blockchain

A consortium blockchain is a stem hybrid that is managed by a limited number of organizations, having shared control over the network. Unlike the fully public blockchain, access is limited yet trust is dispensed with. This enables transparency and collaboration while increasing security, as well as eliminating the pitfalls of full decentralization.

6 The Importance of Blockchain Technology in Hospitality

Within the hospitality sector, processes are made more efficient through the adoption of blockchain technology:

6.1 Payments

Fraud is minimized when business is conducted through the use of cryptocurrency. For instance, Blackbird enables users to pay for their meals using an app that utilizes blockchain technology. Payments made by diners are settled through blockchain. Identity Verification: The instant verification of guest identities through a Blockchain-based digital ID could be a reality. Without the necessity of extensive security checks, guests can be processed in a matter of seconds.

6.2 Loyalty Programs

Guests can redeem loyalty points with tokenized rewards for various services easily. Points in blockchain-based loyalty programs are accessible and manageable within the program.

6.3 Supply Chain Management

The transparency attained during the sourcing and procurement processes increases the trust in the quality of service. Blockchain technology is capable of confirming the supply chain's provenance and path of various products, guarantees their quality, authenticity, and ethical sourcing [20, 26] (Table 3).

By utilizing blockchain technology, the hospitality industry has been enabled to offer more secure, effective, and customer-oriented services. This has improved confidence in the industry.

7 Enhancing Guest Experience Using Blockchain Technology

There has been a transformational change in the hospitality and tourism sector and the application of blockchain is greatly helpful in improving the overall experience. Personalized services from companies, ease of travel, safe business dealings—all these are made possible by the security, transparency and efficiency Blockchain offers. The following applications provide further explanation on how blockchain technology is changing guest experience in modern hospitality [1].

7.1 Blockchain Technology for Personalized Guest Services

Guests personal attention is one factor that has to be taken into account and which is highly likely to improve their satisfaction levels. Hotels and travel agencies can deploy blockchain-based technology to create and store secure and de-centralized guest profiles containing preferences, booking history, and requests. Unlike traditional customer relationship management (CRM) systems which are siloed and can easily be hacked, file guests database is protected through encryption, made practically tamper proof, and is only made available with the user's consent. For example, a luxury hotel chain might want to adopt blockchain for maintaining a guest preference database across all their locations. Any guest with pillow preferences, room decor requirements, and meal choices would have such information saved on a blockchain profile. The user who books accommodation with any hotel in the chain is sure to receive the correct service without having to complete any forms. Accessing their profile is all the staff needs to do. A network business traveler can now have a seamless stay experience in the hotels within the network. For example, the room configuration, entertainment device settings, and preferred foods are already set based on the user's highly sensitive preferences saved in the blockchain ledger [29–30].

7.2 Enhanced Travel Experience

Traveling often involves multiple reservations and arrangements for flights, accommodation, rental cars, activities, and so on. This fragmentation may result in confusion and possible mistakes. With Blockchain technology, several service providers can be connected and share important data, greatly reducing the risk of double bookings, missing reservations, or delays. Imagine a travel ecosystem on the blockchain where purchased algorithms are interconnected like a traveler's options which include the itinerary for a flight, a hotel reservation, and a car rental reservation. In this example, the system can warn the hotel of a delay in check in delay as well as adjusting the car collection time when a required a flight is late. An airline that has a blockchain booking system can, without human intervention, change the status of a passenger who most likely missed a flight because of bad weather and then update the associated hotel with the adjusted time for check in while also rebooking car rentals without the need of the user calling in many times. Feedback and Reviews in Real Time: Stopping Fraudulent Reviews with Legit Blockchain Integrated Feedback.

Blocked reviews and testimonials are common for the hospitality industry, giving guests false information while hurting business' reputations. Blockchain improves the system, allowing only verified guests to leave reviews, preventing dishonest reviews. Every transaction such as checking into a hotel, going to a restaurant are all put as transactions on the blockchain so that only people who have actually used the services are able to leave the reviews. Imagine a traveler visits a resort, and when asked to provide input from their visit, the system verifies whether their stay details are genuine by checking if the hotel has transactions on the blockchain. This enhances the reliability process since only actual guests can provide feedback, making the review process more secure. Online review platforms are prone to manipulation due to fake or paid reviews, making them unreliable. The incorporation of restrictions that allows denial of modification or deletion bolsters authenticity for blockchain-based reviews [31–34].

8 Challenges and Barriers to Adoption

Implementing blockchain in hospitality has its own difficulties. Aside from the higher levels of transparency, security, and reduction of intermediaries, these challenges need to be settled before the full benefits can be obtained.

8.1 High Initial Investment and Infrastructure Costs

The most critical issue in blockchain technology adoption is also one of its most costly components. Setting up a blockchain based protocol requires an investment in hardware and software as well as skilled personnel. Blockchains, in comparison to classical databases, are more resource intensive, and more importantly, they need a special set of equipment for efficient transaction processing. Challenges regarding the expansion of a blockchain and concerns regarding the pace of transaction processing seem to go hand in hand. Blockchains, such as the public chains for Ethereum and Bitcoin, take longer to process transactions than the traditional centralized databases because of certain consensus mechanisms like Proof of Work (PoW) or Proof of Stake (PoS) based systems.

8.2 Varied Compliance and Regulatory Issues Per Region

Responsibility for enforcing blockchain technology varies from region to region, which makes it potentially problematic for international hospitality chains. Added obligations from GDPR in Europe or CCPA in the US along with data sovereignty laws in India and China make it increasingly complicated.

8.3 Resistance from Established Traditions Within the Hospitality Industry Towards Technological Evolution

Many systems within the hospitality industry, such as hotel ownership, travel agencies, and other staff, are proficient with old systems making it harder to automated processes, hence making it difficult for the industry to adopt change at a faster pace.

8.4 Between Different Blockchain Technologies

Cross border or cross chain communication becomes extremely difficult between platforms due to differences in the systems' consensus, smart contract languages, and data structure systems. The current existence of Ethereum, Hyperledger Fabric and Binance Smart Chain makes it difficult for interoperability to function (Table 4).

9 Conclusion

The hospitality industry stands to benefit from blockchain technology, but its application presents some obstacles. Solving problems related to scaling, regulatory compliance, costs, and interfacing are critical to enabling adoption on a larger scale. Companies that take the initiative to modernize their training programs, embrace hybrid business models, and utilize interoperable systems will have the most to gain from the disruptive nature of the blockchain. The proposed work discuss case studies of TUI Group, Webjet, and Winding Tree with successful implementation of blockchain and takeaways. It also discusses significant hurdles like high deployment costs, regulatory issues and interoperability, and presents potential solutions like hybrid blockchain architectures and layer 2 scaling solutions. The proposed research illustrates the conceptual framework of blockchain in hospitality and accommodation in along its benefits, applications and challenges. Finally, the chapter discusses future developments such as blockchain-based digital identities and smart contracts, which have the potential to transform hospitality operations.

References

1. Khanna, A., Sah, A., Choudhury, T., Maheshwari, P.: Blockchain technology for hospitality industry. In: Information Systems: 17th European, Mediterranean, and Middle Eastern Conference, EMCIS 2020, Dubai, United Arab Emirates, Proceedings, vol. 17 (pp. 99–112). Springer International Publishing (2020)

2. Kishore, D., Raju, S.S., Rana, V.: Exploring the impact of emerging technologies on the hospitality industry. In: Marketing Technology-Infused Hospitality: Upskilling Frontline Employees for Competitiveness (pp. 337–364). IGI Global Scientific Publishing (2025)
3. Zeqiri, A.: From traditional to digital: the evolution of business models in hospitality through platforms. Platforms **2**(4), 221–233 (2024)
4. Mitra, A., Jain, V.: AI and IoT applications in the hospitality industry: a study of their potential in transforming the Indian online travel market by 2030. In: Technology and Luxury Hospitality (pp. 151–164). Routledge (2024)
5. Fedosova, K., Katunian, A.: The use of artificial intelligence in the restaurant business. Turystyka i Rozwój Regionalny **22**, 43–54 (2024)
6. Avi, M.A.R., Sardar, S.: Application of innovative technologies in the tourism and hospitality industry of Bangladesh: challenges and suggestions. In: Technology Application in the Tourism and Hospitality Industry of Bangladesh, pp. 369–379 (2021)
7. Ma, H.: Development of a smart tourism service system based on the Internet of Things and machine learning. J. Supercomput. **80**(5), 6725–6745 (2024)
8. Elkhwesky, Z., Elkhwesky, E.F.Y.: A systematic and critical review of internet of things in contemporary hospitality: a roadmap and avenues for future research. Int. J. Contemp. Hosp. Manag. **35**(2), 533–562 (2023)
9. Nur Muharam, I., Tussyadiah, I.P., Kimbu, A.N.: A theoretical model of user acceptance of blockchain-based peer-to-peer accommodation. Curr. Issue Tour. **27**(7), 1008–1025 (2024)
10. Inayatulloh: A blockchain model in the hospitality industry to increase traceability and transparency of transactions. In: 2023 International Seminar on Application for Technology of Information and Communication (iSemantic) (pp. 58–62). IEEE (2023)
11. Kapil, S., Kapil, K.N.: Blockchain in hospitality and tourism industry way forward. Int. J. Bus. Econ. (IJBE) **6**(2), 289–298 (2021)
12. Filimonau, V., Naumova, E.: The blockchain technology and the scope of its application in hospitality operations. Int. J. Hosp. Manag. **87**, 102383 (2020)
13. Hole, Y., Hole, S., Cavaliere, L.P.L., Nair, B., Hasyim, M., Bapat, H.B.: Blockchain usages in hospitality management. In: 2023 3rd International Conference on Advance Computing and Innovative Technologies in Engineering (ICACITE) (pp. 2798–2801). IEEE (2023)
14. Shah Alam, S., Ahsan, M.N., Masukujjaman, M., Kokash, H.A., Ahmed, S.: Adoption of big data analytics and artificial intelligence among hospitality and tourism companies: perceive performance perspective. J. Qual. Assur. Hosp. Tourism 1–35 (2024)
15. Tran, T.H.H., Nguyen, T.M.N., Nguyen, T.T.C., Luu, T.T.D., Nguyen, M.T.: Big data capabilities, strategic flexibility and competitive performance: evidence from Vietnam tourism and hospitality. Asian J. Bus. Res. **14**(3) (2024)
16. Shu, B., Ruan, X., Li, P.: Enhancing hotel efficiency and environmental health with biosensors and big data analytics. Mol. Cell. Biomech. **21**(1), 336–336 (2024)
17. Theodorakopoulos, L., Theodoropoulou, A.: Leveraging big data analytics for understanding consumer behavior in digital marketing: a systematic review. Human Behav. Emerg. Technol. **2024**(1), 3641502 (2024)
18. Rana, V.S.: Enhancing guest experience and data security in hospitality. In: Exploring the World with Blockchain Through Cryptotravel (p. 161) (2024)
19. Dadkhah, M., Rahimnia, F., Filimonau, V.: Evaluating the opportunities, challenges and risks of applying the blockchain technology in tourism: a Delphi study approach. J. Hosp. Tour. Technol. **13**(5), 922–954 (2022)
20. Radic, A.: Blockchain technology and internet of value in the hospitality and tourism industry. In: Exploring the World with Blockchain Through Cryptotravel (pp. 79–89) (2024)
21. Satpute, P.R., Kanitkar, K., Patil, M., Deokar, A.: Use of blockchain technology for management of guest room keys
22. Ghorbian, M., Ghobaei-Arani, M.: Key concepts and principles of blockchain technology (2025). arXiv preprint arXiv:2501.11707
23. Atlam, H.F., Alenezi, A., Alassafi, M.O., Wills, G.: Blockchain with internet of things: benefits, challenges, and future directions. Int. J. Intell. Syst. Appl. **10**(6), 40–48 (2018)

24. Durneva, P., Cousins, K., Chen, M.: The current state of research, challenges, and future research directions of blockchain technology in patient care: systematic review. J. Med. Internet Res. **22**(7), e18619 (2020)
25. Zeba, S., Suman, P., Tyagi, K.: Types of blockchain. In: Distributed Computing to Blockchain (pp. 55–68). Academic Press (2023)
26. Rana, V.S.: Enhancing guest experience in hospitality through blockchain. In: Exploring the World With Blockchain Through Cryptotravel, p. 283 (2024)
27. Dogru, T., Mody, M., Leonardi, C.: Blockchain Technology & Its Implications for the Hospitality Industry (p. 207). Boston University (2018)
28. Dhiraj, A., Kumar, S., Rani, D., Grima, S., Sood, K.: Blockchain payment services in the hospitality sector: the mediating role of data security on utilisation efficiency of the customer. Data **8**(8), 123 (2023)
29. Barkel, C., Kurgun, H., Groen, B.: Blockchain in the Hospitality and Tourism Industry (p. 4). University of South Florida (USF) M3 Publishing, 17(9781732127593)
30. Balasubramanian, K., Kunasekaran, P., Konar, R., Sakkthivel, A.M.: Integration of augmented reality (AR) and virtual reality (VR) as marketing communications channels in the hospitality and tourism service sector. In: Marketing Communications and Brand Development in Emerging Markets Volume II: Insights for a Changing World, pp. 55–79. Springer International Publishing, Cham (2022)
31. Zuhroh, N.F., Suwitho, S., Mildawati, T., Pahlawan, M.R., Baihaqy, A.: Augmented and virtual reality in tourism marketing: a systematic literature review of adoption factors and visitor engagement. In: International Conference of Business and Social Sciences (pp. 887–904) (2024)
32. Solihin, A., Azizah, S.N., Sutomo, S.: Implementation of virtual reality (VR) and augmented reality (AR) in marketing and service development to increase customers in the hospitality sector. Int. J. Manage. Sci. Inform. Technol. **4**(2), 450–458 (2024)
33. Habib, G., Sharma, S., Ibrahim, S., Ahmad, I., Qureshi, S., Ishfaq, M.: Blockchain technology: benefits, challenges, applications, and integration of blockchain technology with cloud computing. Future Internet **14**(11), 341 (2022)
34. Prados-Castillo, J.F., Torrecilla-García, J.A., Andraz, G., Guaita Martínez, J.M.: Blockchain in peer-to-peer platforms: enhancing sustainability and customer experience in tourism. Sustainability **15**(22), 15968 (2023)

Blockchain in Hospitality and Accommodation

Seema Kharod and Jasneet Chawla

Abstract Hospitality and Accommodation Sector is a million-dollar industry that significantly impacts the global GDP. It is a diverse and dynamic sector that includes hotels, lodges, resorts, serviced apartments and other lodging options. The industry is experiencing several internal and external challenges that hampers the good customer experience. This can lead to decline in business revenues, and even deteriorating a brand impression on the customers. As the digital transformation elevates, many businesses have adopted the technological based solutions for addressing the challenges. Blockchain is a special type of digital system that distributes its functions across a decentralized and distributed network. It works without a central authority controlling it and spreads the information across many computers in a network. It is secured and transparent digital system that records transactions in a way that cannot be changed or tampered with. All transactions are permanently recorded and easily verified due to decentralized network. In the hospitality and accommodation industry, it can help by ensuring security, increasing transparency, and making processes more effective to the customers. The technology further provides full transparency, enabling the stakeholders to build trust because every transaction remains accessible for hotels, travel agencies and customers. The data is secured using advance cryptographic techniques. It makes difficult for hackers to access important information and ensures the correctness of the records. The objectives of the chapter are detailed exploration of the existing challenges in the hospitality and accommodation industry along with a thorough investigation of the potential solutions with the aid of blockchain technology. Along with the numerous industrial case studies, the chapter describes the legal and regulatory challenges faced by the technology for its adoption. Further, it concludes with the future direction highlighting the potential aspects in the domain which can be transformed.

Keywords Accommodation · Blockchain · Cryptocurrency · Decentralization · Hospitality · Security · SmartContracts · Transparency

---

S. Kharod · J. Chawla (✉)
Department of Computer Science and Engineering, Chandigarh University, Mohali, India
e-mail: erjasneetchawla@gmail.com

A. Kumar et al. (eds.), *Blockchain in the Tourism Industry: A New Era of Secure and Transparent Travel Solutions*, Information Systems Engineering and Management 57,
https://doi.org/10.1007/978-3-031-95341-5_4

1 Introduction

Hospitality and lodging industry consists hotels, resorts, and vacation rentals as part of it. The Integration of technological revolution has drastically driven the industry, blockchain technology plays an important role in the economy by supporting businesses, driving tourism, and creating jobs to peoples. To work well, it needs good customer relationship and make smart use of data [1]. The technology offers a decentralized and digitized ledger leading to a more secure, transparent, as well as an effective manner to handle transactions and manage with data, thus shaping it as an absolute way for enriching the guest experience, along with enhancing the operational efficiency, and inculcating the faith within the industry.

In today's era, the hospitality industry is growing fast, playing an important and crucial role in the economy, but still, it faces varied challenges. Data breaches, payment frauds, and cyberattacks pose significant threats to the customer privacy and business operations. Along with this, guests and service providers often struggle with trust issues, particularly in context with peer-to-peer lodging related services. Due to the increase in the rate of cyber threats, the centralized databases are more exposed to cyber risks during large and voluminous transactions [2]. Traditional payment processing systems charge high transactional fees, affecting the profitability for service providers. Fake reservations and unauthorized sales are some other challenges experienced by the industry.

The Blockchain is secured and transparent digital way that records the transactions in a manner that cannot be changed or tampered. All the transactions are permanently recorded and easily verified due to the decentralized nature of the network. The decentralization reduces the reliance on intermediaries, cutting the costs and increasing the efficiency whereas the immutability ensures the authenticity of transactions. The smart contracts act as the self-executing contracts used to automate the transactions based on certain predefined conditions. The feature of cryptographic security, reduces the risk of fraud and hacking. In the hospitality and accommodation industry, blockchain can assist by ensuring security, increasing transparency, and making processes more effective to the customers and thus, optimizing the business processes [3].

The Blockchain is a special type of digital system that distributes its functions across a decentralized and distributed network. It works without a central authority controlling it and spreads the information across many computers in a network, making it more secure and transparent. All the nodes in the network share a single ledger which makes blockchain operate as a distributed system. For validation of all blockchain transactions, consensus algorithms are employed. This ensure that all copies of the ledger remain same during the entire process [4]. Further, blockchain technology uses some encryption techniques such as Proof of work (PoW) and Proof of Stake (PoS) to provide security to data. Every transaction receives security protection by using advanced security techniques such as Cryptography security protection. Combination of Digital signatures and hashing is used to make sure that no one can alter the data without being noticed [5].

The Blockchain technology has some relevant potential that enables its practical use throughout the hospitality and accommodation industry such as for safe and secure transactions. By using blockchain, hotels maintain confidentiality in customer data and protect it against data breaches. Since the technology works without any central authority, this decreases the chances of fraudulent activities. The recorded data related to the customer bookings, payments and other relevant details remain unchanged, ensuring accuracy and reliability. The technology further facilitates secure and seamless financial transactions, offering prevention from risks for illegal transfers and reducing the transactional fee as well. It is more beneficial for the cross-border transactions. The blockchain based payment methods like cryptocurrency can lower the transactional costs associated with traditional payment methods to enhance the security. With the help of blockchain technology, the booking as well as payment process is streamlined, as the intermediate agencies are eliminated, the transactional cost is lowered.

Blockchain also enhances loyalty programs for better customer engagement and satisfaction. The technology has simplified the reward earning and point redeeming process through digitally distributed token system. This allows users to acquire rewards securely and transferable across multiple platforms to enhance customer experience. Also, in supply chain management, the technology supports in the management of resources and inventory well to ensure high standards of service. Blockchain technology provide full transparency, enabling the stakeholders to imbibe trust as every transaction remains accessible for hotels, travel agencies and customers. Blockchain technology protected data by using some cryptographic techniques. It makes difficult for hackers to access important information and ensures that records stay accurate.

The hospitality industry handles operations smoothly and securely by using blockchain infrastructure. This makes operations faster and reduce costs by eliminating dependency on third parties like booking agencies or payment processor as much. Furthermore, the technology offers Tamper-Proof Records by preventing against fraud and ensuring accuracy. Once information like bookings, transaction records or guest details are recorded to the blockchain, it cannot be altered and becomes immutable [6]. With the use of strong encryption techniques, the technology provides more protection to data. It also ensures that customers personal data and financial transactions handled securely and safely. Thus, blockchain can assist the hospitality and accommodation industry to solve major challenges. As it can provide faster and safer check-ins to customers by using Digital IDs, visitors can utilize their digital identities instead of presenting the original IDs. In this way the unnecessary paperwork can be avoided, in turn improving visitor security and speeding up the check-in process. The digital tokens can enhance the client interaction and can aid with payments. Guests can use digital token to make payment, earn reward and redeem it in different platforms [7].

Numerous companies like TUI and LockChain have adopted blockchain solutions in creative ways. The former, a huge tourism company of the world, has used the blockchain for management of its inventory allowing real-time price adjustments and for optimization of demand assessment, thus enhancing the operational efficiency. Whereas the latter, has provided a decentralized booking system with zero commission fee, this allows the hotels to manage their places and properties directly, eliminating the middle-men. This demonstrates the potential of the technology in the transformation and shaping of the hospitality industry.

In spite of the advantages of the technology, the adoption of blockchain in hospitality and accommodation industry faces numerous challenges. The first and the foremost is the regulatory concerns for ensuring the compliance of complicated laws enforcement and integration of blockchain. Lack of standardization makes it challenging to incorporate the technology in to the existing systems as it poses some significant technical setbacks.

With the evolution of the industry and increasing market competition, the adoption of technology has become crucial for meeting the better customer experience. The integration of blockchain technology other advancing technologies like the Artificial Intelligence and the metaverse can revolutionize the customer experience. Artificial Intelligence can enhance the customized used experience by analyzing the customer preferences and understanding the previous experiences for prediction of preferences. With metaverse, the guests can be offered impressive as well as secure physical and digital experience to guests.

The blockchain technology provides an ever-changing landscape of opportunities for the hospitality and accommodation industry to revolutionize the operational strategies, evolve the trust among customers and deliver more customized experiences. The technology has the potential to aid and enhance the brand reputation and meet customer expectations. By handling the challenges, the industry can foster and use the cutting-edge technology to take the lead. The blockchain technology holds the potential to overcome as the possible challenges experienced by the hospitality and accommodation industry. The integration and adoption of the technology will significantly shape the user experiences. Figure 1 describes the architecture of blockchain technology in numerous application areas.

In this chapter, the applications, advantages and the future scope of the blockchain technology in hospitality and accommodation industry, identifying the technology's impact on the operations, enhancement of customer experience and innovation will be discussed. With the investigation of case studies, elaborating the challenges and focusing on the future directions, thorough insights of the blockchain technology in shaping the industry will be provided.

2 Related Work

The blockchain technology has been gaining significant attention in hospitality and lodging industry in recent times due to its huge potential to cope up with the challenges faced by the industry. The related work section has explored the

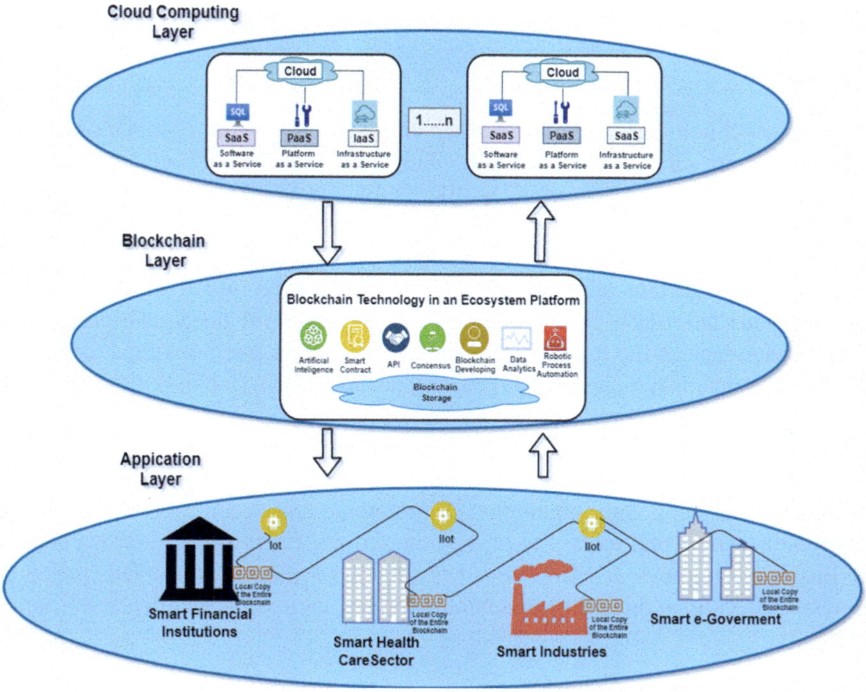

Fig. 1 Describes the architecture of blockchain technology in numerous application areas [8]

different aspects of the industry that have been significantly influenced by the technology. Figure 2 presents the numerous applications of the blockchain technology in Hospitality and Accommodation Industry.

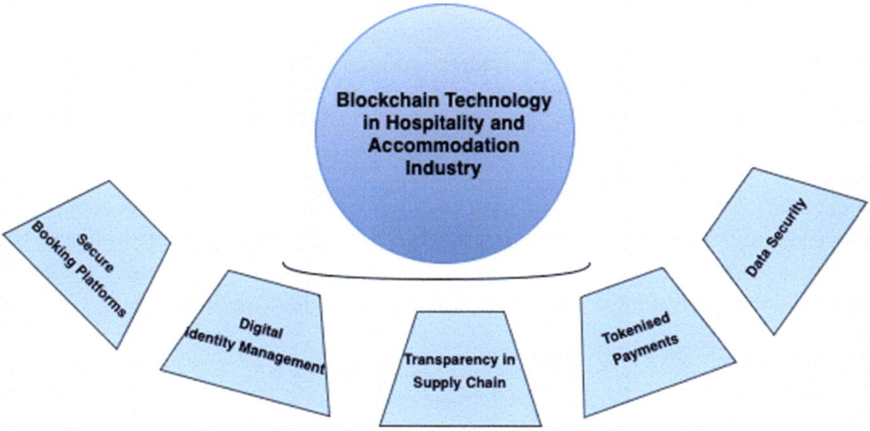

Fig. 2 Applications of blockchain technology in hospitality and accommodation

2.1 Blockchain for Secure and Transparent Booking Platforms

Blockchain technology makes hotel and accommodation booking system more transparent, and more efficient. In Present booking scenarios, for an online booking of a hotel, a third-party service providers like Online Travel Agencies (OTA) charge high fees and sometimes problems like fake reservations, hidden costs, or delay in booking confirmations may occur. By using blockchain technology the middlemen can be skipped and thus allowing guests to do reservation directly with them [9]. This eliminates the risks of fraud, and provide more security to financial transactions. Blockchain enhances the transparency and recorded data like bookings, payments and customer details remains unchanged, stay accurate and reliable. With blockchain the customers will experience:

1. Secure Transactions

The blockchain technology maintains an immutable and decentralized ledger that enhances the safety and reliability of the transactions. Every transaction is chained and linked to the previous transaction, thus making it tamper-proof. This prevents the system from unauthorized fraudulent activities, increasing security at the same time.

2. Decentralized Booking System

With blockchain enabled booking systems, hotel guests can book hotel rooms directly from the hotel through online mode without using third party such as online travel agencies. This has several key points:

- Lower Costs: Hotels save money by avoiding high commission fees charged by OTAs, making bookings more affordable.
- More Transparency: All transactions are recorded on a blockchain, making pricing and policies clear without hidden charges.
- Prevents Fraud: Since blockchain records cannot be changed or deleted, it helps prevent fake bookings, payment scams, or identity theft.
- Builds Trust: Both guests and hotels can check and verify bookings anytime, ensuring fair and secure transactions [10].
- Transparent Record Keeping

Blockchain technology permits the parties involved in the transactions access the history openly. This builds the environment of mutual trust among the users by ensuring accurate and up-to-date information. In the conventional booking systems, the middlemen parties have all the transaction details that can be misused.

4. Smart Contract-Based Reservations

Customers can use blockchain-based smart contracts as automated agreements for hotel bookings, financial transactions and booking cancellations. The automated system increase efficiency, no delay in bookings and making the booking experience

Table 1 Depicting the benefits of various features associated with Secure and Transparent booking platforms

Feature	Benefits
Secure transactions	Reduced frauds and enhanced security
Decentralized booking system	Reduced costs and increased user experience
Transparent record keeping	Trust strengthening and error reduction
Smart contract-based reservations	Enhanced efficiency and decreased disagreements

easy for both customers and hotels. Smart contracts make payment process more secure, clear and stress free without using middleman [4].

Blockchain technology improves hotel booking by using smart contracts as automated agreements for hotel bookings that run on their own. The hotel booking system improves with automation through the implementation of smart contracts. Smart contracts store booking data and maintains all offered conditions like payment dates and check-in requirements. This automation eliminates the third-party involvement during operations. The system performs operations timely and smoothly because it carries out payments when the needed requirements fulfilled [11]. The process of refunds and cancellations becomes streamlined because smart contracts handle these functions automatically. Smart contracts give Tamper-Proof Records by preventing fraud and ensures accuracy. Once blockchain recorded information like bookings, transaction records or guest details to the system, it cannot be changed. The benefits of Smart Contracts in Booking are discussed as.

- Fewer Mistakes: Automating reservations reduces errors like overbooking or incorrect charges.
- Better Customer Experience: Guests get a fast, smooth booking process with clear policies.
- Trust and Reliability: Hotels and guests both have a secure record of the transaction, preventing misunderstandings [12].

With blockchain, all booking information are secure and cannot be changed and eliminate the risk of fraud or scams. Table 1 elaborates.

2.2 *Blockchain for Digital Identity Management for Guests*

The hospitality sector needs a secure process to manage guest identities as well as improving operational efficiency. Blockchain provides tamper proof records and provide better services to customers, reduce fraud risks and provide information security. Guests can use digital identities for their verification and no need to carry original documents everywhere. It uses three key concepts such as reusable digital identities, reduced identity theft risks, and blockchain-based KYC (Know Your Customer) to provide more security [13]. The key services include.

1. Smart check-in systems

To manage and store authenticated credentials guests can use digital identity wallets. By using digital wallet guests don't need to repeatedly show ID during various hospitality interactions [14]. This may help in:

- Minimized Repetitive KYC Checks: Once a guest's identity is verified, they can reuse their credentials without undergoing fresh verification at every hotel.
- Streamlined Check-In Process: Instant identity authentication allows for faster and more convenient guest onboarding [15].
- Secure Identity Access

Blockchain use KYC (Know Your Customer) policies to improve security and guest convenience in the hospitality industry. Guest don't need to verify their identity repeatedly, they only need to do it once, as their data is securely stored in the blockchain system. The key advantages include:

- Fraud Prevention: Cryptographically secured identities mitigate risks related to identity theft and data breaches.
- Self-Sovereign Identity (SSI): Guests retain full control over their personal data, deciding what information to share with service providers.
- Tamper-Proof Identity Records: Once stored on the blockchain, identity data cannot be altered or forged, ensuring data integrity [16].
- Smart check-in systems

A blockchain-based KYC process enables hotels and hospitality providers to process check-ins in fast and secure way. It increases cost efficiency and better guest convenience by preventing repeated KYC procedures. This aids in:

- Operational Efficiency: Hotels can significantly cut down on time and resources spent on repeated KYC verifications.
- Expedited Guest Onboarding: With pre-verified credentials, guests can check in quickly without presenting physical documents each time.
- Global Usability: A guest's digital identity can be used across international hospitality chains, simplifying travel and stay formalities [17].
- Personalized services

The guest preferences, likings and disliking are securely stored, this can help the business to excel by providing personalized and improved user experience.

Table 2 depicts the advantages of various features aligned with Digital Identity Management for Guests.

Table 2 Depicting the benefits of various features associated with digital identity management for guests

Feature	Benefits
Digital identifier	Digitally signed credentials,
Secure identity access	Enhanced security and convenience
Smart check-in system	Reduced identity verification time, improved efficiency
Personalized services	Improved guest experience

2.3 *Blockchain for Transparency in Supply Chain and Ethical Sourcing*

The hospitality industry can achieve better accountability and secure all financial transactions by using blockchain. The features of blockchain like maintenance of a decentralized ledger, tamper-proof record keeping and with the integration of smart contract, the blockchain has enabled transparency in supply chain management. Since the technology provides the traceability, companies can track the movement of items from source till the destination, thus ensuring that all products come from ethical sources. The transparency ensures the sourcing process follows the ethics and helps to builds trust with customers along with reducing the risk of unethical practices. Additionally, the blockchain records cannot mutated, hotel industries can depend on clear and verifiable transaction histories, reducing fraud risks and improving accountability. Through blockchain a business can verify product identities which decreases the entry of fraudulent products into supply networks. The hospitality industry can build an ethical resolution with reliable security through blockchain technology in its supply chain.

1. Transparency in sourcing

The blockchain technology helps to generate records of every transaction that allows the customers as well as investors to keep a track of the products from the point of production to the final delivery. This further helps to enforce the norms for better and informed decisions.

2. Traceability

With the record keeping of every transaction of the network, the blockchain technology provides the ability to track the movement of materials from node to node till the final destination has arrived. This permits the agencies to cohere to the ethics like fair labor practice and environmental sustainability.

3. Responsible sourcing network

The blockchain powered responsible sourcing network, traces and records the product from its generation till outsources, keeping all the minute details secured at every level. This helps to build a transparent relationship of trust and security among the users.

Table 3 Depicting the benefits of various features associated with supply chain management

Feature	Benefits
Transparency in sourcing	Better and informed decision making
Traceability	Ensures ethical sourcing
Responsible sourcing network	Increase reputation and user trust

Table 3 elaborates the advantages of using blockchain for management of supply chain aspects of hospitality and accommodation industry.

2.4 Blockchain for Tokenized Payments and Loyalty Initiatives

Blockchain technology transforms hospitality payments while customer reward management through its efficiency improvements and cost reduction and flexibility enhancement. Cryptocurrency payments help travel services together with hotels to execute swift transactions at reduced costs by removing the restrictions that typical banking system imposes. Under tokenized loyalty programs hotel guests receive blockchain-based incentives which they can earn and use across different hospitality network affiliates. The interoperability in reward systems allows different service providers to join together easily so customers can use loyalty points across multiple platforms for improved engagement. A blockchain adoption by the hospitality sector enables better payments and rewards functionality with reduced costs and improved efficiency [18]. For an instance, the big market brands like Marriott International as well as Shangri-La Resorts have been using the blockchain platform for enhancing the customer experience by enabling the guests to earn and redeem points.

1. Tokenized Payment

The tokenization of payments with the use of blockchain technology has changed the landscape of traditional approaches. In this, the traditional currencies are converted into the digital tokens which can be used for transactions.

Due to this payment methods, the security has increased, thus leading to decreased chances of fraudulent activities and globally accessible payment solutions that permit seamless transborder transactions transparently.

2. Loyalty Tokenized

The loyalty programs offered by the businesses for improved customer satisfaction and better customer experience are tokenized as the digital assets using the blockchain technology. By digitization of the tokens, the redemption has been seamlessly possible through different platforms, thus making it inter operatable. The smart contracts allow the automation of rewards on the basis of predefined conditions.

Table 4 elaborates the benefits of incorporating the blockchain technology to payment and loyalty management system.

Table 4 Depicting the benefits of using blockchain in payment and loyalty management

Feature	Benefits
Tokenized payment	Decreased cost and fraud rate
Loyalty tokenized	Enhanced customer experience

2.5 Blockchain for Data Security and Compliance with Regulations

Blockchain technology enables managers to tackle regulatory problems as well as privacy concerns in the hospitality industry. Hotel businesses resolve their business challenges through blockchain implementations that enhance security together with regulatory compliance and transparent operations. Decentralized data storage ensures safe protection of guest information. A distributed network stores information which minimizes data vulnerability in centralized systems. Customers benefit from enhanced GDPR compliance through blockchain because it gives them control over their personal data. People need to control their personal data through blockchain because it enables them to manage access requests and consent rights for their information. The blockchain platform allows for regulatory compliance through its capability to support essential changes. Businesses maintain clear visible records which operate as tamper-evident logs to guarantee regulatory compliance [19].

1. Encryption and Access Control

The blockchain technology implements numerous encryption algorithms for the enforcement of data protection. Smart contracts are used in order to implement the access control mechanism to monitor the authorized parties for view and edit rights.

2. GDPR Compliance

General Data Protection Regulation Compliance helps the hotels to ensure the basic principles of compliance is followed and the data is processed in accordance with this. Smart contracts are programmed to implement the regulatory automatically, thus reducing the risk of non-compliance.

3. Audit Trials

The immutable feature of blockchain technology ensures the audit trail is maintained securely and transparently. This aids in demonstrating the compliance with requirements and maintaining trust among users (Table 5).

Table 5 Elaborates the benefits of using blockchain in data security and compliance

Feature	Benefits
Encryption and access control	Minimize unauthorized access
GDPR compliance	Regulatory compliance
Audit trails	Audit compliance

3 Result and Discussion on Industry Adoption and Case Studies

The industry has been continuously adopting blockchain technology to solve major challenges like security risks, lack of transparency, and inefficiencies. It enhances protection, prevents fraud, and ensures clear and trustworthy transactions by using blockchain decentralized system. It mainly focusses to secure guest identity management, automated bookings and transactions, gives clarity and reducing fraud risks. These improvements will help hotels and travel businesses operate more efficiently, build trust with customers, and enhance overall security. This will lead the utilization of blockchain technology within the hospitality sector as:

1. **AI Enhanced Smart Contracts**

Smart contracts equipped with AI automation will process verification checks and scheduling operations as well as payments to deliver streamlined booking services without human failure or labor. The system uses AI to examine historical meetings along with visitor choices which enables businesses to customize services that foster customer satisfaction and loyalty. These smart contracts will enable dynamic pricing functionalities thus allowing instant pricing adjustments based on changes in demand together with market patterns and guest activities. The approach ensures maximum revenue generation and ensures competitive market pricing [20].

2. **Unified Digital Identity Systems**

Blockchains establish an interoperable solution to address the persistent hotel industry issue of identity management. Blockchain implements a dual advantage of user-friendly security through its ability to provide users a single point of access for all their digital services [17]. The decentralized system of identity verification lets visitors safely store their login credentials which eliminates the frequent need for identification checks at accommodation locations and air travel businesses and other travel venues. The improved security combined with instant authentication methods and decreased waiting periods will lead to seamless check-in processes through this innovation. The implementation of one unified digital identity system makes it possible to achieve worldwide interoperability and simpler business cooperation between airlines and immigration officials and international hospitality providers. By integrating blockchain technology into its operations the hospitality industry would attain improved productivity together with enhanced security and better visitor experiences.

3. **Decentralized Autonomous Organizations (DAOs)**

DAOs as Decentralized Autonomous Organizations act as a framework that enables businesses to shift authority toward groups of stakeholders resulting in substantial governance reforms in the hospitality sector. Blockchain-based voting systems let guests and stakeholders take part in operational decisions as well as service development and pricing plans thus supporting democracy through inclusion [21].

Throughout every transaction process DAOs enhance openness while building trust since members can examine all financial and governance records with absolute certainty. Licenses of control through decentralization enables DAOs to eliminate middlemen functions that decrease operational costs while distributing profit equitably across all stakeholders. The hospitality industry will build a better system that combines community leadership and openness through this modification [22].

4. **Expansion of Tokenization**

Through tokenization the hospitality business can experience a complete transformation by advancing its payment systems and loyalty rewards and property management practices. The hospitality industry now welcomes cryptocurrency payments through hotels and travel agencies which enables travellers to conduct payments without bank attendance and currency conversions. The system decreases expenses while speeding up the processing time for payments. Loyalty programs based on blockchain technology free customers from point restrictions by enabling them to redeem points with any airline or hotel chain as well as multiple other services. Through tokenization people obtain valuable rewards which maintain their customer commitment. The system allows numerous investors to purchase luxury real estate properties even with smaller investment amounts [23].

Although the adoption of blockchain based solutions have transformed the landscape of hospitality and accommodation industry, there are certain resistances and challenges as well. The transactions in public blockchain networks decrease the overall scalability of the networks. Payment services along with fast hotel reservations encounter operational issues that lead to rising running expenses when their processing standards degrade [6]. Each geographical area adopts independent regulatory guidelines which monitor cryptocurrency activities and smart contracts together with digital identity management. Every Travel and hospitality companies has their own rules and regulations, it become more difficult for global companies to meet all requirements to adopt blockchain in the hospitality. To implementing blockchain systems requires update in software, staff training and system upgrades. To fulfil these requirements, it take long time and increases system complexity too. Due to the implementation cost many companies can't afford new system adoption. This is the main reason that blockchain adoption in the hospitality industry is slow and limited. Table 6 elaborates the observations and findings of key contributions of blockchain technology in hospitality and accommodation industry.

The below discussed are certain case studies of Industry Adoption where the technology has significantly impacted and influenced the various aspects of hospitality and accommodation industry. The hospitality industry maintains a growing trend of blockchain implementation because it achieves operational excellence through security and transparency. Blockchain implementation in various businesses demonstrates its functional advantages to stakeholders.

The instances of industry adoption of the technology has been evident from the case studies like Travala, TUI-Bed Swap, Winding Tree.

Table 6 Observations and findings of key contributions of blockchain in hospitality and accommodation industry

Source	Year	Key findings	Adoption challenges
[1]	2024	Enhanced supply chain management, improved guest experience, integration with emerging technologies	Resistance to change, lack of technical expertise, scalability issues
[5]	2024	blockchain adoption rate, reduced operational costs, loyalty programs, sustainability practices	High implementation cost, regulatory issues, technological complexity, lack of awareness
[15]	2024	Data security, loyalty programs, improved transparency and better quality control, baggage control, personalized guest experience	High adoption cost, regulatory constraints, limited apprehension
[9]	2024	Peer-to-peer accommodation, improved customer experience, sustainable system	Trust and security concerns, scalability concerns, restricted awareness
[24]	2023	Metaverse integration, non-fungible tokens, central bank digital currencies, decentralized finance, internet of value	Security concerns, environmental impact, technical complexity
[25]	2023	immutability, transparency can lead to better user experience, auditability and other organisational factors	Technological complicates, network latency concerns, lack of interoperability
[11]	2022	Decentralized hotel booking systems, economic benefits, enhanced customer satisfaction	Regulatory concerns, lack of awareness, scalability concerns

1. **Travala**

Travala is a Blockchain-Powered Booking Platform that has transformed the way customers can reserve travel services with integration of blockchain based payment mechanisms. Through blockchain technology Travala enables customers to make bookings for hotels flights and accommodations using cryptocurrency alternatives. Founded in 2017, the mission of the company was to connect the cryptocurrency with travel industry worldwide. The implementation of blockchain technology enables Travala to cut out middlemen and achieve lower fees with exclusive security in booking transactions. Through automated reservations Smart contracts reduce the chances of carousel troubles between suppliers and consumers. Travala has established a user reward program that employs its cryptocurrency to promote both participant involvement and customer loyalty [10].

Through blockchain technology Travala enables customers to make bookings for hotels flights and accommodations using cryptocurrency alternatives. The implementation of blockchain technology enables Travala to cut out middlemen and achieve lower fees with exclusive security in booking transactions. Through automated reservations Smart contracts reduce the chances of carousel troubles between suppliers and consumers. Travala has established a user reward program that employs its cryptocurrency to promote both participant involvement and customer loyalty [10].

2. **TUI-Bed Swap**

The worldwide travel and tourism leader, the TUI Group utilizes the blockchain technology. This project is designed to optimize the inventory management and distribution process in the industry. This personalized blockchain based inventory management system allow the company to shift between different points of sale based on the real time demands. This integration ensures the appropriate and enhanced utilization of resources thus leading to cost saving and revenue generation. Further, the technology has assisted to improve its booking management procedures. The migration of TUI hotel inventory to private blockchain operations helps the company cut third-party reliance for bookings while decreasing operational costs and improving customer-to-business contact. The decentralized system allows hotels to have direct transactions and instant updates with guests which results in improved operational efficiency and business profit. TUI shows that blockchain technology can optimize the operations of large hospitality businesses [19].

3. **Winding Tree**

Through its decentralized framework Winding Tree operates as an open-source travel marketplace assisting the connection of service providers with consumers in direct transactions. Winding Tree reduces business expenses through direct connectivity between service providers and travellers thus enhancing travel cost transparency. Through smart contracts Winding Tree provides transaction automation and security which makes the booking process more efficient [26]. The platform was founded in 2017, the system operates on Ethereum cryptocurrency. It has eliminated the intermediates thus reducing the cost. With the help of smart contracts, the direct contacts between the buyers and the sellers, thus promoting transparency.

4. **TravelbyBit**

TravelbyBit is a Binance powered, blockchain based booking platform. This allows customers to book travel and accommodations with the help of cryptocurrencies. The platform provides a decentralized system for payments that eliminates the role of middle men, lower the transactional costs and increase the data and transactional security. This solution is partnered with Brisbane Airport Corporation for creating a digital currency-friendly airport terminal [27].

5. **Trippki**

Trippki is a blockchain powered decentralized loyalty and reward management platform. It offers a transparent and efficient system for revolutionizing loyalty programmes and enhance customer experience. It uses the concept of TRIP tokens that are earned as rewards from various hotel services and can be redeemed for future services [27]. It uses “proof of stay” to complete the verification task. Smart contracts are used to automate the reward distribution process. By reducing the dependency on OTAs, the commission fees was lowered and the profit margin was increased.

6. **Japan's Hospitality Blockchain Platform**

The blockchain powered platform is revolutionizing initiative used to enhance security and efficiency in the hospitality and accommodation industry. This platform provides a secure mechanism for sharing and store the customers critical information related to hotel and travel suggestions, transport choices etc. for better and enhanced personalized experience while their visit to Japan. The immutable ledger ensures transparent audit trail, that helps to inculcate the trust among all the stakeholders [27]. The Japan's blockchain powered model has portrayed as a base model for other countries and is ready to experience global adoption.

4 Conclusion and the Future Direction

Blockchain technology is changing the landscape of hospitality and accommodation industry by implementing operations more securely, effectively and in customer friendly manner. It uses automated system to protect financial transactions, reduce fraud and protect customer personal information from online dangers. It makes hotel bookings, transactions and identity verification faster by using automated process. Blockchain technology boost us economy by supporting businesses, driving tourism, and creating employment to peoples. To work well, it needs good customer relationship and smart use of data. To ensure smooth operations, the hospitality industry truly depends on three aspects such as trust, honesty, and security. The hospitality industry can take full advantage from blockchain technology by implementing these strategies. To ensure operations smoothly, companies must provide training to staff and properly implement the technology [28].

From the enhancement of security and elimination of intermediate parties, the organizational services have improved leading to enhanced customer satisfaction. Customers have enhanced their experiences through personalized services, loyalty schemes, supply chain management processes, digital identity verification and many other services. Blockchain powered platforms like LockChain allow the users to manage the bookings without the intervention of middle parties. This in turn enhance the efficiency of operations and increase the safety.

Despite the advantages offered by the technology, there are certain challenges like scalability issue, technical complexities and other regulatory constraints. The adoption of the technology has become a concern as the regulatory compliance has to be obeyed following the clearly stated guidelines to give direction to the businesses. The technical adoption needs awareness about the technology along with expertise to benefit the applications. Network speed and efficiency are some other aspects for consideration.

As the technology will evolve more with time, it will play a pivotal role in shaping the future of hospitality and accommodation industry. The integration of blockchain technology with other evolving technologies like AI, IoT, metaverse can raise the potential to another level to create a magnificent customer and operational experience.

The future considerations must explore to strengthen the regulatory compliance and focus on developing scalable and interoperable technological solution with reduced costs and better efficiency.

In conclusion, the blockchain technology has been committed towards shaping the hospitality and accommodation industry with its potential. Sustainable, innovative and customer-oriented solutions have been promoted with the aid of the technology. As this industry serves as a crucial part of the economy, ethical practices and environmentally-conscious approaches act as a catalyst for the promotion of integration of technology to the industry.

References

1. Nur Muharam, I., Tussyadiah, I.P., Kimbu, A.N.: A theoretical model of user acceptance of blockchain-based peer-to-peer accommodation. Curr. Issue Tour. **27**(7), 1008–1025 (2024)
2. Khanna, A., Sah, A., Choudhury, T., Maheshwari, P.: Blockchain technology for hospitality industry. In: Information Systems: 17th European, Mediterranean, and Middle Eastern Conference, EMCIS 2020, Dubai, United Arab Emirates, November 25–26, 2020, Proceedings 17 (pp. 99–112). Springer International Publishing (2020)
3. Önder, I., Gunter, U.: Blockchain: is it the future for the tourism and hospitality industry? Tour. Econ. **28**(2), 291–299 (2022)
4. Willie, P.: Can all sectors of the hospitality and tourism industry be influenced by the innovation of blockchain technology? Worldwide Hosp. Tourism Themes **11**(2), 112–120 (2019)
5. Muharam, I.N., Tussyadiah, I., Kimbu, A.: Unlocking the potential of blockchain-based sharing economy in hospitality and tourism: a systematic review. Int. J. Hosp. Manag. **122**, 103863 (2024)
6. Barkel, C., Kurgun, H., Groen, B.: Blockchain in the Hospitality and Tourism Industry (p. 4). University of South Florida (USF) M3 Publishing, 17(9781732127593) (2021)
7. Acikgoz, F., Stylos, N.: Blockchain technology in tourism and hospitality industry. In: Blockchain for Tourism and Hospitality Industries (pp. 7–14). Routledge (2023)
8. Kitsantas, T., Chytis, E.: Blockchain technology as an ecosystem: trends and perspectives in accounting and management. J. Theor. Appl. Electron. Commer. Res. **17**(3), 1143–1161 (2022)
9. Ghasemi Hamedani, I., Bashokouh Ajirlo, M., Rahimi Kolour, H., Seifollahi Anar, N.: Providing a peer-to-peer accommodation service acceptance model based on blockchain in the tourism industry. Sci. Technol. Policy Lett. **14**(3), 28–53 (2024)
10. Morozova, M.A., Stepanov, Y.G., Burlov, D.I.: Innovations in tourism and hospitality through modern information systems and blockchain technologies. Compon. Sci. Technol. Prog. **65**(11), 42 (2021)
11. Tepe, Y.: Decentralisation of hotel and tourism industry: effects of blockchain technology in hotel booking systems (2022)
12. Demirel, E., Karagöz Zeren, S., Hakan, K.: Smart contracts in tourism industry: a model with blockchain integration for post pandemic economy. Curr. Issue Tour. **25**(12), 1895–1909 (2022)
13. Irannezhad, E., Mahadevan, R.: Is blockchain tourism's new hope? J. Hosp. Tour. Technol. **12**(1), 85–96 (2021)
14. Hsieh, M.Y., Weng, T.H., Wang, P.W., Kao, C.H.: Recommendation service for hotel applications on blockchain. Int. J. Comput. Sci. Eng. **25**(6), 629–642 (2022)
15. Youssef, A.B., Dutta, P.K., Doshi, R., Sajnani, M. (Eds.): AI, Blockchain, and Metaverse in Hospitality and Tourism Industry 4.0: Case Studies and Analysis. CRC Press (2024)
16. Erceg, A., Damoska Sekuloska, J., Kelić, I.: Blockchain in the tourism industry—a review of the situation in Croatia and Macedonia. In: Informatics (Vol. 7, No. 1, p. 5). MDPI (2020)

17. Nair, B.K.: Perception of hotel staff towards blockchain integration in tourism and hospitality sector. Uluslararası Ekonomik Araştırmalar Dergisi **7**(3), 1–12 (2021)
18. Jain, P., Singh, R.K., Mishra, R., Rana, N.P.: Emerging dimensions of blockchain application in tourism and hospitality sector: a systematic literature review. J. Hosp. Market. Manag. **32**(4), 454–476 (2023)
19. Prados-Castillo, J.F., Torrecilla-García, J.A., Andraz, G., Guaita Martínez, J.M.: Blockchain in peer-to-peer platforms: enhancing sustainability and customer experience in tourism. Sustainability **15**(22), 15968 (2023)
20. Halkiopoulos, C., Antonopoulou, H., Kostopoulos, N.: Integration of blockchain technology in tourism industry: opportunities and challenges. In: International Conference of the International Association of Cultural and Digital Tourism (pp. 353–368). Springer International Publishing, Cham (2022)
21. Onder, I., Acikgoz, F. (Eds.): Blockchain for Tourism and Hospitality Industries. Taylor & Francis (2023)
22. Calvaresi, D., Leis, M., Dubovitskaya, A., Schegg, R., Schumacher, M.: Trust in tourism via blockchain technology: results from a systematic review. In: Information and Communication Technologies in Tourism 2019: Proceedings of the International Conference in Nicosia, Cyprus (pp. 304–317). Springer International Publishing (2019)
23. Rashideh, W.: Blockchain technology framework: current and future perspectives for the tourism industry. Tour. Manage. **80**, 104125 (2020)
24. Radic, A.: Blockchain technology and internet of value in the hospitality and tourism industry. In: Exploring the World with Blockchain Through Cryptotravel (pp. 79–89) (2024)
25. Acikgoz, F., Stylos, N.: Blockchain technology in tourism and hospitality industry. In: Blockchain for Tourism and Hospitality Industries (pp. 7–14). Routledge (2023)
26. Kumar, D., Shandilya, A.K., Kumar, N.: Comprehending the multifaceted realm of blockchain within the hotel industry: science mapping approach. In: Hotel and Travel Management in the AI Era (pp. 203–218). IGI Global (2024)
27. Aiazbekov, A.: Cryptocurrency as a method of payment in the tourism sector. Financ. Internet Quart. **19**(1), 57–65 (2023)
28. Rana, R.L., Adamashvili, N., Tricase, C.: The impact of blockchain technology adoption on tourism industry: a systematic literature review. Sustainability **14**(12), 7383 (2022)

Decentralized Tourism: The Transformative Role of Blockchain in the Travel Industry

Utsav Krishan Murari and Hemlata Parmar

Abstract The tourism sector is at the cusp of adopting new transformative solutions based on blockchain technology, which shows promise to solve two major issues: transparency, among other problems such as security vulnerabilities, high transaction costs and difficulties created by intermediaries. In this research paper, the first part explores blockchain as a Decentralized system that can lead to impact through a systematic review of literature and qualitative case studies of industry leaders with blockchain initiatives: Winding Tree, Travala and IATA's Travel Pass. The results demonstrate that blockchain is indeed an effective way of bringing efficiency and transparency to transactional processes, given that the ledger becomes immutable and decentralised; it eradicates fraud to a great extent and builds consumer trust. The advancement of security offered by blockchain includes a robust identity verification system and other innovations, such as non-fungible tokens (NFTs), to reduce counterfeiting risks. Above all, blockchain-enabled smart contracts, cryptocurrencies, and Decentralized finance (DeFi) significantly improve operational efficiencies through streamlined transactions, reduced intermediary costs, and increased consumer satisfaction. However, there are still issues such as scalability, regulatory abyss, and consumers' resistance to the technological complexity under which blockchain is adopted. A verdict is reached that future research should concentrate on technological scalability improvement, standardized regulatory platforms, user-friendly interface, and the integration of blockchain with previously arising technologies such as artificial intelligence, the Internet of Things, and the Metaverse so that blockchain may day serve as the backbone of a robust, protected, open, and innovative planetary tourism ecosystem.

U. K. Murari
Journalism and Mass Communication, Manipal University Jaipur, Jaipur, Rajasthan, India

H. Parmar (✉)
Department of Computer Science and Engineering (AIML), Manipal University Jaipur, Jaipur, India
e-mail: hemlata.parmar@jaipur.manipal.edu

A. Kumar et al. (eds.), *Blockchain in the Tourism Industry: A New Era of Secure and Transparent Travel Solutions*, Information Systems Engineering and Management 57,
https://doi.org/10.1007/978-3-031-95341-5_5

Keywords Blockchain · Decentralized tourism · Smart contracts · Transparency · Security · Cryptocurrencies · Decentralized Finance (DeFi) · Tourism industry · NFTs · Metaverse

1 Introduction

Rapidly, Blockchain technology has become a revolutionary innovation across many sectors, such as Finance, Supply Chain Management and Health, with tourism being added to the list. Because of the possibilities blockchain offers to tackle age-long challenges in tourism and security, transparency, and intermediary costs, blockchain has gained much attention in the tourism industry. Tourism is one of the most significant and fastest-growing economic sectors worldwide, significantly contributing to GDP, employment and cultural exchanges [29]. However, it has numerous setbacks, such as fraud involving transactions, intermediaries' inefficiencies, data breaches, and lack of transparency in service availability and pricing [16].

In other words, Blockchain technology was initially conceived with cryptocurrencies such as Bitcoin in mind, which is why it has the characteristics of decentralization, immutability, and transparency [23]. Studies have explored the possibility of its use in the tourism context, especially in booking platforms, identity verification systems, loyalty programs and financial transactions [26]. By nature, blockchain is a Decentralized ecosystem, and data is spread over various nodes from multiple geolocations, thereby reducing the risks of centralized data storage and controlling the characteristics of the present tourism systems [37].

A significant benefit of blockchain is that it can enhance transactional transparency. Typically, other than the airline, there are many intermediaries: online travel agencies (OTAs), global distribution systems (GDS), and other payment processors that all have fees and commissions, thereby adding to the expenses of the consumers [12]. Besides, hidden charges and opaque pricing approaches make consumers dissatisfied and mistrustful. Blockchain addresses these problems through a set of real-time, transparent and immutable records of transactions that any relevant stakeholder may access, which assures fairness and clarity in value calculating and availability [20].

The second is that blockchain provides enhanced security within the tourism industry. The industry includes thousands of travellers whose information is being stored and significantly handled within the system, so it is very susceptible to cyber-attacks and identity fraud. One of the most attractive to cybercriminals [2] is the centralised databases owned by an airline, hotel, or any other travel agency. Blockchain's architecture is decentralised, therefore abrogating these risks as well. Furthermore, blockchain-based digital identity solutions also enable secure Decentralized storage of traveller credentials, facilitating safer and expedient traveller credentials verification at different checkpoints, including airports and hotel receptions [14]. These solutions streamline customer experience by shortening time duration and focusing on data privacy and control for users [11].

Blockchain technology also simplifies and secures the process of financial transactions in the tourism industry. In traditional financial transactions, such as international ones, there is a long chain of intermediaries, and these intermediaries generate higher transaction fees, extended processing time and currency conversion complexities [7]. Blockchain-based solutions like cryptocurrencies and stablecoins provide a much better solution by allowing faster, cheaper, secure and borderless payments than the traditional banking system. Furthermore, this approach lowers costs and provides financial inclusion, especially to travellers and small tourism businesses in developing regions [35].

Similarly, blockchain can facilitate innovative Decentralized finance (DeFi) forms adapted solely for tourism industry use. With smart contracts as coded self-executing contracts defining preconditions for various types of automation, including insurance claims, refunds and loyalty program distributions, DeFi removes the administrative burdens and operational inefficiencies [5]. With smart contracts, automatic transaction execution happens when real-time verifiable conditions are achieved to reduce disputes and foster trust between service providers and consumers [17].

Furthermore, tourism is a significant example of innovation in using blockchain in loyalty programs. Fragmented, opaque, and incompatibility between service providers make traditional loyalty schemes frustrating and ineffective. Blockchain solves such challenges by enabling tokenised reward systems that are secure, transparent and interoperable, taking tokenised rewards made within such systems, and users can use them in live time over many platforms. This innovation is a significant payback to the benefit of both the customers and the businesses, as it has significantly increased customer engagement, customer satisfaction, and customer retention [15].

Although blockchain technology has many benefits, integrating such technology into tourism is limited by several significant barriers. There are substantial challenges with technological limitations, including inclining towards scalability and speeds of slow transactions for real-time applications in high-volume tourism contexts [18]. It also presents considerable barriers to regulatory uncertainty, making it challenging to implement blockchain across borders, specifically regarding data protection and financial issues [34]. The complexity and perceived risk of widespread adoption by users without user-friendly interfaces and clear value propositions are critical [21].

Blockchain is saturated and closely linked to its incorporation into technologies such as AI, IoT, and virtual reality, which are used in the latest Metaverse projects. With such integration, unparalleled personalisation, immersive experience, improved operational efficiencies, and significant improvement in security measures are the promises. For instance, using blockchain and AI-based analytics, personalised recommendations and proactive fraud detection are possible. IoT integration is realised with real-time authentication of traveller credentials or efficient operation management of hotels and airports [6].

Thus, it is concluded that blockchain technology has much transformative potential for the tourism industry and the capability to navigate security, transparency, and cost-efficiency challenges. According to him, blockchain's further adoption and effectiveness would be enabled through its continued evolution, which current and future advances in regulatory frameworks and technology scalability would support.

In a tourism ecosystem with blockchain applications, blockchain will be a building block for becoming more secure, transparent, and decentralised.

2 Literature Review

Blockchain technology has become incredibly revolutionary in the tourism industry and can solve problems like security vulnerabilities, transactional transparency and high intermediary costs. In this paper, we explore the possibility of blockchain technology adoption and integration in the tourism sector to see how it can add to the tourism sector by adding to its transparency, security and decentralization. The research also examines case studies and real-world applications on how blockchain can address fraud, double bookings, and data breaches if it does not streamline operations via Decentralized finance (DeFi) and cryptocurrencies.

The scholarly interest in blockchain technology has been ever-increasing, as it has an enormous impact across many industries, especially tourism, leading to increased transparency, improved security and operational efficiency, and it is indeed encouraging to witness this growth in knowledge acquisition. Trust and accountability across the stakeholders are increased by blockchain, which makes them more transparent via immutable, publicly verifiable transaction ledgers [32]. According to [20], the use of the blockchain is transformative since intermediaries are eliminated and fairness and transparency are introduced to travel-related transactions.

Where blockchain affects the most is in diminishing the need for intermediaries. To minimise the presence of intermediaries, such as online travel agencies (OTAs) and payment processors, blockchain brings costs and efficiency benefits, as mentioned by [35]. Dolgui et al. [9] also show that decentralized travel booking systems eliminate intermediaries, which allows subsequent reductions in transaction costs and increases in transparency in pricing and availability.

The other important domain of application for blockchain in tourism is security enhancement. As per [13], the security vulnerabilities in traditional centralized databases, like data breaches and identity theft, are substantially reduced in blockchain, where data is stored in a Decentralized format and cryptographic security features. In more detail, explain the part that blockchain plays in building robust digital identity systems, which enable the creation of robust, secure identity verification without being dependent on centralised authorities [27].

To illustrate blockchain in confirming digital identity, the International Airline Transport Association (IATA) provides its practice example of the application of blockchain in securely managing COVID-19 test results and vaccination credentials [10]. As Treiblmaier [28] points out, blockchain's potential to prevent ticketing and reservation fraud through these nonfungible tokens (NFTs) is to create unique, verifiable, and impossible-to-copy tickets.

Chen et al. [7] also view blockchain use in smart contracts and highlight the efficiency of blockchain in automating transactions such as refunds, cancellations and loyalty point distribution to eliminate disputes and improve customer satisfaction. In their article, Lee et al. [17] provide examples of how blockchain can make loyalty programs more secure and transparent and improve user experience and engagement.

This also has promising applications when integrating cryptocurrencies and DeFi, short for 'Decentralized Finance.' In this system, as introduced by [22], cryptocurrencies like Bitcoin are based on a concept of a blockchain underlying Decentralized financial systems, which have the potential to automate financial transactions completely friction between two countries. Chen et al. [7] further examine the DeFi tourism solution of microloans in travel and suit insurance, spotlighting how blockchain-based micro-loans and travel insurance can significantly improve both the transaction costs and fraud associated with financial activity in tourism.

Its potential for other emerging technologies such as Artificial Intelligence (AI), the Internet of Things (IoT), and Metaverse has ranked highly. Chen et al. [6] also explore how blockchain leverages AI to aid in customising service delivery, real-time data analytics, and improved security monitoring in travel interaction. By combining IoT with blockchain, real-time tracking and verification of travellers and luggage is possible, significantly reducing the operation's inefficiency. Patel and Gupta [25] highlighted the integration of IoT in blockchain for real-time tracking and verifying travellers and luggage.

Existing literature is also reviewed, and blockchain adoption challenges are critically reviewed. I think that barriers to the use of blockchain in the tourism industry are brought up in many conversations regarding the use of blockchain, and one of those is the technology constraints, such as blockchain scalability limitations and the time required for a transaction timeline. It also becomes a challenge of regulatory uncertainty. The complexity of regulatory compliance, as described by [34], is manifest in the absence of the global uniform standards of cross-border transactions, impeding the adoption of the blockchain in tourism globally.

Additionally, consumer trust is an important element for adopting blockchain in tourism. Moyano and Ross [21] address the challenges of interacting with the blockchain, such as managing private keys and permanently deemed irreversible transactions that can discourage users from accepting the same. Utilising user-friendly blockchain interfaces and having a clear value proposition will go a long way to boosting consumer trust, as highlighted by [19].

Overall, the reviewed literature suggests that blockchain has multi-faceted benefits and huge potential for significant improvement in transparency, security and efficiency in the tourism industry while facing significant challenges to its widespread adoption.

3 Research Methodology

A systematic literature review and case studies have been used as the research methodology for this study. In addition, this mixed method approach to the topic—Blockchain applications in the tourism sector—provides a rigorous exploration of this topic as a consequence of combining theoretical insights from academic literature with practical observations from industry implementations.

Step 1: Systematic Literature Review

The first step will be secured by a systematic literature review based on the recent five-year span (2019–2024) of academic works, industry reports, and technological white papers published. The purpose of this literature review is to find out the current state of blockchain technology application in the tourism sector, starting from theoretical models and key trends, technologies applied by the tourism sector, and challenges that have been identified on the online and mobile applications utilising blockchain technology in the sector.

Databases utilised for this extensive search will include:

- Web of Science
- Scopus
- IEEE Xplore
- SpringerLink
- Google Scholar

The search criteria will utilise combinations of key phrases such as "blockchain technology," "tourism," "peer-to-peer marketplaces," "smart contracts," "cryptocurrency integration," "Decentralized finance (DeFi)," and "blockchain security." The search will be complemented by backward and forward citation tracking to ensure comprehensive coverage and the inclusion of seminal articles [4, 26].

Step 2: Selection and Evaluation of Literature

Following the initial retrieval of articles, a careful screening process based on predefined inclusion and exclusion criteria will be applied:

- **Inclusion criteria**:
  - Peer-reviewed journals, books, and conference proceedings.
  - Relevance to blockchain technology applications, specifically within the travel and tourism industry.
  - Empirical studies, theoretical frameworks, and detailed case studies.
- **Exclusion criteria**:
  - Studies published before 2019, except landmark articles.
  - Non-peer-reviewed sources or opinion articles without substantial empirical backing.
  - Studies focused primarily on blockchain applications outside tourism and hospitality.

Evaluating literature will ensure methodological rigour, relevance, and clarity of contributions to the topic [33].

Step 3: Qualitative Case Study Analysis

Specific organizations that have successfully implemented blockchain technology in the tourism sector will be reviewed, and then a qualitative case study will be developed based on the literature review. Therefore, the choice of qualitative approach via the case study is made because of the rich context, contextual insights, and understanding of complex phenomena such as blockchain adoption in real-world settings [33].

Cases to be analysed include organizations such as:

- Winding Tree
- Travala
- Blockchain-based loyalty programs (e.g., Singapore Airlines KrisFlyer)

Data for these cases will be obtained from publicly available sources, including company whitepapers, annual reports, press releases, and credible online articles.

Step 4: Analytical Framework

The analysis will employ an analytical framework grounded in specific performance criteria, enabling a structured comparison across cases. The framework includes evaluating:

- **Degree of Decentralization**: Reduction of intermediaries and the impact on transaction efficiency [9].
- **Transparency and Trust**: Improvements in transparency, verifiable transactions, and consumer trust-building mechanisms [3].
- **Security Enhancements**: Effectiveness of blockchain-based solutions in preventing fraud, data breaches, and counterfeit activities [1].
- **Economic and Operational Efficiency**: Cost reduction, increased speed of transactions, and operational improvements [30].
- **Consumer Adoption and Satisfaction**: Customer engagement, ease of use, and consumer feedback post-adoption [38].

Step 5: Comparative and Interpretive Analysis

A synthesis of the insights derivable from the literature review and case studies is done through a comparative interpretive analysis. It will highlight commonalities, differences, best practices, and important challenges in blockchain implementation in multiple contexts in tourism. It will also provide more information for recommendations and discussion on ways to increase blockchain adoption in the tourism sector in the future.

3.1 Justification of Methodology

This research methodology is deliberately structured as a mixed-method approach, combining systematic literature reviews and qualitative case studies. It is justified due to:

- **Comprehensiveness**: Ensuring coverage of theoretical and practical dimensions [4].
- **Depth of Understanding**: Qualitative case studies provide deep insights into complex and dynamic phenomena, such as blockchain integration, which quantitative methods might overlook [33].
- **Triangulation**: The combination of literature review and case analysis strengthens the validity of findings through method triangulation, reducing potential biases inherent in relying on a single approach.

3.2 Anticipated Outcomes of Methodology

Expected outcomes are a strong, broad understanding of how blockchain could affect tourism, how to recognise key issues and opportunities and recommendations on using blockchain technologies strategically or involving blockchain technologies in the travel business.

4 Results

This section details the key results of applying blockchain technology in the tourism sector through a comprehensive methodology, including a systematic literature review and qualitative case studies. The results are presented considering year-wise technological progress based on an analytical framework of Degree of Decentralization, Transparency and Trust, Security Enhancements, Economic and Operational Efficiency, Consumer Adoption and Satisfaction.

4.1 Degree of Decentralization

Blockchain technology substantially enhanced decentralization, and its role significantly reduced the intermediaries involved, such as traditional online travel agencies (OTAs), Global Distribution Systems (GDS), and centralised financial institutions. Decentralized implementation pioneered by companies such as Winding Tree and Travala proved successful. Ethereum-based smart contracts are a good example of using existing technology to achieve direct interaction between consumers (people) and service providers (companies) without intermediaries, which traditionally claim commissions from 10 to 30%. It was a fair marketplace with lower costs and more price competitiveness for small to medium enterprises [9, 12].

Finally, similar to this, the Decentralized travel platform from Travala, through blockchain-based transactions, provided direct bookings at much lower booking fees and less intermediary costs that benefited both the service providers as well as the consumers with fewer transaction costs [38].

4.2 Transparency and Trust

Integrating blockchain became a thing of transparency that was prominently seen. However, the records of transactions, bookings and service availabilities were all immutable and trusted by each consumer. Well, simply put, blockchain-based reputation systems like Travala's mechanism of feedback reduced fraudulent and manipulated reviews by directly linking the verified transactions to the people who provide feedback, almost guaranteeing that consumers place a significant amount of trust in online travel reviews [3].

In addition, innovative contract implementations simplified travel insurance claims and claims. Looking at studies where blockchain systems guarantee straightforward, fast, and condition-based transactions, we observed that automated, transparent contracts reduced the instances of consumer dispute and increased customer satisfaction [7].

4.3 Security Enhancements

The most significant improvement of Blockchain was in its security—it addressed common problems, including identity fraud, ticket counterfeiting, and security breaches, among many others. A standard example of this is the blockchain digital identity solutions developed by the International Air Transport Association (IATA) in the form of a "Travel pass." The system applied blockchain to secure verification and storage of travellers' COVID-19 testing and vaccination data with forgery-proof and ultra-strong security against unauthorized access, showing the great potential of blockchain in the aspect of secure identity management within global travel [14].

There were indeed a couple of blockchain applications utilising non-fungible tokens (NFTs) that were proven effective in ticketing and reservation security. In event ticketing sectors, NFTs offered a unique digital representation of tickets, thereby preventing ticket counterfeiting, unauthorized resale, and duplication; the significant potential of adoption in this area was exhibited in hotel and airline reservations.

4.4 Economic and Operational Efficiency

Within the tourism industry, blockchain significantly improved economic and operational efficiency. This increased the efficiency of cross-border financial transactions through blockchain-based cryptocurrencies such as Bitcoin, Ethereum, and Stablecoins (such as USDC). Both overlapped similarly by drastically reducing transaction times from days to minutes and lowering currency conversion and transfer fees, especially in international travel situations [30].

More operationally, Decentralized platforms and smart contracts ensured direct operational aspects such as double booking and fraudulent chargebacks went moot simply by uniting the booking, cancellation and refund within a single contract, thereby transforming the operation. The studies showed that with blockchain automation, companies such as Travala and Winding Tree experienced better operation performance in terms of fewer manual contact points and reduced administrative weight, which in general helps improve the overall operational efficiency [4, 7].

4.5 Consumer Adoption and Satisfaction

Consumer adoption of blockchain implementation was optimistic but slow, cautious, and experienced. Decentralized loyalty programs such as Singapore Airlines's Kris-Flyer were highly successful, leveraging blockchain tokenization to seamlessly interoperate the rewards of its partners and ensure real-time redemption within the consumers' reward spend [8].

However, adoption by consumers suffered from complications in using technology—technological complexity and people not being familiar with blockchain interfaces. Paramount to adoption was continuous user education and a simplified user experience that shows clearly what the product does [24]. This indicates the need for user-friendly blockchain interfaces and coherent communication of value benefits.

4.6 Challenges and Barriers to Adoption

They further analysed the challenges significantly limiting the broader adoption of blockchain. However, blockchain scalability was constrained by technological factors, precisely scalability issues like high latency and energy-intensive aspects of consensus on blockchain, such as Proof of Work. In addition, impediments consisted of regulatory uncertainty: regular chaos of inconsistent international standards and ongoing discussions of cryptocurrency legality and privacy in data complicating global adoption [18, 34].

Furthermore, consumer resistance was considerable, mainly because consumers deemed blockchain technology complex and insecure, indicating the need for heavy consumer education, a more transparent regulatory landscape, and less complex technology integration to make it more mainstream [24].

4.7 Year-Wise Technological Progress and Implementation

The progressive adoption and refinement of blockchain applications within tourism were observed as follows:

- **2019–2020**: The initial phase involved foundational blockchain implementations, predominantly Decentralized travel marketplaces like Winding Tree and Travala. Ethereum-based smart contracts and basic cryptocurrency integrations marked early-stage blockchain adoption, primarily aimed at reducing transaction costs and intermediaries [9].
- **2021–2022**: Significant advancements included robust identity management systems leveraging blockchain for secure verification of travellers' credentials, notably demonstrated by IATA's Travel Pass initiative. This period also introduced biometrics integrated with blockchain, enhancing secure identity validation and operational efficiency at border checkpoints [14].
- **2023–2024**: More advanced blockchain applications, such as Decentralized finance (DeFi) solutions and tokenization in loyalty programs, gained prominence. Notably, stablecoins and NFTs became integral parts of tourism payments and ticketing systems, significantly enhancing operational transparency and security while reducing costs and fraud [8, 30].

The cumulative results underscore blockchain's transformative potential and illustrate successes and critical areas requiring further development and regulatory clarity to facilitate broader implementation in the tourism industry.

5 Discussions

Blockchain technology greatly benefits the tourism industry in terms of transparency, security, operational efficiencies, and consumer satisfaction. Amongst the improvements in this study are significant improvements in these critical areas and important challenges that hinder widespread adoption. These findings are elaborated in the following discussion and placed in the context of the extant research and industry practice.

5.1 Enhancing Transparency and Trust

Out of all things, transparency became one of blockchain's most influential contributions to tourism. The nature of blockchain technology is Decentralized and immutable, which means that every transaction is verifiable, which makes it difficult for fraudsters, manipulators, and manipulators to manipulate. Blockchain-based reputation systems significantly reduce the problem of fraudulent reviews on traditional online platforms, such as customer feedback being linked directly to verified transactions [2, 17]. Consistent with previous research that has stressed transparency as a core element of consumer trust in online travel bookings [20], these findings indicate one way to improve both travel bookers' trust.

Further trust provides automation and immutability in smart contracts → clarify conditions for refunds, etc., regarding insurance claims. Unlike the manual processes and the delays and disputes that often lead to blockchain-driven smart contracts that provide timely and transparent execution of agreed terms, they give customers more confidence [6]. Transparency of this nature not only places consumer satisfaction high but also equips blockchain with the position to act as a tool of trust required for future tourism growth.

5.2 Improving Security in Travel Transactions

Blockchain provides a solution to some of the long-standing vulnerabilities like data breaches, identity theft, counterfeit bookings and fraudulent activities that affect security in the tourism sector. The Decentralized architecture substantially reduces risks from the centralised data storage systems that may fail on a single point and be susceptible to cyberattacks. As an example, blockchain-based identity verification systems such as IATA Travel Pass, which demonstrated they mitigate the risk of document forgery and unauthorized access alike, provide an excellent role of blockchain security in the practice of secure global travel [31, 36].

Furthermore, this can be applied using non-fungible tokens (NFT) to integrate blockchain into the ticketing and reservation processes to deal with counterfeit tickets and resale by unlicensed agents. NFTs create a unique and immutable verifiable proof of each ticket to ensure there is no fraudulent duplication of tickets on such event-based forms of tourism. This correlates with earlier observations by [12] that blockchain potentially provides this radical security improvement to tourism operations.

5.3 *Operational and Economic Efficiency*

Decentralization and reduced dependence on intermediaries were the ideal aspects of blockchain integration, bringing operational efficiency. Travala and Winding Tree are some platforms that provide evidence of how blockchain can streamline processes, reduce overhead operational costs, and reduce intermediary costs. Cryptocurrencies and stablecoins are also further helping blockchains "promote transactional efficiencies" such as low cross-border fees, faster payment settlement and lighter administrative burdens on international financial transactions [17].

While much efficiency has been achieved, scalability constraints inherent to blockchain technology lower these benefits. However, although current systems continue to experience performance limitations, notably transaction latency and network congestion than peak use in public blockchain networks like Ethereum, the possibility of enabling a high volume transaction environment in global tourism operations [31] is not immediate. The scalability of these issues will remain essential for long-term sustainable operational integration.

5.4 *Challenges to Blockchain Adoption*

Although the tourism field gained advantages from blockchain adoption at this stage, there are some critical issues. Barriers to scalability, with technological scalability remaining the predominant one, except for some experimental note cryptography systems, significantly impact transaction speed and the number of concurrent transactions such networks can handle. These scalability issues impose substantial limitations to the practical implementation where, for example, one needs to receive instantaneous booking confirmation or instantaneous payment processing in real-time tourism services [20, 36].

It should also be noted that regulatory compliance is a significant issue. Jurisdictions have different regulatory frameworks relating to blockchain and cryptocurrencies, making it challenging to adopt across borders. Data protection, privacy compliance and anti-money laundering regulations require further clarification from the regulator. The lack of consistent international standards further impedes the deployment of blockchain solutions in global tourism, making stakeholders uncertain and hindering the adoption of such a technology at scale.

Consumer trust and adoption represent additional barriers, primarily due to perceived technological complexity and the nascent nature of blockchain applications. However, it surprises no one that blockchain is not commonplace as most user concerns revolve around its technical usability, private key security, and other pretexts of not having one. For successful consumer adoption, user-friendly interfaces, extensive consumer education and strong consumer support match those prior literature recommendations [31, 36].

5.5 *Future Directions and Technological Integration*

The identified challenges can or require further research focusing on overcoming them. Addressing current scalability concerns can be significantly expedited by technological progress, mainly where Proof-of-Stake, a more affordable and energetic scaling method, can replace energy-rigorous Proof-of-Work. It also means that intuitive blockchain interfaces and user education campaigns would significantly decrease consumer adoption in this case.

As technological developments happen, the timing of policy frameworks must also adapt to be more explicit about what constitutes the policy and more standardised policy frameworks. Coherent, mutually beneficial, sustainable blockchain support policies based on placing technology providers, tourism industry stakeholders, regulatory bodies and end consumers on the same table to encourage multi-stakeholder dialogues on the issue will be encouraged.

Lastly, while blockchain finally converges with other nascent innovative technologies, such as Artificial Intelligence (AI), the Internet of Things (IoT) and the three environs of the Metaverse, they bring the spectres of additional potential changes and innovations in the tourism value chain. AI-enhanced Blockchain systems can lead to predictive analytics for offering personalised tourism and better fraud detection. Luggage tracking from a real-time source could be further integrated with an IoT to track luggage, automated check-in at the hotel, personalised tourist experience, etc. Virtual travel experiences based on metaverse enabled by blockchain digital identity to enable safe and secure transactions and offer considerably newer tourism market and services [6].

Blockchain technology has great promise to transform the tourism industry at the foundational level to become more transparent, secure, and operationally efficient. However, those challenges must still be overcome with technological innovation, clarity in regulators and prising consumers closer. Future research and efforts by the industry will help determine how far blockchain will go in supporting the resilient, innovative tourism industry as a central pillar of a technological infrastructure.

6 Conclusion

Finally, blockchain technology significantly disrupts the tourism industry's challenges, such as security, security gaps, and high intermediary costs. Blockchain platforms greatly diminish reliance on traditional intermediaries through decentralized platforms and improve operations' transactional transparency and efficiency. However, an application of blockchain in the real world includes organizational presence through Winding Tree, Travala, and IATA's Travel Pass, which helps establish consumer trust via immutable transaction records and secure digital identity.

Additionally, financial processes are streamlined, costs have been reduced by eliminating transaction fees, and there is a lack of a middleman to make the interaction between consumers and service providers as smooth and trustworthy as possible.

One can hardly speak about what ICs have achieved without at the same time mentioning the many significant barriers. However, the spread of this adoption has been hindered by technological constraints, namely, scalability and real-time processing ability, as well as regulatory and complexity resistance from consumers. These limitations should be considered in future research and development to get a better technological idea through a more transparent regulatory framework and user-friendly interfaces. In addition, the combination of blockchain with its new technology, like artificial intelligence, Internet of Things (IoT), and Metaverse, offers new pathways for developing a secure, personalised, and easy tourism ecosystem in the global community. To a certain extent, blockchain's implication in decentralizing tourism will depend on sustained innovation and partnerships with stakeholders.

References

1. Al-Fuqaha, A., Dinh, T.N., Tsai, W.T.: Blockchain and security: a deep dive into decentralized technologies. J. Cybersecurity Stud. **8**(1), 23–42 (2022)
2. Alonso, S., Borras, J., García, C.: J. Travel Res. **59**(5), 873–890 (2020)
3. Benčić, F., Podnar, Ž: Blockchain-based reputation systems: enhancing transparency in tourism. Tourism Rev. **76**(3), 455–470 (2021)
4. Casino, F., Dasaklis, T.K., Patsakis, C.: A systematic literature review of blockchain-based applications: current status, classification, and open issues. Telematics Inform. **36**, 55–81 (2019)
5. Chen, Y., Liu, X., Wang, Z.: J Tourism Technol. **10**(3), 145–162 (2022)
6. Chen, H., Zhang, Y., Wang, L.: Int. J. Travel Technol. **29**(1), 55–71 (2023)
7. Chen, X., et al.: Tourism Manage. **87**, 104358 (2021)
8. Dinh, H., Tran, K., Pham, J.: Int. J. Hosp. Manag. **95**, 102545 (2022)
9. Dolgui, A., Ivanov, D., Sokolov, B.: Decentralized travel booking systems: Blockchain's role in disintermediation. Int. J. Prod. Res. **58**(12), 3756–3771 (2020)
10. IATA: Travel Pass Initiative. International Air Transport Association (2021). Retrieved from www.iata.org
11. International Air Transport Association (2023). Available from www.iata.org.
12. Iwamura, M., Kitamura, Y., Matsumoto, T., Saito, K.: Technol. Soc. **58**, 101247 (2019)
13. Janssen, M., et al.: Gov. Inf. Q. **37**(3), 101522 (2020)
14. Jiang, S., Liu, Y., Huang, J.: IEEE Trans. Blockchain **7**(2), 89–106 (2021)
15. Kim, J., Oh, S.: J. Digit. Bus. **14**(2), 87–102 (2023)
16. Kim, M.J., Lee, C.K., Jung, T.: Exploring consumer behavior in virtual reality tourism using an extended stimulus–organism–response model. J. Travel. Res. **59**(1), 69–89 (2020). https://doi.org/10.1177/0047287518818915
17. Lee, C., Huang, R., Wang, T.: Tourism Rev. **78**(4), 202–218 (2023)
18. Li, J., Greenwood, D., Kassem, M.: Autom. Constr. **111**, 103004 (2020)
19. Mekdad, Y., et al.: J. Hosp. Tourism Technol. **13**(2), 263–284 (2022)
20. Morabito, V.: Blockchain and Smart Contracts: Disruptive Technologies in Tourism. Springer Nature, Cham (2017)
21. Moyano, J.P., Ross, O.: Bus. Inf. Syst. Eng. **59**(6), 411–423 (2017)
22. Nakamoto, S.: (2008). Bitcoin.org
23. Park, J., Lim, E.: J. Policy Tour. **14**(3), 77–95 (2022)

24. Parmar, H.: Blockchain-enabled food supply chain management: enhancing traceability, transparency, and trust through IoT and smart contracts. In: Food in the Metaverse and Web 3.0 Era: Intersecting Food, Technology, and Culture. IGI Global (2025). https://doi.org/10.4018/979-8-3693-9025-2.ch003
25. Patel, A., Gupta, R.: Int. J. Smart Tourism **17**(4), 122–140 (2022)
26. Saberi, S., Kouhizadeh, M., Sarkis, J., Shen, L.: Blockchain technology and its relationships to sustainable supply chain management. Int. J. Prod. Res. **57**(7), 2117–2135 (2019)
27. Sharples, M., Domingue, J.: Educ. Technol. Soc. **19**(4), 247–261 (2016)
28. Treiblmaier, H.: J. Bus. Res. **100**, 574–590 (2019)
29. World Tourism Organization: Tourism investment report 2022. UNWTO. https://www.unwto.org/investment/tourism-investment-report-2022 (2022)
30. Xu, J., Lee, M.: Decentralized finance and cryptocurrency adoption in the tourism industry. J. Financ. Innov. **28**(3), 187–205 (2022)
31. Xu, X., Weber, I., Staples, M.: Architecture for Blockchain Applications. Springer, Cham (2019)
32. Xu, X., et al.: Springer (2021)
33. Yin, R.K.: Case Study Research and Applications: Design and Methods. Sage Publications, Thousand Oaks (2018)
34. Zavolokina, L., Dolata, M., Schwabe, G.: Electron. Mark. **30**, 447–462 (2020)
35. Zheng, Z., et al.: Inf. Manage. **59**(6), 103471 (2022)
36. Zheng, Z., Liu, Y., Xie, B.: Tourism Manag. Rev. **14**(2), 56–73 (2020)
37. Zheng, Z., Xie, S., Dai, H., Chen, X., Wang, H.: Int. J. Web Grid Serv. **14**(4), 352–375 (2018)
38. Zhou, R., Park, J., Li, X.: Decentralized booking platforms and blockchain: the future of transparent travel transactions. J. Hosp. Travel Technol. **11**(4), 201–219 (2021)

Blockchain-Powered Airlines: Enhancing Security, Transparency, and Efficiency in Aviation

Anupam Bonkra, Pummy Dhiman, Amandeep Kaur, Gaganpreet Kaur, and Sandeep Sharma

Abstract Security remains one of the biggest critical success factors affecting the consolidation and growth of the airline business, operations, and sales transparency, as well sas compliance with aviation laws. Centralized systems worked on databases and several layers of go-between services are not only prone to fraud and hacking but also time and cost-consuming. Blockchain solution can be defined as a decentralized and immutable distributed ledger system that provides security to and increases efficiency in the performance of the system and builds trust among the users. The application of blockchain is also expected to transform ticketing through the smart contracting feature, sharing of data in real-time, and record-keeping that cannot be altered easily, baggage tracking and management of aircraft as well as supply chain. Furthermore, the use of blockchain technology in various solutions improves passengers' identification, fights ticketing fraud, as well as develops a more secure and transparent loyalty program. Nevertheless, challenges are realized when introducing this technology in the aviation sector, such as legal restraints, expansion factors, and structures. In this chapter, the subject matter focuses on the possibilities of using blockchain in the airline industry, the opportunities and difficulties related to the implementation of blockchain and possible future developments which may contribute to the improvement of security, efficiency and transparency of the airline industry.

Keywords Blockchain · Aviation security · Smart contracts · Airline operations · Baggage tracking · Supply chain management · Loyalty programs · Data transparency · Cybersecurity · Digital identity

---

A. Bonkra (✉) · A. Kaur
Department of Computer Science and Engineering (CSE), MM Engineering College, Maharishi Markandeshwar (Deemed to Be University) Mullana, Ambala, Haryana, India
e-mail: anupambonkra@gmail.com

P. Dhiman · G. Kaur
Chitkara University Institute of Engineering and Technology Chitkara University, Punjab, India

S. Sharma
Department of Computer Science Engineering, Lovely Professional University, Phagwara, Punjab, India

A. Kumar et al. (eds.), *Blockchain in the Tourism Industry: A New Era of Secure and Transparent Travel Solutions*, Information Systems Engineering and Management 57,
https://doi.org/10.1007/978-3-031-95341-5_6

1 Introduction

The airline industry is a diverse and a fast-growing industry, which involves several different participants to include passengers, airlines, airports, regulatory agencies and supply chain and value chain partners [1]. Thus, the industry has problems of security, transparency, and efficiency of operations today, even with constantly progressing technological background. In light of this, Blockchain technology comes out as an innovation that can be adopted to solve these basic challenges affecting the airline sector. In this chapter, a brief look at the problems that exist in the airline industry is made, and the significance of digitalization is discussed to understand the role of blockchain in improving aviation services. The airline industry plays its operations across the world, catering to millions of passengers, cargo, and other administrative tasks on daily basis [2]. But different factors affect its efficiency and effectiveness of its growth. Issues of security and fraud are considered key challenges due to passengers' identification theft, ticketing fraud, and cyber threats. These risks are compounded by the absence of safe and clear processes of sharing information. Furthermore, mishandled baggage also causes customer discontent and losses since centralized tracking systems for conveyance lacks updates on the status of the baggage, making it easy for it to go missing. The analysis of the structure of airline ticketing systems where tickets may be sold by an intermediary such as an agency or a payment processor introduces elements of potential fraud, unfair prices, and poor or unclear refunding of money to customers [3]. Also, the logistic and quality of aircraft maintenance is not sufficient and safe because what is being used now are paper logs or records or digital record systems that are separated from each other which does not allow to verify whether the maintenance or records are original or fake, and tracing the spare parts are also a problem. Another issue to note is the matters of regulations and compliance because aviation is one of the industries which are most strictly regulated with strict adherence to national and/or international standards. Monitoring the data quality and having to adhere to the regional requirements of different countries is still a laborious process [4]. Also, there is limited trust and customer involvement regarding pricing policy, unreliable extra charges, and different assessment of travellers, which makes confidence in airlines deficient. Resolution to these issues can only be achieved through implementation of solution that proactively promotes security, openness and operation efficiency in this industry.

1.1 The Need for Digital Transformation in Aviation

Given the challenges mentioned above, digital transformation has become a priority for the airline industry. Traditional centralized systems are proving to be inadequate in addressing modern-day complexities, necessitating a shift towards more advanced technological solutions.

1.1.1 Enhancing Operational Efficiency

Efficiency is an important factor in the airline business since it helps lower the expenses and offer upgraded customer service. Digital technologies help to automate various operations in an airline as well as make operation decisions based on collected data [5]. A traditional method of ticketing in airline industry take much longer time, involve more than one intermediate party and is expensive, prone to fraud. This has led to direct and efficient communication between the airlines and passengers eliminating any form of fraud as well as falsely inflated ticket prices. Likewise, mishandled baggage continues to be a major issue, which affects the travellers and airline companies economically. There are potential solutions for implementing block chain and IoT [6] based tracking systems help in real time update and reduce chances of misplacing the baggage. Maintenance management is also a complex factor in the safety and efficiency of aircrafts and aircrafts are most of the time not maintained efficiently due to traditional methods of record keeping that separates records and data. This paper also demonstrates how blockchain-based maintenance records offer well-organized records and lower paperwork whereas improving spare parts tracking, compliance along with decreasing aircraft downtime. Also, to mention, those industries involve a vast supply chain for fuel, spare parts, catering, and other necessary services, making it necessary to enhance transparency. Blockchain application in procurement and distribution helps in increased transparency, cutting of cost, and counterfeiting check on aviation parts, and overall supply chain management.

1.1.2 Improving Security and Data Integrity

Since risks of cyber threats are steadily increasing, the need to have a secure and impenetrable system for storing airline data is paramount. Thus, by using blockchain technology, there is the decentralization of records with improved security of information. Counterfeit tickets and unauthorized sales also continue to be common in the air travel industry as actual airline tickets, consequently posing a threat towards the security and financial stability of an airline company [7]. Smart contracts established on the blockchain platform increases the security of tickets because ticket purchases and access cannot frauds and unauthorized people. Also, it important for the passenger's data and privacy especially because airlines deal with substantial personal data. Privacy of the users is protected by encrypting the passenger information and only the relevant parties are allowed to access such information within the prescribed legal requirements like the GDPR [8]. In addition, airlines constitute main threats to cyber criminals mainly because their databases contain large quantities of information. Many conventional systems detected to have singular points of control, so the chance of hacking is always high as compared to blockchain security models which distribute records and therefore decreasing the likelihood of hacking. The Fig. 1 illustrates how blockchain enhances security and data integrity through decentralization, smart contracts, immutability, cryptographic security, and transparency.

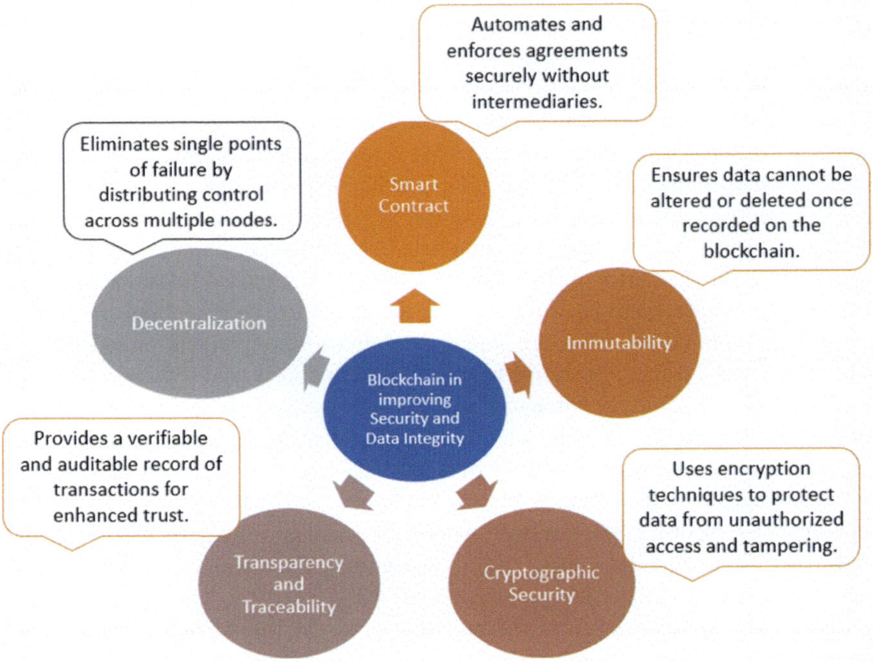

Fig. 1 Blockchain's role in enhancing security and data integrity

1.1.3 Ensuring Seamless Passenger Experience

Modern travellers expect efficiency, convenience, and transparency throughout their journey. Digital transformation enables airlines to enhance the customer experience through personalized and automated services [9]. Lengthy check-in and security procedures often result in delays and inconvenience for passengers, making efficient identity verification a necessity. Blockchain-powered digital identities provide a secure and seamless method for passenger authentication, significantly reducing processing times at airports while enhancing security. Additionally, ticketing and refund processes are often slow and inefficient, leading to customer dissatisfaction. Smart contracts streamline these operations by automating refund policies and ticket modifications, ensuring instant and transparent transactions without the need for intermediaries [10]. Moreover, frequent flyer programs face challenges such as reward fraud and complicated redemption processes, limiting their effectiveness. Blockchain-based loyalty programs enhance transparency and security, allowing passengers to track and redeem rewards instantly, fostering greater customer engagement and satisfaction.

1.1.4 Achieving Regulatory Compliance

Today's passengers demand speed and the freedom to manage their own affairs throughout the trip. With the help of digital options, the whole idea of flying is being recreated, making it possible for consumer expectations to be delivered and met through reinforced and streamlined services [11]. Time-consuming check-in and security processes mean unnecessary time wastage which is why identification processes are crucial. This is a secure and efficient method of authenticating people since via blockchain technology, ensures faster processing of passengers especially at the airports while at the same time increasing security. Also, ticketing and refund policies are generally weak and cumbersome, which create discontent among the customers. Smart contracts also enhance refund policies and modification of tickets in that it simplifies such processes to instant transactions without third parties [12]. Furthermore, frequent flyer programs have issues such as fraud concerning the awards as well as a complex system of award redemption. Loyalty programs that use the blockchain are more secure and offer transparency hence the passengers can quickly track and easily redeem the gift-vouchers hence customer satisfaction.

1.1.5 Enabling Sustainable Aviation Practices

Aviation is perhaps one of the most controlled and monitored industries in term safety and general operations. By using digital transformation, one is able to fulfil compliance because it can then be proven through documented records [13]. Thus, compliance with regulations in the field of aviation is a highly time-consuming and comprehensive process, which has to consider international safety and operational norms. Blockchain ensures that compliance operations are handled automatically and entails records of aircraft inspections, maintenance records, and acts safety audits. Complexity and errors on the traditional manual paperwork exposes vulnerability to negative occurrences such as misplacement and alterations of records [14]. All jobs that have to do with Maintenance can easily record all maintenance work done and Aircraft part changes in block chain technology, making it easier for the regulators to have instant access to information as it provides only the verified information. Also, it is noted that the airlines work across the borders and therefore there is the need to engage with several agencies from across the world. Blockchain makes it easier for the airlines, airports and other aviation stakeholders to share information and conform to the set regulatory requirements easier and more effectively [15].

1.2 Role of Blockchain in Aviation

Blockchain also provides an innovative solution to all problems of the airline industry [16]. Given that it is a distributed database and can create an unchangeable and credible record of transactions, blockchain facilitates transparency and reduces the risk

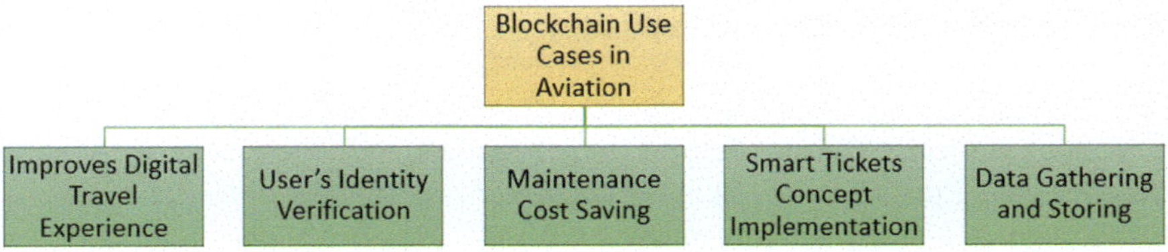

Fig. 2 Blockchain use cases in aviation

of fraud. The main argument is that the potential of the blockchain technology in scope of bringing the new waves of changes to the airline industry is enormous as it will make this industry more open, safe and fast [17]. One of the focus areas is the airline ticketing that is beneficial when it eliminates the middlemen resulting in a reduction in fake ticket selling and is still helpful in cases of better ticket pricing. Smart contracts can also add improvement to refund processes to be executed automatically and without any delay on the client's side. Also, the bag tracking systems built on blockchain technology ensure that passengers and airlines can track baggage movements in real-time, enhancing records for cases of lost baggage and fast and easy processing of claims. In aircraft maintenance, the use of blockchain enables the security of maintenance logs and reduction of possible instances where a substandard product was serviced and passed through the system due to counterfeits. Thus, SCM receives great benefits by addressing issues in real-time tracking such as aircraft component, fuel and spare parts supply, and minimizing the unauthentic material usage through the implementation of blockchain technology [18]. It also offers a higher level of security to the assets of the frequent flyer program and also provides more efficient strongly urged tokenization of the reward points in the block chain platform to enable peer to peer transaction to more involve the customers and reduce the aspect of fraud. In addition, the bureaucracy of compliance is eased because blockchain-generated compliance reports are easily created and easily submitted to regulatory bodies to monitor safety and operational standards that are recorded in the blockchain. This is because blockchain can be implemented in the various focal areas to help airlines work more efficiently, increase customer confidence, and promote better accountability in the aviation sector. The Fig. 2 illustrates various applications of blockchain in aviation, including improving digital travel, identity verification, cost savings, smart ticketing, and secure data storage.

2 Fundamentals of Blockchain Technology

Blockchain technology is built upon several fundamental principles that make it a powerful tool for enhancing security, transparency, and efficiency across industries, including aviation.

This Table 1 provides a structured comparison of the fundamental blockchain concepts and their applications in the airline industry.

Table 1 Key concepts of blockchain technology

Concept	Description	Benefits in aviation
Decentralization	Data is distributed across a network of nodes rather than being stored in a single central authority. Each participant in the network has a copy of the ledger, ensuring transparency and security	– Reduces risk of fraud and cyberattacks – Eliminates reliance on intermediaries in ticketing and payments – Enhances data sharing between airlines, regulators, and airports
Immutability	Once a transaction is recorded on the blockchain, it cannot be altered or deleted. Each block is cryptographically linked to the previous one, ensuring a tamper-proof system	– Ensures secure and verifiable maintenance logs – Prevents data manipulation in ticketing and loyalty programs – Enhances auditability and compliance with aviation regulations
Smart contracts	Self-executing contracts with predefined rules and conditions stored on the blockchain. These contracts automate transactions and processes without the need for intermediaries	– Automates ticketing, refunds, and baggage tracking – Reduces disputes and delays in airline operations – Improves efficiency in supply chain management and loyalty programs

2.1 Types of Blockchain: Public Versus Private Blockchains

Blockchain technology can be categorized into two primary types: Public Blockchain and Private Blockchain, each with distinct characteristics and use cases.

2.1.1 Public Blockchain

A public blockchain is an open ledger that can be accessed by the public and where equal decisions are made by various members of the network [19]. Being a decentralized platform, it works without the influence of an individual or a team performing the work as a middleman. However, it is necessary to note that the transactions

Fig. 3 Decentralized network security in blockchain

conducted in the blockchain are not checked by third parties or banks but are verified through consensus models such as PoW or PoS [20]. This is secure and easy to track but may have limitations since its processing time is several times longer than the time necessary for validation. Public blockchains can be seen as being utilized efficiently in cryptocurrencies [21], such as Bitcoin or Ethereum, etc., or applications such as Decentralized Finance or Non-Fungible Tokens, etc. The Fig. 3 represents a decentralized blockchain network where multiple connected devices ensure secure and transparent data exchange without a central authority.

2.1.2 Private Blockchain

Private blockchain, also called as the permission blockchain is a type that has access control to participation through approval from the network's administration [22]. Furthermore, private blockchain is not open to anyone, it is owned by an organization/several organizations where certain rights to access and validation of blocks, and transactions are controlled by the owners. Some of these blockchains utilize consensus solutions that are also fast such as Practical Byzantine Fault Tolerance (PBFT) as well as Proof of Authority (PoA) [23]. Private blockchains are implemented to a larger degree in the corporate environment, like supply chain management, financial systems and applications, and the sharing of organizational data, and put more value in data privacy and usability rather than decentralization. The Fig. 4 illustrates a centralized network where all connected devices rely on a single central authority for data security and communication.

Fig. 4 Centralized network security

2.2 *How Blockchain Works in Secure Data Management*

Decentralization, encryption techniques, and constant record updates are the fundamental principles of the blockchain that assuage the problem of data security [24]. Every entry is protected by cryptographic hashing (for example, SHA-256) and stored in blocks so that no one can make adjustments to the information [25]. As compared with typical databases where all the information is stored in one place by an administrator, blockchain system involves distributing of information inputs, processing, and storage across the nodes minimizing the effect of hacking and centralized points of failure. All transactions or data modifications are to be encrypted by consensus protocols such as the Proof of Work (PoW) or Proof of Stake (PoS) and can only alterations the original records by allowing only validated ones. Once data is recorded in block and entered in chain, it cannot be changed or removed and therefore makes the complete record highly secure for all the transactions. In the same spectrum, blockchain can incorporate flexible smart contracts to enable information access to specific data as set by specific conditions before specific people could be allowed to retrieve or alter information. While in Public blockchain, all information is open to the various users, the private blockchain way allows limited visibility to specific details seen as highly sensitive such as industrial health, finance, and supply chain. Blockchain together with its features such as, encryption, decentralization, consensus, immutability makes it strategically safe to manage present and future data.

3 Enhancing Security in Airlines with Blockchain

3.1 Preventing Ticket Fraud and Identity Theft

The need to IoT Security in airlines cannot be overemphasized since it is instrumental in tackling two major threats in the flying business; they include ticket fraud, and identity theft [26]. The old school ticketing system involves several brokers through whom an authentic ticket may not reach the buyer and most of the tickets that are in the market are fake or those that have been procured through illicit means. They are avoided in block chain based ticketing since smart contracts and decentralized ledgers are used to validate tickets on real time basis. In particular, every issued ticket is saved on a blockchain net that may hardly be duplicated, altered or sold fraudulently. However, one major threat which currently poses a significant threat to aviation security, particularly for identities of passengers is the identity theft. This because, a group of hackers can easily compromise the appropriate databases and use information found in them for malicious purposes. It is possible to improve the identity verification by applying blockchain and offering passengers' self-controlled identity data storage on the distributed ledger. To avoid impersonation or identity theft as parts of conventional methods of identification, the blockchain-based digital identity technique incorporates bio metrics and cryptographic hash to validate the traveling individuals [27]. This helps prevent identity theft, fastens the process of check-in and boarding, and improves general passengers' confidence. Moreover, blockchain eliminates a centralized approach to data storage in terms of passengers, and therefore, the hackers cannot easily break into the system. Through the use of blockchain, the airlines will be able to design an immobile and secure platform that will help in enhancing the security in the ticketing operations, protect passengers' identity and other aspects of aviation security.

3.2 Securing Passenger Data and Privacy

Security is one of the biggest issues affecting the passengers' information especially in the aviation industry due to the emerging issues such as cyber threats and data breaches [28]. The information that is held in traditional airline databases might include personal information such as passport details, payment information, and travel history and, therefore, good targets for hackers. Blockchain technology presents a rather strong solution since it distributes data, so no concentration of data occurs in one place. Instead of hosting the application and passenger data on a central server which can be hacking, blockchain implements the concept of distributed ledger and thus makes it difficult for the unauthorized people to access this data. In addition, through the use of the key, the data in the Blockchain is secured thus enhancing the privacy of the passengers and protection of their identity from any form of identify theft [29]. Regarding the data privacy, smart contracts can be also used to establish strict rules for passengers' data access, and therefore, decide who can gain access to the information and for what purpose. It also meets requirements outlined in data protection laws such as GDPR since passengers can be informed on how their information will be used, thus affording a chance to record how the information is utilized. Also, applying blockchain smart technology in the recognition of passengers' identities will help the airline minimize cases in which the passengers provide the same information with other parties, thereby jeopardizing their security. In this way, blockchain would help airlines to improve passengers' confidence, make identification much easier and provide vastly improved security measures for handling personal data in the aviation industry.

3.3 Mitigating Cybersecurity Threats in Airlines

Protection of such airline business from cyber threats is crucial because it involves the protection of key infrastructure, passengers' information and operational technology systems [30]. The flight industry has emerged as common target for hackers mainly due to the extensive use of it based IT system for ticketing, flight operations and customer data processing. The conventional centralized database systems are, thus, vulnerable to ransomware attacks, loss of data, and a break down of the systems which may lead to the risk of passengers' lives and huge losses [31]. Cyber threats security risk is a significant benefit offered by blockchain technology since it has a decentralized architecture that is highly resistant against hacking attempts. A proper set of data is shared across the network, and this makes it almost impossible for hackers to compromise or corrupt the data since it is not stored in one area that is easily accessible. Also, through the use of advanced cryptographic encryption, blockchain guarantees that the passenger and operation related details are protected from easy infiltration by hackers. Smart contracts also improve the efficiency and security because the ticket validation, identity check and/or compliance with rules and regulation are automatically carried out minimizing the chances of experiencing a vulnerability in the system. The following are some of the reasons why use of block chain enabled security measures are implemented; the airline is also able to prevent fraudulent works such as unauthorized access to their systems and payment fraud. Also, blockchain enables data to be shared between the airlines, airports, and other regulatory bodies with the privacy of such information from hacker attacks. Thus, through the utilization of blockchain technological solutions based on decentralization and cryptographic security, as well as using verifications of the results of their activity, airlines can enhance the level of security in their work and increase the probability of safe and reliable functioning in the context of the constant digitalization of the aviation industry.

4 Transparency and Trust in Airline Operations

4.1 Blockchain for Real-Time Data Sharing

Real time information exchange is vital in improving accountability, increase effectiveness and credibility of airline operations by enabling airline, airport, regulatory authorities and maintenance service providers to have an easy access to the information they require. Previous data-sharing methodologies are based on central databases that triggers inconsistency issues, loss of data, and even time delay which cause inefficiencies in operations and communication failures [32]. A decentralized system does not have such a problem since it offers a safe, non-adjustable, and centralized data-sharing network, which would need to ensure that the proper records attained and updated by all the relevant actors. It enables passing of flight schedules, passengers' information, maintenance history records as well as cargo tracking records in real time; hence, minimizing duties such as mistakes, time wastage, and fraud. Further, it facilitates operations coherency as well as resource management since real-time commodity availability assists the airlines and airports to manage disruptions and challenges concerning baggage handling and passenger services. This also enhances the aspect of compliance with the regulations because authorities can directly access the confirmed flight and maintenance records without much hassle involved in bureaucratic procedures, safety audits are also achieved in less time. Depending on the cases, passengers want to see prices on tickets being more transparent and clear rules for receiving a refund, as well as the usage of loyalty programs helps create confidence in airline services. Thus, secure, efficient, and verifiable exchanges of data can benefit the aviation industry by increasing its reliability security and customer satisfaction of its stakeholders in order to introduce a work-flow-oriented trust.

4.2 Enhancing Trust Between Airlines, Regulators, and Passengers

Better relations between the airlines, the authorized regulatory bodies, as well as the passengers are quintessential for the healthy operation of this sector. Some of the historical complaints that have demoralized passengers include hidden fees, ticket prices that are hard to decipher, inconsistent refund policies and obscured schedules on matters such as flight delays and regular maintenance records. In the same way, the regulatory organisations are equally in a fix due to scattered data and inadequate reporting systems. These issues are well catered by a decentralized and an immutable system that would ensure that all the parties in the system get the correct information in real time without necessarily going through third party influencers. For the regulator, the verification of compliance becomes almost as smooth because logs of maintenance and other records are maintained, flight safety reports, as well as operational information can be kept secure and quickly reviewed for authenticity. They get more cooperation from the authorities and that leads to getting approvals faster, less compliance issues slow down the working of the airlines and so more efficiency. Customers benefit because of an increase in transparency in issues such as the pricing system, refund policies, frequent flyer programs, and even flight timings and better security in i.d. checks and in baggage management [33]. The elimination of data manipulation risks fosters accountability across the industry, leading to better service standards, improved passenger experiences, and a more secure and compliant aviation ecosystem.

4.3 Case Studies on Transparent Airline Operations

The Table 2 highlights how major airlines are leveraging blockchain technology to enhance transparency and efficiency across various operations. From ticketing and baggage tracking to carbon emission monitoring and fraud prevention, blockchain ensures secure, real-time data management, improving customer trust and regulatory compliance.

5 Improving Efficiency in Airline Management

5.1 Optimizing Baggage Handling and Tracking

A significant subset of improving efficiency when it comes to management of an airline service entails the optimization of baggage handling and tracking. Conventional approaches of the baggage systems involve the use of centralized databases and limited tedious efforts to manage them, resulting in lost bags and slow delivery time. This way, the utilization of decentralized and automated method also improves accuracy and offers real time control over every luggage from check-in to delivery. Using IoT sensors and RFID tags in particular, it is possible to provide the existence of one single logistics chain that links each stage in baggage movement [44]. Travellers can easily receive information on the whereabouts and status of their luggage, and are no longer having to worry or stress over the situation. Further, through electronic documentation of the baggage, one cannot manipulate the luggage and additionally, there is a record of how the luggage was checked in when handling a case. In P2P partnerships, smart contracts can be used as a quality remedy when baggage is lost, delayed or damaged in transit to determine the amount to be paid to the affected party [45]. Concrete enabler: better coordination between airlines and ground handlers: Through better coordination of the activities of airlines, ground handlers, and airports, some common operational pains such as constriction in the baggage claim areas and long turnaround time for the flights are eliminated. In this

Table 2 Case studies of airlines operations

Case study	Airline/organization	Key implementation	Impact on transparency
Blockchain-based ticketing [34]	Lufthansa	Implemented a decentralized ticketing system	Reduced ticket fraud and enhanced pricing transparency
Baggage tracking with blockchain [35]	British Airways	Used a decentralized ledger to track luggage in real time	Minimized lost baggage cases and improved passenger trust
Smart contracts for refunds [36]	Air Canada	Automated refund policies with smart contracts	Improved customer satisfaction with faster refunds
Maintenance record transparency [37]	Emirates	Stored aircraft maintenance logs on a blockchain ledger	Ensured regulatory compliance and reduced tampering risks
Secure identity verification [38]	Delta Airlines	Implemented blockchain for passenger identity verification	Enhanced security and reduced check-in times
Supply chain management [39]	Singapore Airlines	Used blockchain to track and authenticate aviation spare parts	Reduced counterfeit parts and improved safety standards
Loyalty program enhancement [40]	Cathay Pacific	Tokenized frequent flyer miles using blockchain	Provided real-time tracking and secure point redemption
Digital flight data sharing [41]	Qatar Airways	Shared real-time flight operation data with aviation authorities	Improved regulatory compliance and flight efficiency
Fraud prevention in ticket sales [42]	Etihad Airways	Deployed a blockchain-based ticket verification system	Eliminated fake ticket sales and enhanced customer trust
Carbon emission tracking [43]	KLM Royal Dutch Airlines	Utilized blockchain to monitor carbon footprints	Increased transparency in sustainability efforts and offset programs

regard, the airline companies stand to benefit from operational expatriate because it leads to lesser damages to customer baggage, few postponements of flights and enhanced passenger satisfaction hence operational efficiency.

5.2 Blockchain-Based Smart Contracts for Ticketing and Refunds

New possibilities of ticketing and refund functions are implemented through the use of blockchain smart contracts. The elements of self-ticketing are as follows, traditional airline ticketing is a complex process involving several middlemen who charge hefty service charges, they take a long time to process the tickets and the transactions could also be fraudulent. Many passengers have also complained that refunds are so slow and time-consuming for the purpose of verifying so many details [46]. Using smart contracts, major airlines are managing to get rid of such inefficiencies today. It refers to how contracts can be placed directly into code so that any transaction can happen only as long as certain conditions have been met regarding a ticket purchase. This helps to demy the sale of fake tickets as well as bookings of the same ticket by different persons. Also, passengers can be issued immediate refunds for cancellations, delays or change in policy since the system autonomously determines the admissibility of the request and authorize refunds. The decentralized nature of the system ensures that all transactions are recorded immutably, preventing disputes and unauthorized modifications. Moreover, airlines can customize smart contracts to include dynamic pricing, upgrade options, or loyalty program integrations, further enhancing the passenger experience. By reducing reliance on third parties and minimizing paperwork, this approach improves operational efficiency, builds customer trust, and streamlines financial transactions in the airline industry.

5.3 Streamlining Aircraft Maintenance and Supply Chain Management

The management of aircraft and the supply of engines and other diagnosed instruments is critical in demanding safer aircraft for flying as well as reducing operational costs and sustaining efficiency of operation in airlines. Current methods of maintenance recording involve manual record keeping in paper logs or multiple isolated record keeping systems in a computer attached to a plant, and this it delays it, contain errors and results to inefficiencies when recording the occurrence, part replacement and compliance checks. Having a decentralised and immutable record of maintenance also helps in improving the record of checks, repairs, and replacement of any part of an aircraft making the records more credible. Real-time tracking of spare inventories eliminates delays arising from lacks of components or use of fake parts and also enhances aircraft availability hence cutting down the operational time. Other technologies that can be interfaced with IoT include using sensors and other devices coupled with the use of Artificial Intelligence in analyzing the performance of the aircraft and in anticipating the failure occur [47]. This prevents untimely breakdowns which results in preventing most of the time the need to fix the rail Tracks lowering the maintenance cost as well as improving the safety of passengers. A transparent system also increase co-ordination between manufacturer and supplier and regulatory bodies because the end product has to conform to certain quality and safety standard and the suppliers must not supply substandard or blacklisted parts. This particular issue entailing a non-optimal manner of scheduling and provisioning of aircraft maintenance and spare parts can be costly to airlines as it negatively affects the direct operating cost, service reliability, and fleet durability.

6 Blockchain in Airline Loyalty Programs

6.1 Revolutionizing Frequent Flyer Programs

Deregulation and decentralization of frequent flyer programs make these common travelling programs more transparent, secure, and customer-oriented. The traditional method of the same fails in many ways, including a mechanism for accumulating obscure points, restricted remedies, and fraud. This best can be evidenced by the fact that passengers usually struggle to know their rewards balance, clarify different points programs with the partners, or transfer credits to other people. With decentralization, it is possible for the airlines to bring together all the loyalty points, which would then exist in a digital form and would be recorded on the blockchain. It allows tracking of reward points in real-time as well as instant redemption across the airlines or non-airlines partners without experiencing an issue with data falsification or unauthorized transactions. Passengers fully possess their rewards wherein they can buy, transfer or redeem points without going through a middleman. Also, several contracts are self-executing smart contract through the use of IT which guarantees that point redemption, transfer of points or even the expiry of these points will be implemented instantaneously and without bias through contractual terms. This is beneficial to the customer as it increases the level of convenience as well as cutting down airline overheads. Thus, airlines can build triangular customer loyalty and devise more effective and appealing rewards for the representatives of their flying programs.

6.2 Secure and Transparent Reward Redemption

It has emerged clearly that secure and transparent means of redemption of the rewards is a vital factor that needs to be enhanced for the sustenance of the loyalty programs. The traditional consumers frequent flyer programs have a number of drawbacks, which include: intricacy in the redemption process, restrictions on the redemptions, the presence of fees apart from a specific reward and fraudulent activities. Disadvantages of using such programs may include availability of the rewards, blackouts on flights, frequent changes in the value of the points and compatibility or lack of it between airlines and its partners. By so doing, there will be a way of recording the loyalty points in a decentralized and secure manner hence allowing the passengers to check them in real-time. This has the advantage of enabling the users to have flexibility on how to trade the miles so that he or she can be able to use them on services such as upgraded flights, hotel accommodations, car leases, and goods from partner stores. Automated smart contracts [48] provide better redemption features to execute the transactions with the help of set terminology by themselves and without any time lag. It helps to avoid such problems as expired points or unauthorized operations as well as to provide passengers with more transparent system. The benefits of real-time auditing occur in the fact that every transaction can be reviewed and tracked, which in turn, increases the trust of people, especially which this process is visible to passengers, and they can see when and how the necessary bonuses were added. Also, the systems can be integrated to allow a traveller to coordinate different programs and point systems and use them with different airlines and partners.

6.3 Preventing Fraud in Loyalty Programs

Every airline needs to implement mechanisms for avoiding, or at least minimizing, fraud when issuing rewards in relation to loyalty programs since consumers mostly spend money with the expectation that the promised rewards will be compensated. It has been reported that the traditional frequent flyer programs have been mainly exposed to a variety of fraudulent activities such as fake account creation, unauthorized point transfers, reward hacking, and account takeovers. There are those who take advantage of the relatively poor security, hack the program tables to change the balance and use other people's account numbers to cash in on the points, to the detriment of the airlines and much inconvenience to genuine customers. A decentralized and tamper-proof system increases the level of security by providing safe storage of transactions that are associated with the loyalty points. For every issuance, transfer, or redemption, each point is recorded, making it impossible to perform any form of fraud such as double spending or performing any unauthorized transaction. The smart contracts are fixed and self-executing mechanisms that only allow valid transaction, thus eradicating human bias and cheating. Also, Account Authentication and identification enhancement through the use of multi-factor authentication and identity card encryption make accounts securer than before to prevent unauthorized and identity theft. The other benefit is that AI [49] has the potential of incorporating real-time ways of identifying frauds using data analysis. Indeed, the system makes assessments automatically based on the frequency of the transaction, the number of points accumulated and redemeption or used in a given period of time, the terminals or stores from which the fraud is attempted. This helps in not only protect passengers from being subjected to accounts' breaches, but also helps reduce losses incurred by airlines. It seems that the truthful, safe, and technical means of rewarding the loyalty program can decrease fraudulent incidents and increase the level of customer satisfaction for a particular airline as well the airline industry in general.

7 Challenges and Limitations of Blockchain Adoption in Aviation

7.1 Regulatory and Compliance Issues

Regulatory and compliance issues present significant challenges to the adoption of blockchain in the aviation industry. Given that airlines operate across multiple jurisdictions, they must comply with a complex web of international, national, and local regulations governing data security, financial transactions, safety protocols, and passenger privacy. The introduction of decentralized ledger technology brings uncertainty regarding how existing aviation laws apply and whether new regulations are required to accommodate this innovation. One of the primary concerns is data privacy and protection laws, such as the General Data Protection Regulation (GDPR) in Europe and similar laws in other regions. Blockchain's immutable nature makes it difficult to modify or erase personal data, which may conflict with the right to be forgotten and other privacy regulations. Airlines must find a way to balance data transparency with compliance, possibly through hybrid blockchain models that allow selective access and data anonymization. Another regulatory challenge involves smart contracts and their legal recognition. Many aviation transactions, including ticketing, refunds, and aircraft leasing, could be automated using smart contracts, but legal frameworks for enforcing these agreements remain unclear in several jurisdictions. Disputes over contract terms or liability in the event of errors could create legal complications. Additionally, compliance with aviation safety and security regulations must be ensured when using blockchain for critical functions like aircraft maintenance and supply chain tracking. While blockchain can enhance record-keeping and regulatory audits, aviation authorities, such as the Federal Aviation Administration (FAA) and the International Civil Aviation Organization (ICAO), must approve its implementation to ensure it meets safety and operational standards. Collaboration between airlines, regulators, and technology providers is essential to establish clear guidelines for blockchain adoption. Without standardized regulations, the widespread implementation of blockchain in aviation may face delays, limiting its potential to improve efficiency, security, and transparency in the industry.

7.2 Scalability and Integration Challenges

Blockchain technology faces numerous challenges regarding its scalability and integration in the aviation industry and its widespread adoption. The huge number of transactions that occur during an airline's day-to-day operations like ticket bookings, baggage checking, aircraft servicing, and supply chain activities would require the blockchain networks to be able to complete transactions quickly. Much of the existing blockchain architecture, especially public architecture, has no scalability due to low transaction per second (TPS) rates and intensive computation requirements. The speed of how transactions can be completed is one of the key concerns. For instance, Ethereum and Bitcoin blockchains process transactions in a linear manner, which leads to congestion where there are a lot of users. The ability to process information in real time is extremely crucial for the aviation industry during functions such as baggage handling, aircraft arranging and scheduling, and even during the confirmation and boarding procedures for the passengers. The restriction of efficiency may serve as a problem for consumer satisfaction as well as interfere with the operating schedule for the aircraft. There are methods and technologies like layer-2 scaling solutions, sharding, and off-chain frameworks that can help with improving the overall performance of block based chains at aviation systems transitions with digital blocks, but the implementation of these solutions in aviation is still very limited. Another significant obstacle is integration with current airline systems. Airlines rely on complex, legacy IT infrastructures that manage reservations, flight operations, loyalty programs, and compliance reporting. Since these legacy systems were not initially intended for decentralized data processing, they must work seamlessly with blockchain. Retrofitting blockchain into these systems may require significant investments in infrastructure upgrades, middleware solutions, and staff training. Moreover, different airlines, airports, and regulatory authorities operate on various digital platforms, making standardization a critical issue. Without a unified approach to blockchain implementation, interoperability between different blockchain networks and traditional databases may become difficult, limiting the technology's effectiveness. Establishing industry-wide standards and ensuring compatibility with existing IT frameworks is essential for successful blockchain adoption. The aviation sector must make research and development investments, work with blockchain solution providers, and test pilot projects prior to full-scale adoption in order to address scalability and integration issues. For airline operations, hybrid blockchain architectures that blend public and private networks could provide a more flexible and scalable option.

7.3 Cost and Infrastructure Requirements

Airlines and other stakeholders face considerable obstacles as a result of the substantial infrastructure and financial requirements associated with the use of blockchain technology in the aviation sector. In order to achieve a smooth integration into current aviation frameworks, implementing a decentralized system requires not only financial investment but also technology restructuring, training, and regulatory adaptations. The initial outlay for blockchain infrastructure is one of the main financial issues. Airlines and airport authorities must allocate substantial financial resources to set up blockchain-based systems, which include developing or integrating a blockchain platform, upgrading existing IT infrastructure, and ensuring interoperability with legacy systems such as airline reservation databases, baggage tracking systems, and maintenance logs. Additionally, securing partnerships with blockchain developers and technology providers adds to the overall expenditure.

Ongoing operational costs also pose a challenge. Unlike traditional centralized databases, blockchain networks require significant computational power, particularly in permissionless (public) blockchains that rely on consensus mechanisms such as Proof of Work (PoW) or Proof of Stake (PoS). Maintaining decentralized ledgers, ensuring data storage efficiency, and continuously validating transactions demand high processing capacity and energy consumption, which can lead to increased operational costs for airlines? Infrastructure requirements also include the need for skilled personnel capable of managing and maintaining blockchain systems. Airlines must invest in employee training programs to equip their workforce with knowledge of decentralized technologies, smart contract management, and cybersecurity protocols. Recruiting blockchain experts may further drive up costs, as professionals with expertise in this field are in high demand. Another consideration is regulatory compliance, which may require additional investment in legal consultation and compliance management to ensure blockchain implementations align with aviation industry standards, data protection laws, and international regulations. Airlines must also work closely with aviation regulatory bodies to address concerns related to data security, governance, and cross-border operations. Despite these challenges, airlines can explore cost-effective solutions such as hybrid blockchain models, consortium-based blockchain networks, and cloud-based blockchain services to reduce infrastructure costs while benefiting from the advantages of decentralized systems. Collaborative efforts between industry players, regulatory bodies, and technology providers can also help develop standardized and cost-efficient blockchain solutions tailored to the aviation sector.

8 Conclusion and Future Scope

Blockchain technology brings a revolutionary opportunity to the aviation sector by addressing critical problems concerning security, transparency, efficiency, and compliance. Its decentralized nature minimizes fraud, enhances the security of passenger information, and apropos central databases enhances the operating systems. Smart contracts automate processes such as supply chain dealings, ticket sales, and refund issurance, all while cutting expenses and ameliorating business processes. Moreover, health and safety abnormalities can be lessened by bettering the reliability of real-time monitoring through blockchain integration in luggage handling and aircraft maintenance. Although promising, the aviation sector is still hindered by the lack of acceptance owing to the overwhelming regulations, difficulties in scaling, and costly enactment. To surpass these challenges, airlines, regulative bodies, and the technology sector need to collaboratively form low-cost solutions and common frameworks. Nonetheless, in light of the hurdles, blockchain has the potential to revolutionize air travel through a more secure, transparent, and efficient aviation ecosystem making it an appealing technology for the aviation industry. The future of blockchain in aviation is promising, with continuous advancements expected to enhance its adoption across various airline operations. Predictive maintenance, flight optimization, and real-time baggage monitoring may all be further enhanced by combining blockchain technology with artificial intelligence (AI) and the Internet of Things (IoT). Combining public and private blockchains, hybrid blockchain models might provide scalable solutions for the aviation industry. Furthermore, by streamlining passenger processing, the growth of decentralized identity verification might lessen airport traffic and security check hold-ups. It is anticipated that regulatory frameworks will change, making it easier to integrate blockchain-based smart contracts for financial transactions, aircraft leasing, and inter-airline settlements. More cooperation and standardization will spur innovation as more airlines, airports, and industry partners investigate blockchain technology, opening the door to a more technologically sophisticated, safe, and transparent aviation sector.

References

1. Gillen, D., Gados, A.: Airlines within airlines: assessing the vulnerabilities of mixing business models. Res. Transp. Econ. **24**(1), 25–35 (2008)
2. Borenstein, S., Rose, N.L.: How airline markets work… or do they? Regulatory reform in the airline industry. In: Economic Regulation and Its Reform: What Have We Learned?, pp. 63–135. University of Chicago Press (2014)
3. Waterson, M.: Independent review of consumer protection measures concerning online secondary ticketing facilities (2016)
4. Crochemore, L., Isberg, K., Pimentel, R., Pineda, L., Hasan, A., Arheimer, B.: Lessons learnt from checking the quality of openly accessible river flow data worldwide. Hydrol. Sci. J. **65**(5), 699–711 (2020)
5. Dias, C.A.S.: Smart airports: analysing digital transformation and technological advancements in aviation. Doctoral Dissertation (2024)
6. Bonkra, A., Dhiman, P.: IoT security challenges in cloud environment. In: 2021 2nd International Conference on Computational Methods in Science & Technology (ICCMST), pp. 30–34. IEEE (2021)
7. Getaye, T., Ababa, A.: Current practices, challenges and prospects of using credit cards as a payment method: the case of Ethiopian airlines enterprise (2015)
8. Benyahya, M., Kechagia, S., Collen, A., Nijdam, N.A.: The interface of privacy and data security in automated city shuttles: the GDPR analysis. Appl. Sci. **12**(9), 4413 (2022)
9. Talukder, M.B., Hoque, M.: Navigating the digital horizon: transforming travel agencies in the digital era. In: Perspectives on Digital Transformation in Contemporary Business, pp. 383–410. IGI Global Scientific Publishing (2025)
10. Bhattacharya, A., Sharma, V., Mishra, R.R., Sonker, J.: Blockchain technology in hospitality: enhancing security and streamlining operations. In: Hotel and Travel Management in the AI Era, pp. 149–178. IGI Global (2024)
11. Laurier, E., Lorimer, H., Brown, B., Jones, O., Juhlin, O., Noble, A., Weilenmann, A.: Driving and ‘passengering’: notes on the ordinary organization of car travel. Mobilities **3**(1), 1–23 (2008)
12. Staples, M., Chen, S., Falamaki, S., Ponomarev, A., Rimba, P., Tran, A.B., Zhu, J.: Risks and opportunities for systems using blockchain and smart contracts. Data61. CSIRO, Sydney (2017)
13. Johnson, C.W.: A handbook of incident and accident reporting. Fail. Safety-Crit. Syst **1**, 1–1000 (2003)

14. Xu, T., Zhou, Y.: Systems approaches to tackling configuration errors: a survey. ACM Comput. Surv. (CSUR) **47**(4), 1–41 (2015)
15. Ahmad, R.W., Salah, K., Jayaraman, R., Hasan, H.R., Yaqoob, I., Omar, M.: The role of blockchain technology in aviation industry. IEEE Aerosp. Electron. Syst. Mag. **36**(3), 4–15 (2021)
16. Lopes, D.P., Rita, P., Treiblmaier, H.: The impact of blockchain on the aviation industry: findings from a qualitative study. Res. Transp. Bus. Manag. **41**, 100669 (2021)
17. Rashideh, W.: Blockchain technology framework: current and future perspectives for the tourism industry. Tour. Manage. **80**, 104125 (2020)
18. Gulmesoff, I.: A study of aircraft engine counterfeit parts mitigation through the application of blockchain technology. Doctoral dissertation, American Public University System (2024)
19. Cerf, M., Matz, S., Berg, A.: Using blockchain to improve decision making that benefits the public good. Front. Blockchain **3**, 13 (2020)
20. Baliga, A.: Understanding blockchain consensus models. Persistent **4**(1), 14 (2017)
21. Tiwana, P.S., Singh, J.: Load optimization accessions, ramification the QoS in software-defined networking. In: 2023 International Conference on Artificial Intelligence and Smart Communication (AISC), pp. 1013–1017. IEEE (2023)
22. Henry, R., Herzberg, A., Kate, A.: Blockchain access privacy: challenges and directions. IEEE Secur. Priv. **16**(4), 38–45 (2018)
23. De Angelis, S., Aniello, L., Baldoni, R., Lombardi, F., Margheri, A., Sassone, V.: PBFT vs proof-of-authority: applying the CAP theorem to permissioned blockchain. In: CEUR Workshop Proceedings, vol. 2058. CEUR-WS (2018)
24. Juliet, A.H., Kumar, K.S., Kirubakaran, D., Dinesh, M., Thilagam, K.: The future of data protection: empowering data security with identity-centric blockchain technology. In: 2024 2nd World Conference on Communication & Computing (WCONF), pp. 1–7. IEEE (2024)
25. Michail, H.E., Athanasiou, G.S., Theodoridis, G., Gregoriades, A., Goutis, C.E.: Design and implementation of totally-self checking SHA-1 and SHA-256 hash functions' architectures. Microprocess. Microsyst. **45**, 227–240 (2016)
26. Makhdoom, I., Lipman, J., Abolhasan, M., Challen, D.: Science and technology parks: a futuristic approach. IEEE Access **10**, 31981–32021 (2022)
27. Ahmed, M.R., Islam, A.M., Shatabda, S., Islam, S.: Blockchain-based identity management system and self-sovereign identity ecosystem: a comprehensive survey. IEEE Access **10**, 113436–113481 (2022)
28. Sampigethaya, K., Poovendran, R.: Aviation cyber–physical systems: foundations for future aircraft and air transport. Proc. IEEE **101**(8), 1834–1855 (2013)
29. Zhang, N., Zhong, S., Tian, L.: Using blockchain to protect personal privacy in the scenario of online taxi-hailing. Int. J. Comput. Commun. Control **12**(6), 886–902 (2017)
30. Ukwandu, E., Ben-Farah, M.A., Hindy, H., Bures, M., Atkinson, R., Tachtatzis, C., Bellekens, X.: Cyber-security challenges in the aviation industry: a review of current and future trends. Information **13**(3), 146 (2022)
31. Usama, M., Ullah, U., Sajid, A.: Cyber attacks against intelligent transportation systems. In: Cyber Security for Next-Generation Computing Technologies, pp. 190–230. CRC Press (2024)
32. Hafiz, M.M., Zaman, A.K.M., Shaf, M.S.: Preventing data loss using Raft consensus algorithm in a decentralized database system. Doctoral dissertation, Department of Computer Science and Engineering, Islamic University of Technology, Board Bazar, Gazipur-1704, Bangladesh (2023)
33. Mishra, D., Mishra, A.: Improving baggage tracking, security and customer services with RFID in the airline industry. Acta Polytechnica Hungarica **7**(2), 139–154 (2010)
34. Riechmann, J.M.: Blockchain takes to the skies: an assessment of blockchain applications in the airline industry. Master's thesis, Universidade Católica Portuguesa (2020)
35. Nadeem, S.N.: Can blockchain disrupt the traditional airline distribution for the better? If so, what are the benefits of this new technology, and how can it be implemented? Doctoral dissertation, University of Westminster (2018)
36. Alqaisi, A.: Trustworthy decentralized last mile delivery framework using blockchain (2023)

37. Chaterera-Zambuko, F.: Enterprise content management system implementation: insights from the National Library and Archives of UAE. Inf. Dev. (2023). https://doi.org/10.1177/02666669231211539
38. Varma, T.M.: Responsible leadership and reputation management during a crisis: the cases of Delta and United Airlines. J. Bus. Ethics **173**, 29–45 (2021)
39. Low, J.M., Yuan, X.M., Tang, L.C.: Performance linkages between an airport and the air cargo supply chain—evidence from Hong Kong and Singapore. In: 2008 6th IEEE International Conference on Industrial Informatics, pp. 1722–1728. IEEE (2008)
40. De Boer, E.R.: Strategy in airline loyalty. Springer (2018)
41. Alshamsi, M.A., Sipos, A.: The legal implications of the aviation industry' entrance to the metaverse. Access Just. Eastern Eur. 285 (2024)
42. Ancy, G.: Airline industry—Special reference with Emirates Airline. Doctoral dissertation, College of Management and Economics Studies, UPES, Dehradun (2015)
43. Port, U.: Exploring the relationship between the estimated cost of carbon emissions in passenger flights and the reported cost of compensation: a case study of KLM passenger flights from Schiphol Airport (2023)
44. Alghamdi, A., Alansari, S., Alshamsi, R.: A proposed solution for tracking air baggage handling system using IoT platform. In: 2024 Advances in Science and Engineering Technology International Conferences (ASET), pp. 1–5. IEEE (2024)
45. Sinha, D., Roy Chowdhury, S.: Blockchain-based smart contract for international business–a framework. J. Glob. Oper. Strateg. Sourc. **14**(1), 224–260 (2021)
46. Tiwana, P.S., Singh, J.: Software-defined networking's adaptive load optimisation technique for multimedia applications with multi-controller utilisation. In: 2023 9th International Conference on Signal Processing and Communication (ICSC), pp. 167–172. IEEE (2023)
47. Tiwana, P.S., Tiwana, H.S., Sharma, R., Gupta, A.: An scrutiny of run-time ramification for 5-proviso busy beaver proving empirical composition. Int. J. Innov. Technol. Explor. Eng. **8**(9), 726–732 (2019)
48. Dhiman, P., Kaur, A., Iwendi, C., Mohan, S.K.: A scientometric analysis of deep learning approaches for detecting fake news. Electronics **12**(4), 948 (2023)
49. Dhiman, P., Bonkra, A., Kaur, A., Gulzar, Y., Hamid, Y., Mir, M.S., Elwasila, O.: Healthcare trust evolution with explainable artificial intelligence: bibliometric analysis. Information **14**(10), 541 (2023)

Blockchain Applications in Aviation Securing Transactions, Streamlining Operations, and Improving Passenger Experience

Inam Ul Haq, Vandita Nandal, and Himani Uppal

Abstract The aviation industry works within a very complex group of stakeholders which includes airlines, airports, regulatory authorities, and passengers. This system faces lots of challenges and difficulties related to secure transactions, streamlined operations, and improved customer experience. Blockchain has come up as a revolutionary solution, offering decentralized, transparent, and tamper-proof data management. The paper reviews the evolution of the blockchain from the time of Bitcoin until today. Some important applications include secure ticketing using smart contracts, passenger identity management using biometric blockchain solutions, real-time baggage tracking with IoT integration, and reward and loyalty program automation. In addition, blockchain ensures data authenticity, real-time tracking of parts, and predictive maintenance models in aircraft maintenance and safety records. By using traditional decentralized ledgers, blockchain reduces operational inefficiencies, improves security, and builds trust among aviation stakeholders. Its adoption in supply chain management further ensures transparency, counterfeit prevention, and seamless collaboration across the industry. Blockchain is revolutionizing the aviation industry by enhancing transparency, security, and efficiency. Case studies of SITA's Flight Chain, Lufthansa's blockchain loyalty program, and British Airways' flight data experiment demonstrate improved data exchange, fraud prevention, and operational accuracy. These implementations highlight blockchain's potential to streamline airline operations and enhance passenger experience. Yet, among these merits is an array of complications that blockchain adoption will bring, the most telling of which are integration complexities, scalability limitations, and regulatory uncertainties, coupled with high deployment costs. Most crucially, the bulk of aviation legacy systems are not designed to interact with blockchain networks, demanding considerable investment in updating those systems as well as staff training. Furthermore, concerns for data privacy and requirements for standardization at the industry level would act as constraints for the easy adoption of blockchain solutions. Whatever those challenges are, the future of blockchain in aviation looks bright when

I. U. Haq (✉) · V. Nandal · H. Uppal
Department of CSE, FEAT SGT University, Gurugram, India
e-mail: mirinam525@gmail.com

A. Kumar et al. (eds.), *Blockchain in the Tourism Industry: A New Era of Secure and Transparent Travel Solutions*, Information Systems Engineering and Management 57,
https://doi.org/10.1007/978-3-031-95341-5_7

viewed along with the integration of other developing technologies such as Artificial Intelligence (AI), the Internet of Things (IoT), and 5G networks. With AI-based smart contracts and predictive analytics, decision-making will also improve, while IoT-enabled devices will make operational efficiency and accurate asset tracking possible. Speedy adoption of 5G technology will take care of latency challenges to realize real-time transactions and secure data sharing across global aviation networks. This article depicts the transformational potential of blockchain technology for the aviation sector, which could improve process efficiency, heighten safety, and increase stakeholder trust. Although challenges have yet to be resolved, the future will be bright for blockchain use for modern air travel, since research is still ongoing, industry stakeholders collaborate, and regulatory initiatives spring up.

Keywords Blockchain · Airlines · Tourism

1 Introduction

A network full of complexities can be seen through the various stakeholders that usually depend on the governments and international organizations, which include: airlines, airports, passengers, and regulatory authorities. Among these changes, there is a demand for very secure transactions; a faster, streamlined process making it easy to do business with one another; and improved customer experience. To address these challenges blockchain is emerged as a promising technology.

1.1 Evolution of Blockchain Technology

Blockchain technology has traveled a lot since the time of its inception. It all started in 2008 through some anonymous entity that is popularly known as Satoshi Nakamoto. A digital currency which is known as Bitcoin was introduced. More precisely, Bitcoin was into the blockchain, which introduced a publicly distributed ledger to record transactions in a secure, transparent, and tamper-proof way. In the beginning, blockchain simply was the backbone of different cryptocurrencies, but it later became clear that it had uses beyond that [1].

So, as industries started exploring what blockchain is, researchers and tech innovators began working on improving the technology. The first generation of blockchain (Bitcoin) focused completely on decentralized digital transactions. Like that, this generation focused uncompromisingly on simple transactions only only some limitations, such as problems regarding scalability, actually being unable to support complex operations. In 2015, smart contracts were incepted by the second generation, Ethereum being the prototype of its kind. Smart contracts are termed self-executing contracts which do not require a third party to mediate [2]. There came a wider range of potential applications-from finances to healthcare and supply chain

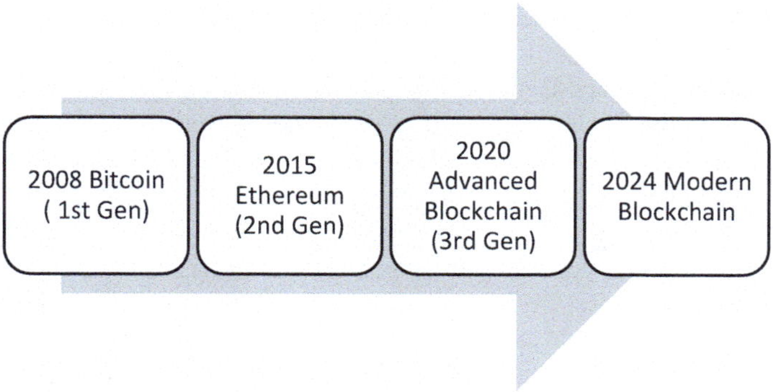

Fig. 1 Blockchain evolution stages

management. Because of the flexibility of Ethereum, blockchain did no longer find itself as a mere enabler of cryptocurrency. At its present stage of development, third-generation blockchain aims at addressing those problems of the past, like slow transaction speeds, excessive energy consumption, and non-interoperability. Latest platforms, like Polkadot, Cardano, and Hyperledger create more scalable, sustainable, and enterprise-friendly blockchain networks. With such advancements, blockchain has gained popularity in sectors like aviation, where it helps to improve customer experience while streamlining operations and security [3].

This means that from a simple ledger for digital currency, it turned into a powerful weapon to support developing secure, decentralized applications. It kept growing and changing as industries started considering its potential. Innovations will keep driving the concept toward mainstream adoption and the eventual revolution of business processes around the globe. Figure 1 shows each generation of blockchain represents a step forward in terms of functionality, scalability, and real-world applicability. From the initial focus on digital currency with Bitcoin to the current exploration of advanced features and enterprise solutions, blockchain technology continues to evolve rapidly. The "generation" concept helps us understand this progression and the key improvements and focuses of each stage.

1.2 Blockchain Adoption in Aviation

Blockchain technology is quite amazing, and revolutionary and is resulting in changing systems. In theory, it is a type of distributed database that is characterized by a different feature from previous traditional databases, where data resides on a central system with backup and redundancy issues [4]. As a result of its present organization, the aviation sector is outdated and fragmented because most of its systems developed over time with little or no interconnection with other air traffic control systems. This results in difficulty in the seamless transmission of data throughout the interconnected and complex network of industry actors [5]. These are some of the problems that can be addressed with the use of blockchain technology. A blockchain is a database made across information "blocks" that are "chained together" and that is continually verified and authenticated by independent users using the database [6]. The adoption of Blockchain within aviation has created some buzz among aviation business stakeholders, mostly due to its remarkable benefits. Indeed, according to the International Air Transport Association (IATA), the aviation industry itself handles more than 50 million tonnes of freight annually, so comprising roughly 35% of the global trade value [5].

The aviation sector might be considered a "safety–critical industry" due to its complexity, high level of dynamic activity, and competitive nature. Factors including precision level, data availability, and data consistency affect airlines' use of data-driven decision-making in the aviation sector [7]. Data inconsistency in the aviation industry can affect safety, airworthiness score, and finance. Blockchain can facilitate a unified view of data to reflect the actual airworthiness score, enhance safety, and increase the sharing of information. Maintenance logbooks for aircraft are still stored manually in the aviation sector as a physical ledger in the portfolio of the airline [8, 9]. Thus, missing maintenance logbooks reduce the trust in the aircraft's integrity and reliability and the aircraft owner has to reconstruct the complete maintenance logbooks from the beginning, which is time and resource-consuming. Such issues can be addressed through blockchain technology whereby decentralized storage services will permit immutable storage of aircraft service records on the Blockchain ledger with complete maintenance records so that transactions do not exist without improving transparency, reliability, availability, and Federal Aviation Administration (FAA) high compliance. Each of these cases will be enhanced through the appropriate application of blockchain technology and smart contracts, which automate company rules and resolve conflicts in a far more dependable and trustworthy method. A smart contract is an auto-executing piece of code establishing confidence among the companies involved. It has a significant impact on the airline industry by storing aircraft maintenance records, verifying crew certificates, and monitoring ground handler operations at the airport [10]. Figure 2 shows blockchain offers numerous benefits for the aviation sector, ranging from improving customer experience to enhancing safety and regulatory compliance. While the technology is still in its early stages of adoption, its potential to transform the industry is significant. As blockchain technology matures and more aviation companies explore its applications, we can expect to see even greater innovation and efficiency in the years to come.

1.3 Comparison with Traditional Systems

For decades, the aviation sector has functioned within a framework dominated by centralized databases and third-party intermediaries, orchestrating crucial operations such as ticketing, frequent flyer programs, and logistical coordination. Blockchain technology has opened a new era, creating a paradigm-shifting decentralized scheme

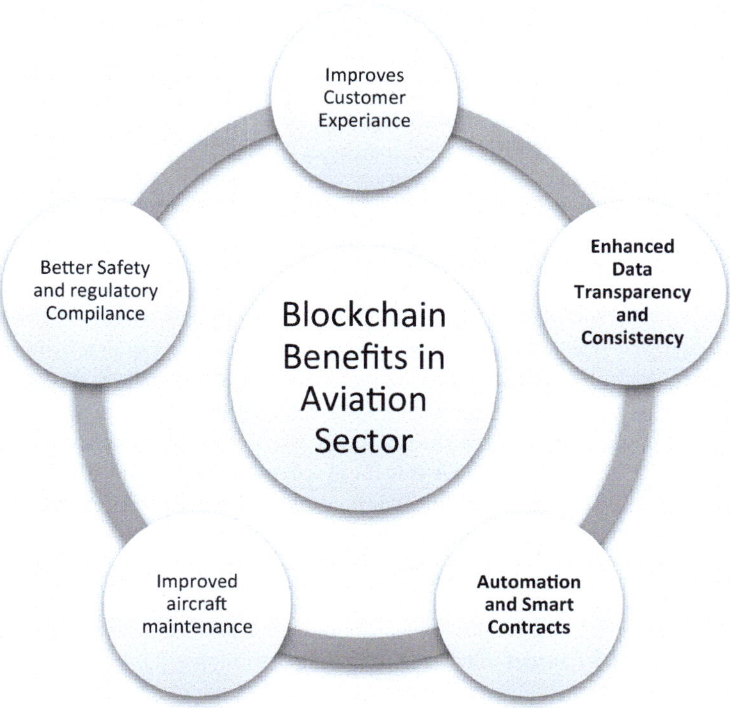

Fig. 2 Blockchain benefits in the aviation sector

in the aviation system, which would efficiently make transparency, high security, and smooth operations across the highly intertwined processes of the sector. Table 1 explains how blockchain improves aviation operations by replacing centralized databases with decentralized ledgers, ensuring data security and immutability. It eliminates third-party intermediaries, reducing costs and delays through peer-to-peer transactions. Transparency is enhanced as real-time traceability replaces fragmented supply chain visibility. Loyalty programs benefit from instant updates, improving customer experience. Additionally, smart contracts automate processes, reducing errors and inefficiencies in-flight data management and operations.

Table 1 Comparison with traditional methods

Aspect	Traditional methods	Blockchain applications
Data management	Centralized databases are more likely to have single points of failure and data leaks	Decentralized ledgers ensure data immutability and enhanced security
Intermediaries	Reliance on third-party intermediaries leads to increased costs and potential delays	Direct peer-to-peer transactions help to lower the intermediary demand
Transparency	Limited visibility across the supply chain, making tracing and verification challenging	Enhanced traceability and transparency across the supply chain, improving stakeholder trust
Loyalty programs	Fragmented systems cause delays in point allocation and redemption	Simplified procedures including real-time updates improve client experience
Operational efficiency	Manual processes and reconciliations lead to inefficiencies and errors	Automated processes through smart contracts, reducing errors and improving efficiency

2 Applications of Blockchain in the Airline Industry

2.1 Secure and Fraud-Proof Ticketing

Blockchain-Based Smart Contracts to Eliminate Ticket Fraud and Scalping: Traditional ticketing platforms suffer losses from fake tickets and unauthorized resale, in turn affecting user trust. These problems can be best solved by utilizing the decentralized structure of blockchain and automated features of smart contracts. The use of smart contracts on the Ethereum blockchain would make ticketing sales more efficient, transparent, and secure. Scams and reselling are effectively curtailed by the system's automation of the entire process of issuing tickets and transferring ownership of tickets [11].

Transparent Ticket Transactions Without Intermediaries: The conventional ticketing methods often contain intermediaries that cause complexity, additional costs, and other possible points of failure. Blockchain technology facilitates the direct interaction between event organizers and attendees, thereby ensuring a smooth ticketing process. In this sense, an entirely decentralized ticketing management called "DeTi" is also based on smart contracts for electronic ticket management. This guarantees that each ticket is unique, traceable, and thereby reducing counterfeiting and regulating any aftermarket sales without any intermediaries [12].

Digital Tokens Replacing Traditional Ticketing Methods to Reduce Operational Costs: Digital tokens in ticketing systems are capable of reducing operational costs substantially since they abolish the need for physical tickets and the logistical hassles involved with distributing them. An analysis of a decentralized digital token architecture for public transport demonstrated that such a system could achieve transaction times comparable to existing methods while potentially lowering operational expenses. Pointing towards much lower operational costs relative to existing options, the study thus goes on to introduce a potential cost-effective ticketing pathway [13].

2.2 Passenger Identity Management

Safe Keeping of Passenger Data: Blockchain's decentralized and immutable ledger is an effective safe storage system that can take on board the passenger's fingerprints in one striped identity verification mechanism in all check posts. Storing data in a network is a system that minimizes dangers associated with central breaches of patient information. Putting information in a decentralized network makes it ambiguous and tamper-proof. The introduction of this secure data management will not only create trust among passengers and aviation authorities but will also make identity verifications an easier process besides reducing administrative time consumed [14].

Enhancing Airport Security with Biometric-Based Blockchain Solutions: Biometric authentication with blockchain technologies can dramatically enhance aspects of airport security while speeding up processing times for all passengers. Biometric identifiers like facial recognition or fingerprints, when securely stored and managed on a blockchain, allow entry to only authorized persons into restricted areas or services. This not only safeguards identity fraud but also enables faster check-in and boarding procedures by permitting swift authentication without repetitive manual verifications. Such an implementation deters fraudulent access, reduces queuing time, prevents lost documents, and turns the passenger into a human carrier of his credentials. Blockchain will keep the biometric data safe and intact, thus leading to a travel experience that is efficient and secure [15].

Empowering Passengers with Self-sovereign Identity (SSI): The Self-Sovereign Identity (SSI) framework has distributed architecture technology designed for blockchain, thus empowering visitors to gain control over their data. Individuals under the SSI model receive their identity credentials under their exclusive management. They may choose when to share data, with whom data should be shared, and for which purpose data is shared. This privacy enhances the devolution of reliance on centralized identity providers. In public transportation, SSI will allow passengers to have the same travel credentials that they can use with various operators in different countries but without compromising the safety of their personal information. This form of decentralization makes the journey easy, but it also harmonizes with modern data protection principles as it grants a person autonomy over his or her personal information [16].

2.3 Baggage Tracking and Management

The implementation of blockchain in baggage tracking and management systems offers significant improvements in real-time monitoring, reduction of mishandled luggage, and overall operational efficiency within the aviation industry.

Real-Time Tracking with Decentralized Ledger: A blockchain technology named decentralized ledger allows for tracking luggage in real-time through multiple checkpoints, resulting in improving accuracy and transparency in the area of baggage handling. An immutable ledger that holds records for every transaction and movement of the baggage allows all the stakeholders—airlines, airports, and travelers—to access the latest information regarding the location and condition of luggage. The additional time and inconsistencies are reduced as opposed to the conventional luggage tracking mechanisms. A study proposed a baggage tracking system utilizing RFID and blockchain technology, ensuring timestamped records of luggage movements and enhancing tracking accuracy [17].

Reduction of Lost or Misrouted Baggage: Using blockchain in baggage management can help to decrease instances of lost or misplaced luggage, hence improving customer satisfaction. The transparency and traceability inherent in blockchain allow for immediate identification of errors or misplacements, enabling swift corrective actions. Singh et al. (2023) described a blockchain-based tracking system to minimize baggage mishandling at smart airports in their research, indicating the capabilities of such technology to increase efficiency in baggage handling and consequently decrease the rate of mishandling [18].

Enhancing Tracking Efficiency through Blockchain and IoT Integration: Integration of IoT with Blockchain ensures increased efficiency in tracking baggage through real-time data acquisition and safe sharing of information. The IoT devices, in particular the RFID tags and sensors, could monitor the baggage conditions and locations continuously and transfer the data into the secure blockchain network for later use by authorized parties. This collaboration secures the integrity and availability of data, making the baggage handling more efficient. Zhang et al. (2021) suggested a blockchain and edge computing-based IoT system for airport luggage tracking, therefore illustrating how such integration may improve security and distribution in baggage management [19].

2.4 Loyalty and Reward Programs

Blockchain technology will transform reward systems and airline loyalty by offering a unified, interoperable system, enhancing liquidity through tokenization, and automating reward distribution via smart contracts. These innovations can address inefficiencies in traditional loyalty programs and improve customer experience.

Unified, Interoperable Loyalty Programs: Reward programs of airlines traditionally have been quite fragmented, making it difficult for consumers to claim points redeemable across different airlines and partners. The decentralized loyalty program could allow passengers to seamlessly exchange reward points with hotels and any other travel-related services across airlines along with hotels. Therefore, an increased amount of interoperability can engage customers better and maximize the worth of loyalty points. Sharma and Singh (2023) state that ecosystems based on blockchain technology will lessen the inefficacies of operational methods while increasing the usability of rewards across brands [20].

Tokenization Enhancing Liquidity and Fraud Prevention: The tokenization of airline miles turns them into a form of digital asset, thereby allowing a passenger to store, trade, or redeem the same on multiple platforms to enhance liquidity. Any traditional loyalty program suffers under various constraints due to the expiration of points and limited redemption options, leading to user dissatisfaction. A blockchain-based token will prevent fraud and double spending, ensuring hugely safe transactions through an immutable ledger. Lee et al. (2022) stated that blockchain and tokenization would be most beneficial in terms of avoiding the consequences of data breaches and unauthorized redemptions on the ecosystems of reward programs.

Smart Contracts Automating Reward Distribution: Smart contracts would automate the issuance and redemption of loyalty rewards upon meeting specified conditions for users. In this way, they eliminate the need for intermediaries and ensure that points are awarded in real-time, as and when customers engage in qualifying activities like flight bookings or hotel stays. As illustrated by Zhang and Li (2021), the use of smart contracts improves the efficiency of program functions and minimizes both administrative costs and errors in the allocation of points [21].

2.5 Aircraft Maintenance and Safety Records

It tremendously benefits operations in predictive maintenance capacities, operational transparency, and data integrity through the application of blockchain technology in aircraft maintenance and safety records. The following summarizes the details of these impacts which are referred to in recent research.

Maintaining Authenticity and Integrity in the Maintenance Logs: Traditional maintenance records, which are commonly paper-based or shared in a central database, are more vulnerable to loss and tampering. A decentralized and immutable ledger would ensure that the data once recorded cannot be altered or deleted by anybody without consensus in the blockchain, and thus it helps to maintain the authenticity and integrity of the maintenance logs. This greatly minimizes chances for inputs into the system of falsified data [22].

Real-Time Tracking of Parts and Service History: Blockchain allows every party-including the operator, airline, and regulator access to real-time data on aircraft parts and their service records. This transparency ensures that every component's life cycle manufacturing through installation and maintenance will have been comprehensively documented and immediately available. Such comprehensive tracking would also significantly increase accountability and give swift responses to possible safety concerns [23].

Enhancing Predictive Maintenance Models through Data Transparency: Predictive maintenance is based on extensive and accurate data to ensure that potential breakdowns are predicted and timely countermeasures can be taken accordingly to repair faulty parts. Blockchain offers a safe and open surrounding for the data so that high-quality input will feed predictive models, resulting in better maintenance schedules and improved safety for the aircraft [24].

2.6 Supply Chain Management in Aviation

With an extensive network of stakeholders and laws, the supply chain of the aviation industry is highly complicated. However, revolutionizing the aviation industry's supply chain through blockchain specifies the solution to the problems of component tracking, counterfeit prevention, and enhancement of trust among parties through transformational solutions.

Tracking Aircraft Components: A decentralized ledger in the case of a blockchain can track aircraft components precisely from its manufacturers to end-users. Such continuous traceability makes it possible to create a history for a part whereby the information can never be altered and made available for maintenance and compliance assessments. For example, it was shown that the inclusion of blockchain in MRO operations could greatly enhance parts optimally monitored in real-time, thus improving inventory management and reducing operational delays [25].

Reducing Counterfeit Parts: Adulterated components can be detrimental to both safety and operational integrity. With the verifiable records that it could provide, blockchain can offer authenticity at every stage of the supply chain. Studies show that blockchain can effectively authenticate parts. This will decrease the distribution of counterfeit items in aviation [26].

Enhancing Trust Among Stakeholders: Trust is a crucial word among suppliers, airlines, and regulators for smooth operations. Blockchain offers such trust with a transparent and tamper-proof data-sharing platform. Researchers proved that using blockchain in developing transparent material supply chains in aviation favors better collaboration and trust among stakeholders [27].

Figure 3 illustrates various applications of blockchain technology in the aviation industry, enhancing security, transparency, and efficiency. Blockchain enables fraud-proof ticketing, ensuring secure and verifiable transactions. It also enhances passenger identity management by leveraging decentralized digital IDs, reducing check-in times and security risks. Baggage tracking and management benefit from

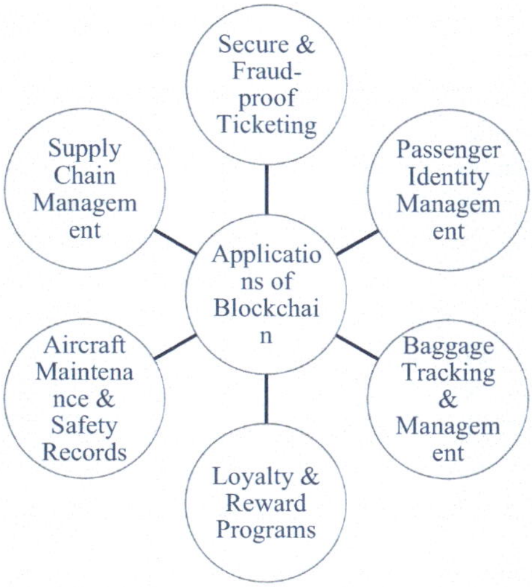

Fig. 3 Applications of blockchain in the airline industry

real-time tracking, minimizing lost luggage. Blockchain revolutionizes loyalty programs by offering transparent, tokenized reward systems. Additionally, aircraft maintenance and safety records stored on an immutable ledger improve compliance and reliability. Finally, blockchain optimizes supply chain management, ensuring the traceability and authenticity of aviation components.

3 Case Studies and Real-World Implementations

The advent of blockchain technology is redefining widely diverse activities in the airline sector-from passenger management to ticketing, supply chain tracking, and operational efficiency. Hence, airlines, aviation service providers, and regulators are looking into blockchain technology to improve transparency, security, and automation in the aviation ecosystem. Several airlines and aviation technology companies have initiated pilot programs and large implementations of blockchain technology to gain increased effectiveness, fight fraud, and streamline information management.

3.1 SITA's Blockchain-Based Aviation Data Exchange

The Société Internationale de Télécommunications Aéronautiques (SITA) is a leading provider of IT solutions to the air transport industry. Recognizing the fragmented and inefficient nature of traditional airline data exchange, SITA launched a blockchain-based initiative to improve transparency, security, and operational efficiency.

The blockchain-based data exchange platform of SITA is a major IT provider for the air transport industry and aims to enhance efficiency and transparency in airline operations. The solution facilitates secure and decentralized data sharing between airlines and their stakeholders, including airports and other aviation players, thus minimizing inconsistencies and data silos. Because of blockchain, SITA has been able to enhance the operational tracking of aircraft, security of passenger data, and accuracy of flight schedules. Furthermore, by creating an immutable ledger of transactions, the technology ensures compliance with aviation regulations, enhancing auditability and traceability in air transport operations.

The SITA Blockchain Initiative aims to enhance real-time data exchange among air traffic control systems, airports, and airlines while ensuring security, data integrity, and operational efficiency. By implementing Flight Chain, a blockchain-powered aviation data exchange, SITA enables automated and verified updates to flight schedules, delays, and cancellations through smart contracts. The immutable blockchain ledger reduces discrepancies in flight data across different airline databases, ensuring all stakeholders access a synchronized, accurate dataset. This enhances passenger experience by providing reliable flight status updates, improves security by preventing unauthorized modifications, and reduces reliance on centralized databases, minimizing operational errors and points of failure. Additionally, it facilitates seamless coordination among aviation entities, streamlining industry-wide networking and collaboration. Blockchain technology can significantly improve data integrity and reduce inconsistencies in flight records by providing a single, immutable version of truth (SITA, 2018) [28].

3.2 Lufthansa's Blockchain Loyalty Program

The Miles & More loyalty program of Lufthansa is one of the largest and most significant airline groups in Europe the first and only one to implement a blockchain. Traditional airline loyalty programs often suffer from complex reward structures, high transaction fees, and susceptibility to fraud. Blockchain has implemented a more secure, flexible, and efficient system for managing airline rewards.

Lufthansa has adapted its Miles and More Loyalty Program to embrace blockchain technology to ensure greater transparency and simplicity in reward redemption for passengers. A decentralized system prevents all types of fraudulent activities-all loyalty points and transactions are safely stored in a tamper-proof ledger. This, in turn, allows passengers to earn and redeem points across multiple partnering companies without the risk of double spending or manipulation. Using smart contracts, Lufthansa's blockchain program automates the transactions, improves efficiency, and lowers administrative costs. All this results in increased levels of engagement of passengers and consequently higher traction of loyalty programs. Thus, there is a lot of improvement that can already be seen in the program.

Traditional airline loyalty programs face several challenges, including limited flexibility in redeeming miles, lack of transparency in point accumulation, fraud risks due to system vulnerabilities, and slow transactions caused by manual processing. To address these inefficiencies, Lufthansa partnered with Winding Tree, a blockchain-based travel platform, to tokenize loyalty points, making them more accessible, transferable, and fraud-resistant. This blockchain-backed system enables travelers to redeem miles instantly without third-party interference, use loyalty tokens across multiple platforms, track points in real time for full transparency, and securely transfer or trade miles. Key benefits of Lufthansa's blockchain loyalty program include fraud prevention via immutable ledgers, real-time point tracking, interoperability with other airline programs, enhanced customer engagement through flexible redemption, and improved transaction efficiency, reducing delays and processing costs.

Blockchain-powered loyalty programs ensure secure, seamless, and fraud-resistant reward transactions, enhancing passenger experience (Lufthansa 2020) [29].

3.3 British Airways Blockchain Experiment for Flight Data

British Airways (BA) experimented with blockchain technology to enhance flight tracking and ensure data synchronization across the different systems of airlines and airports. This was prompted due to challenges that were related to discrepancies in flight schedules, unqualified updates of information, and miscommunication among stakeholders.

British Airways, on the flip side, ventured into the exploration of blockchain technology to enable the enhancement of flight tracking in the lessening of discrepancies in the records of flight data. It also allowed the airline to try out a decentralized ledger, wherein real-time information on flights is going to be stored securely, giving the same accurate and impenetrable data to all stakeholders airport, air traffic controller, and passengers. Such moves resulted in fewer flight delays due to miscommunication and improved overall operational efficiency. Transparency and reliability of blockchain made it easier for regulatory bodies to verify compliance with aviation standards, hence basically reducing operational risks and enhancing customer satisfaction.

Traditional flight data management faces challenges such as inconsistent flight information across airline databases, passenger confusion due to conflicting updates, operational inefficiencies caused by misaligned data, and the absence of a single source of truth, requiring redundant verifications. To resolve these issues, British Airways implemented a permissioned blockchain network that synchronizes flight data between airports, air traffic control systems, and partner airlines. This system ensures real-time updates, maintains secure and immutable records of schedule changes, and eliminates misinformation by providing uniform updates to all stakeholders. The key advantages of this blockchain solution include minimized discrepancies in-flight data, enhanced operational efficiency with fewer delays, improved passenger experience through real-time flight status accuracy, and greater transparency and trust in airline operations.

Blockchain technology has the potential to provide a single version of truth for flight data, ensuring greater transparency and operational efficiency (British Airways 2019) [30].

In addition to these cases, major airlines and industry leaders are continuing to explore blockchain-powered solutions to combat ticket fraud, enhance identity verification, and streamline supply chain management. The adoption of blockchain across various aviation operations highlights its potential to redefine security, efficiency, and trust in the airline industry.

4 Challenges and Limitations

Blockchain technologies bear an incredible promise for changing the airline market. Nevertheless, despite their advantages, many challenges and limitations remain even for their limited-scale adoption. These challenges can be posed technically, regulation-wise, economically, and through full-ended integration into an industry standard.

4.1 Regulatory and Compliance Challenges

Some of the most salient obstacles impeding blockchain adoption in aviation include regulations and regulatory requirements. Since the aviation arena is heavily regulated by different international bodies, the integration of blockchain requires compliance with many different authorities. The role of regulatory bodies in blockchain adoption is critical, as aviation operations must adhere to stringent safety and security standards to ensure passenger and cargo safety [2]. Blockchain technology poses a challenge for regulatory authorities due to its decentralized nature, which may conflict with existing legal frameworks [27].

Aviation stakeholders must also cooperate with entities such as the Federal Aviation Administration (FAA), the International Air Transport Association (IATA), and the European Union Aviation Safety Agency (EASA) to set a common policy on blockchain applications. The current regulatory landscape is not yet well-equipped to handle decentralized technologies, creating uncertainties in the adoption of blockchain in aviation [31]. In the absence of clear regulatory boundaries, the capacity to adopt blockchain is further delayed, because airlines and stakeholders are fearful of facing penalties for non-compliance, as well as operational risks.

4.2 Scalability and Performance Issues

Another barrier to blockchain use in the aviation sector is scalability. The aviation industry has a vast transaction volume and requires fast processing and high-speed rates in transaction processing, passenger data, and logistics management services. Traditional blockchains using proof-of-work consensus mechanisms are slow in processing transactions with wide computational requirements. Blockchain networks such as Bitcoin and Ethereum face significant scalability problems, making them unsuitable for high-frequency transactional systems like airline operations [32].

Blockchain implementations in aviation require improvements in transaction throughput and processing times to support real-time data exchanges [20]. The millions of transactions per day handled by the aviation industry, ranging from ticket bookings to baggage tracking, should be matched by blockchain volumes for efficiency. The scalability challenge makes it difficult for blockchain networks to handle the growing demand for real-time aviation data processing [21].

4.3 Integration with Legacy Systems

Airline companies have invested much in legacy systems over the years. It is quite challenging to replace or integrate the existing systems with blockchain. Airline reservation, passenger management, and cargo handling systems are built on legacy infrastructures that are incompatible with decentralized networks [29].

Integrating blockchain with traditional airline systems requires significant overhauls, leading to high costs and operational disruptions [33]. The technological challenge is coupled with operational as a lot of training will be necessary for employees and stakeholders about the new blockchain-based workflows. This complexity prevents airlines from adopting rapid changes to operate on the blockchain.

4.4 High Implementation Costs

The expense associated with implementing blockchain technology is, without a doubt, one of the most important barriers to adoption in the airline industry. Blockchain systems require a heavy investment in developing, deploying, and maintaining the infrastructure and skills needed for the solution. Airlines face significant cost barriers in deploying blockchain solutions, particularly in terms of hardware, software, and network security investments [34].

Blockchain adoption in airline ticketing systems is financially demanding, requiring companies to allocate substantial resources for research, development, and deployment [35]. Many airlines operate on tight profit margins, and convincing stakeholders to invest in blockchain technology remains a challenge. The return on investment (ROI) for blockchain implementation in aviation is still uncertain, which deters companies from making long-term financial commitments [30].

4.5 Data Privacy and Security Concerns

Blockchain security through encryption and decentralization also offers devices to violate privacy. Transparency, a bedrock attribute of blockchain technology, can be troublesome. Sensitively stored passenger or airline data must be recorded into such a ledger carefully to meet the conditions set forth under existing data protection laws, such as the General Data Protection Regulation (GDPR). Sensitive passenger and airline data once recorded on an immutable ledger, must be handled carefully to ensure compliance with data protection laws. The transparency of blockchain may lead to conflicts with data privacy laws, especially in cases where passenger details and flight records are publicly accessible [36].

While blockchain ensures data immutability, it is difficult to modify or erase sensitive information once it has been recorded, posing challenges in compliance with right-to-be-forgotten policies [37]. Airlines now have to try to balance the transparency offered by the blockchain and the requirement of data privacy. Cybersecurity risks remain a challenge, as attackers may exploit vulnerabilities in smart contracts and blockchain applications to manipulate airline transactions [38].

4.6 Energy Consumption and Environmental Concerns

Some blockchain consensus mechanisms, particularly proof-of-work (PoW) systems, are computationally intensive and consume much energy. Blockchain applications in aviation need to shift towards more energy-efficient models such as proof-of-stake (PoS) or hybrid consensus mechanisms to ensure sustainability [30].

Airlines aiming to integrate blockchain into baggage tracking and theft prevention systems must consider energy-efficient blockchain networks to reduce carbon footprints [39]. As sustainability becomes a priority for the aviation sector, blockchain solutions must align with environmental standards [40]. The aviation sector keeps making efforts towards sustainability, and blockchain solutions must be aligned with the environment. Green blockchain technologies, such as energy-efficient mining protocols, must be developed to mitigate aviation's carbon footprint [41].

4.7 Industry-Wide Collaboration Challenges

For blockchain technology to operate efficiently in the aviation sector, it requires the full collaboration of many different entities, including airlines, airports, regulatory agencies, and service providers [42, 43]. Difficulties arise in reaching a consensus and uniform adoption of the technology across the industry [44, 45]. The lack of standardization in blockchain protocols across different airline networks makes interoperability a major challenge [46].

A unified approach is needed for flight data sharing on blockchain networks, but competition between airlines often impedes such collaborations. The absence of standardization will result in a fragmented rendering of blockchain adoption, thus rendering it less effective [47].

4.8 Smart Contract Limitations

Smart contracting technology; however, carries its limitations. Because of buggy code, smart contract programs may yield unwanted consequences that later cannot be amended since the contract is already enforced. Bugs in smart contract code can result in financial losses or system vulnerabilities, requiring rigorous testing before deployment in airline operations [48].

Smart contracts in air cargo management must accommodate dynamic conditions such as flight delays, regulatory changes, and emergency handling, which require flexible coding structures [7].

While blockchain holds great promise for transforming the airline industry, its adoption is not without challenges. Regulatory compliance, scalability, legacy system integration, high costs, security concerns, environmental impact, and collaboration hurdles all pose significant barriers. As blockchain technology evolves, addressing these challenges will be crucial in determining its long-term success in aviation.

Table 2 presents key challenges associated with integrating blockchain technology in the aviation industry and their potential solutions. Regulatory and compliance issues arise due to blockchain's decentralized nature, which can be addressed by establishing aviation-specific policies through governing bodies like IATA and FAA. Scalability issues affecting transaction speeds can be mitigated using Layer-2 solutions, sharding, or hybrid blockchains. The lack of compatibility with existing airline systems requires middleware solutions for seamless integration. High implementation costs can be managed through public–private partnerships and government incentives. Data privacy concerns, particularly with GDPR, can be handled using permissioned blockchains and off-chain storage. Energy consumption issues from Proof-of-Work (PoW) blockchains can be resolved by adopting Proof-of-Stake (PoS) or energy-efficient alternatives. Industry collaboration challenges due to standardization gaps can be tackled through blockchain consortiums. Finally, smart contract limitations can be mitigated by conducting rigorous auditing before deployment.

Table 2 Blockchain challenges versus possible solutions

Challenge	Description	Possible solutions
Regulatory and compliance issues	Decentralized blockchain conflicts with centralized aviation regulations	Establish aviation-specific blockchain policies via IATA and FAA
Scalability issues	Current blockchain transaction speeds are too slow for aviation demands	Implement Layer-2 solutions, sharding, or hybrid blockchains
Integration with legacy systems	Existing reservation, ticketing, and logistics systems are not blockchain-compatible	Develop middleware solutions for smooth blockchain integration
High implementation costs	Airlines face budget constraints in adopting blockchain technology	Public–private partnerships and government incentives for R&D funding
Data privacy and security	Passenger data stored on immutable ledgers conflicts with GDPR laws	Implement permissioned blockchains and off-chain storage solutions
Energy consumption issues	Proof-of-Work (PoW) blockchain networks require high energy usage	Shift to Proof-of-Stake (PoS) or energy-efficient hybrid blockchains
Industry collaboration challenges	Lack of standardization across airline networks leads to interoperability issues	Establish industry-wide blockchain consortiums for unified adoption
Smart contract limitations	Bugs in smart contracts could cause financial and operational losses	Rigorous smart contract auditing before deployment

5 Future Outlook

Blockchain technology combined with new technologies such as Artificial Intelligence, the Internet of Things, and even 5G holds great potential for transforming airline business processes. International air transport has been increasingly focused on improving the efficiency, security, and transparency within the aviation system; this reliance is expected to promote the ever-increasing adoption of blockchain technology over the next few years. Some futuristic expectations for blockchain in aviation include seamless automation, heightened security levels, and cost-effective operations other words, a new frontier for air travel.

5.1 AI-Powered Blockchain Systems

Optimized operations for airlines through AI-powered blockchain solutions comprise predictive maintenance, ticketing, fraud detection, and personalization of passengers. A large portion of flight and maintenance data can be scrutinized through machine learning algorithms, enabling proactive measures to resolve any emerging issues before they develop into serious problems. Automating procedures relative to smart contracts steers any manual intervention away and further aids the flow of transactions. The combination of AI and blockchain enhances operational efficiency by providing automated decision-making capabilities for airlines, reducing downtime, and improving safety [26]. Another innovation can be AI-supported and blockchain-integrated virtual chatbots functioning to improve customer services with secured and personalized communication in response to encrypted passenger data.

5.2 Integration of IoT and Blockchain

By integrating devices with the Internet of Things with blockchain technology, the organization of logistics and management of assets would be transformed in the areas of baggage handling and cargo management. The real-time data sharing over smart contracts on blockchains will also reduce the incidence of lost baggage, thus improving passenger experiences. The detection of malfunction can be greatly enhanced with the input of real-time sensor data from individual aircraft components into secure databases. It significantly enhances the tracking of wear and tear, improving safety costs in maintenance. Blockchain-powered IoT applications can enhance security and traceability in aviation logistics, ensuring greater transparency across the supply chain [49].

5.3 5G and Blockchain for Real-Time Transactions

The adoption of 5G will be geared toward eliminating another major barrier to the entry speed of blockchain transactions, namely very slow transaction speeds. Airlines will be able to process their transactions on a blockchain in real time, as fast as possible. Operations will include validation of tickets, passenger identification, and fast linkage of communications between airports and aircraft. Further, the 5G technology provides digital identities and e-passports supported by blockchain, as it would speed up check-in at airports while improving their security.

5.4 Increased Acceptability of Decentralized Solutions

More companies are expected to adopt decomposing solutions as airlines have opened up an understanding of the security, cost-savings, and operational efficiency of blockchain. The trials of blockchain-enabled applications such as smart ticketing systems, loyalty programs, and even identity management are being piloted by most airlines and airport authorities. Blockchain-based payment systems will also allow cryptocurrency transactions that enable more flexible and secured payment options for passengers. With growing support from the industry and continuous advancement in technology, the adoption of blockchains in aviation is poised to rise in the coming years, whereas decentralization in airline data management will also mitigate cybersecurity risks associated with centralized databases, thus leading to a more secure airline operating system.

5.5 Sustainability and Green Blockchain Solutions

As concerns about environmental sustainability grow, the future of aviation's blockchain will point toward solutions that consume less energy. Airlines will more likely continue moving toward eco-friendly consensus mechanisms such as Proof-of-Stake and hybrid models as consensus algorithms to reduce energy consumption and carbon emissions. Sustainable innovations in blockchain technology will fit into the side commitments of the aviation industry toward minimizing its environmental footprint. Furthermore, blockchain can facilitate the effectiveness of carbon offset by transparency in emission tracking and carbon credit trading. The airline industry can harness smart contracts built on blockchain technology to automate their authentication processes around sustainable aviation fuel (SAF) use-more so to advance the industry into greener air travel.

5.6 Barriers and Considerations in Future Adoption

While the initial performance of blockchain in aviation looks promising, many challenges remain to be overcome for widespread adoption. Investment, technical expertise, and regulatory compliance all have substantial roles in the deployment of blockchain technology with the present airline infrastructures. Thus, setting up global blockchain standards that ensure interoperability among different airlines and the other stakeholders within aviation will require a collaborative industry effort. Additionally, privacy concerns and cybersecurity issues must be addressed—the utmost concern in establishing confidence in passengers and airlines toward blockchain. Surmounting these hurdles through research and strategic implementation will be critical for the successful maximization of blockchain's potential in the aviation sector.

This scenario continues to progress as the years go by and evolving technologies, such as blockchain, will transform the landscape of the airline industry over time. For instance, converging AI, IoT, and 5G along with blockchain will change the innovation call into overall improved operational efficiencies for airlines as well as enhanced passenger experience. Future predictions are a shift toward fully automated airline processes, enhanced security frameworks, and cost-effective solutions. For the time being, there are still a lot of challenges left to be solved, but as research continues and as the industry collaborates for the advancement of such a technology, widespread acceptance of blockchain in aviation will take place, making air travel safer, more secure, and more efficient in the future.

6 Conclusion

The airline industry is seeing a major revolution through the implementation of the blockchain, which touches almost every aspect of airline operations, from ticket sales to passenger identity management, baggage handling, and aircraft maintenance. A blockchain with its decentralized and immutable ledger system can further ensure the security, transparency, and efficiency of the processes while reducing risks of operations and chances of fraud. The research has mapped out salient applications of blockchain technology in aviation, outlining how the adoption of this innovative technology may enhance regulatory compliance, customer experience, and supply chain management. Blockchain technology certainly has its advantages; however, wide-scale adoption poses significant challenges, such as regulatory uncertainty regarding the use of blockchain technology, issues related to interoperability, and high costs of integrating it with existing airline systems. Concerns also exist regarding the capability of such networks to scale to accommodate the very great number of transactions in the aviation sector. These challenges will require continued research, cooperation between industry players, and the drafting of favorable regulations to smoothen the adoption pathway. This technology shall only be getting more powerful as it merges with AI, IoT, and 5G, thereby enabling real-time data exchange, automation, and greater safety of transactions. There must be cooperation between the airlines and the regulators to arrive at standardized blockchain protocols for broader adoption throughout the sector.

In conclusion, blockchain technology stands to be the paradigm-shifting technology in the airline industry for providing data security, operational efficiency, and building customer confidence. While there are hurdles to be addressed, the enormous opportunities await the sector. Continued research and pilot projects that endorse blockchain viability will only hasten the adoption of the technology in the aviation sector, leading to a more efficient, transparent, and technologically advanced air travel ecosystem.

References

1. Azzi, R., Chamoun, R.K., Sokhn, M.: The power of a blockchain-based supply chain. Comput. Ind. Eng. **135**, 582–592 (2019)
2. Ahmad, R.W., Salah, K., Jayaraman, R., Hasan, H.R., Yaqoob, I., Omar, M.: The role of blockchain technology in aviation industry. IEEE Aerosp. Electron. Syst. Mag. **36**(3), 4–15 (2021)

3. Momoh, M.O., Chinedu, P.U., Nwankwo, W., Aliu, D., Shaba, M.S.: Blockchain adoption: applications and challenges. Int. J. Softw. Eng. Comput. Syst. **7**(2), 19–25 (2021)
4. Ismail, L., Hameed, H., AlShamsi, M., AlHammadi, M., AlDhanhani, N.: Towards a blockchain deployment at uae university: performance evaluation and blockchain taxonomy. In: Proceedings of the 2019 International Conference on Blockchain Technology, pp. 30–38 (2019)
5. Choudhary, C., Haq, I.U., Rather, A.H.: Utilizing dynamic QR codes to enhance secure payment transactions: an approach to secure computer based transactions. In: 2023 International Conference on Communication, Security and Artificial Intelligence (ICCSAI), Greater Noida, India, pp. 552–557 (2023). https://doi.org/10.1109/ICCSAI59793.2023.10421294
6. Li, X., Lai, P.L., Yang, C.C., Yuen, K.F.: Determinants of blockchain adoption in the aviation industry: empirical evidence from Korea. J. Air Transp. Manag. **97**, 102139 (2021)
7. Poleshkina, I.: Blockchain in air cargo: challenges of new World. In MATEC Web of Conferences, vol. 341, p. 00021. EDP Sciences (2021)
8. Schyga, J., Hinckeldeyn, J., Kreutzfeldt, J.: Prototype for a permissioned blockchain in aircraft MRO. In: Hamburg International Conference of Logistics (HICL) 2019, pp. 469–505. epubli GmbH (2019)
9. Efthymiou, M., McCarthy, K., Markou, C., O'connell, J.F.: An exploratory research on blockchain in aviation: the case of maintenance, repair and overhaul (MRO) organizations. Sustainability **14**(5), 2643 (2022)
10. Hewa, T., Ylianttila, M., Liyanage, M.: Survey on blockchain based smart contracts: applications, opportunities and challenges. J. Netw. Comput. Appl. **177**, 102857 (2021)
11. King, O., Goodell, G.: A decentralised digital token architecture for public transport (2020). arXiv preprint arXiv:2012.01382
12. Rafati Niya, S., Bachmann, S., Brasser, C., Bucher, M., Spielmann, N., Stiller, B.: Deti: a decentralized ticketing management platform. J. Netw. Syst. Manage. **30**(4), 62 (2022)
13. Megala, G., Venkatesan, R.: Empowering secure and transparent ticket sales using smart contracts on the ethereum blockchain. In: Applying Internet of Things and Blockchain in Smart Cities: Industry and Healthcare Perspectives, pp. 131–148. IGI Global (2024)
14. Mühle, A., Grüner, A., Gayvoronskaya, T., Meinel, C.: A survey on essential components of a self-sovereign identity. Comput. Sci. Rev. **30**, 80–86 (2018)
15. Sharma, P.K., Singh, S., Jeong, Y.S., Park, J.H.: Distblocknet: a distributed blockchains-based secure sdn architecture for iot networks. IEEE Commun. Mag. **55**(9), 78–85 (2017)
16. Zhu, X., Badr, Y.: Identity management systems for the internet of things: a survey towards blockchain solutions. Sensors **18**(12), 4215 (2018)
17. Mul, S., Philip, A., Correia, M., Gadhikar, L.: Baggage tracking using rfid and blockchain technology. In: 2021 4th Biennial International Conference on Nascent Technologies in Engineering (ICNTE), pp. 1–5. IEEE (2021)
18. Jiang, Y., Tran, T.H., Williams, L.: A surveillance-and-blockchain-based tracking system for mitigation of baggage mishandling at smart airports. J. Airline Oper. Aviat. Manage. **2**(2), 33–51 (2023)
19. Wang, F., Si, P., Sun, E., Su, Y.: Bei-tab: Enabling secure and distributed airport baggage tracking with hybrid blockchain-edge system. In: 2021 IEEE 21st International Conference on Communication Technology (ICCT), pp. 1221–1225. IEEE (2021)
20. Udegbe, S.: Impact of blockchain technology in enhancing customer loyalty programs in airline business. Int. J. Innov. Res. Adv. Stud **4**(6), 257–263 (2017)
21. Pérez-Sánchez, M.D.L.Á., Tian, Z., Barrientos-Báez, A., Gómez-Galán, J., Li, H.: Blockchain technology for winning consumer loyalty: social norm analysis using structural equation modeling. Mathematics **9**(5), 532 (2021)
22. Aleshi, A.: Secure aircraft maintenance records using blockchain (SAMR) (2018)
23. Andrei, A.G., Balasa, R., Costea, M.L., Semenescu, A.: Building a blockchain for aviation maintenance records. In: Journal of Physics: Conference Series, vol. 1781, No. 1, p. 012067. IOP Publishing (2021)

24. Jensen, W.L., Jessing, S., Chiu, W.Y., Meng, W.: AirChain-towards blockchain-based aircraft maintenance record system. In: 2022 IEEE International Conference on Blockchain and Cryptocurrency (ICBC), pp. 1–3. IEEE (2022)
25. Fernández-Caramés, T.M., Blanco-Novoa, O., Froiz-Míguez, I., Fraga-Lamas, P.: Towards an autonomous industry 4.0 warehouse: a UAV and blockchain-based system for inventory and traceability applications in big data-driven supply chain management. Sensors **19**(10), 2394 (2019)
26. Zhang, C.: Research on transparency and optimization of aviation materials supply chain based on blockchain. In: 2024 IEEE 13th International Conference on Communication Systems and Network Technologies (CSNT), pp. 334–339. IEEE (2024)
27. Lopes, D.P., Rita, P., Treiblmaier, H.: The impact of blockchain on the aviation industry: findings from a qualitative study. Res. Transp. Bus. Manag. **41**, 100669 (2021)
28. Anurag, C.C., Thakur, J., Bhardwaj, H., Haq, I.U., Rather, A.H.: Revolutionizing wind energy: a enhanced machine learning perspective on wind turbines fault prediction. In: 2023 2nd International Conference on Automation, Computing and Renewable Systems (ICACRS), Pudukkottai, India, pp. 1763–1769 (2023). https://doi.org/10.1109/ICACRS58579.2023.10404768
29. Riechmann, J.M.: Blockchain takes to the skies: an assessment of blockchain applications in the airline industry. Master's thesis, Universidade Catolica Portuguesa (Portugal) (2020)
30. Cheung, T., Li, B., Lei, Z.: A paradigm shift in the aviation industry with digital twin, blockchain, and AI technologies. In: Handbook on Artificial Intelligence and Transport, pp. 323–346. Edward Elgar Publishing (2023)
31. Revin, V., Shevcheko, Y.: The opportunities and challenges of implementation of blockchain in aviation industry. Молодий вчений **4**(2), 206–209 (2019)
32. Abuayied, S., Alajlan, F., Alghamdi, A.: Implementing blockchain in the airline sector. In: Enabling Machine Learning Applications in Data Science: Proceedings of Arab Conference for Emerging Technologies 2020, pp. 379–388. Springer Singapore (2021)
33. Irannezhad, E., Mahadevan, R.: Is blockchain tourism's new hope? J. Hosp. Tour. Technol. **12**(1), 85–96 (2021)
34. Mahfouz, A., McKey, V.: Blockchain adoption to improve efficacy and transparency within the airline industry: aviation leasing sector. In: Logistics Research Network Conference, pp. 8–10 (2021)
35. Sahoo, A.K., Pasayat, A.K.: Implementing blockchain in airline ticketing system. In: AIP Conference Proceedings, vol. 2852, No. 1. AIP Publishing (2023)
36. Bierrings, M.J., Sivakumar, G., Wunderlich, N.: Blockchain in the aviation industry: a decentralized solution to the transparency issue in baggage handling. In: Business Digital Transformation: Selected Cases from Industry Leaders, pp. 45–72. Springer International Publishing, Cham (2023)
37. Madhwal, Y., Panfilov, P.B.: Blockchain and supply chain management: aircrafts' parts' business case. Ann. DAAAM Proc. **28**, 1051–1056 (2017)
38. Barua, B., Kaiser, M.S.: Blockchain-based trust and transparency in airline reservation systems using microservices architecture (2024). arXiv preprint arXiv:2410.14518
39. Muruganantham, A., Joseph, B.: Smart airline baggage tracking and theft prevention with blockchain technology. Test Eng. Manage. **83**(3), 3436–3440 (2020)
40. Barua, B., Kaiser, M.S.: A next-generation approach to airline reservations: integrating cloud microservices with AI and blockchain for enhanced operational performance (2024). arXiv preprint arXiv:2411.06538
41. Nadeem, S.N.: Can blockchain disrupt the traditional airline distribution for the better? If so, what are the benefits of this new technology, and how can it be implemented. Doctoral dissertation, University of Westminster (2018)
42. Haq, I.U., Hassan, I.U., Shah, H.A.: Machine learning techniques for result prediction of one day international (ODI) cricket match. In: 2023 IEEE 8th International Conference for Convergence in Technology (I2CT), pp. 1–5. IEEE (2023)

43. Ramesh, K., Gupta, P., Singh, A.: 5G-enabled blockchain for real-time airline transactions and security enhancements. IEEE Trans. Transp. Syst. **18**(1), 22–39 (2024)
44. Haq, I.U., Rather, A.H., Kaur, G.: A comparative analysis of machine learning algorithms for the early prediction of cardiovascular disease. In: 2023 2nd International Conference on Edge Computing and Applications (ICECAA), pp. 987–993. IEEE (2023)
45. Valafar, A., Maleki MinBashRazgah, M., Zarei, A., Davood, F.: Identifying and ranking the antecedents of digital marketing development based on blockchain technology(case: aviation industry). Bus. Intell. Manage. Stud. **12**(46), 219–257 (2023)
46. Rather, A.H., Haq, I.U.: Intelligent framework for early prediction of diabetic retinopathy: a deep learning approach. In: International Conference on Paradigms of Communication, Computing and Data Analytics, pp. 377–388. Springer Nature Singapore, Singapore (2023)
47. Zkik, K., Sebbar, A., Nejjari, N., Lahlou, S., Fadi, O., Oudani, M.:. Secure model for records traceability in airline supply chain based on blockchain and machine learning. In: Digital Transformation and Industry 4.0 for Sustainable Supply Chain Performance, pp. 141–159. Springer International Publishing, Cham (2023)
48. Yadav, J.K., Verma, D.C., Jangirala, S., Srivastava, S.K., Aman, M.N.: Blockchain for aviation industry: applications and used cases. In: ICT Analysis and Applications, pp. 475–486. Springer Singapore (2022)
49. Patel, R., Choudhury, A.: Blockchain-powered IoT applications in aviation logistics: security and traceability enhancement. Int. J. Aviat. Technol. **12**(2), 67–85 (2023)

Enhancing Supply Chain Transparency and Efficiency in Tourism Through Blockchain Technology

Khushbu Cheetu, Gurwinder Singh, Anil Manohar Dogra, and Neetu Malik

Abstract The tourism industry operates within a diverse and complex linked supply chain, and connects to hotels, transport services and travel agencies with local businesses and regulatory bodies. While incompletes such as data fragmentation, scam risk and high transaction costs continue to challenge stakeholder confidence along with even business. The traditional tourism supply chain depends on centralized systems, which increases operating costs and data security risk during the delay. To solve these challenges, the blockchain technique provides a decentralized alternative, ensuring more openness, security and automation in tourism operations. Unlike traditional platforms, blockchain removes the need for middlemen using smart contracts and cryptographic verification to automatically act. This change reduces fraud risk, transaction costs and disabilities in economic processing and management of the supply chain. This chapter suggests how blockchain integration can revolutionize tourism supply chains, with special attention to simplify payments, pursue logistics, refine inventory management and raise customer experiences. With real-time tracking, automatic execution of contract and decentralized verification mechanisms installs blockchain a safer, efficient and transparent ecosystem, which lays the foundation for the tourism industry clearly for the future.

Keywords Blockchain · Tourism supply chain · Decentralized ledger · Transparency · Fraud prevention · Logistics management · Digital transactions

K. Cheetu
Department of CSE, Chandigarh University, Punjab, India
e-mail: khushbu.e16075@cumail.in

G. Singh (✉) · A. M. Dogra
Department of AIT-CSE, Chandigarh University, Punjab, India
e-mail: singh1001maths@gmail.com

A. M. Dogra
e-mail: anil.e13431@cumail.in

N. Malik
Department of Mathematics (AIT-CSE), Chandigarh University, Punjab, India

A. Kumar et al. (eds.), *Blockchain in the Tourism Industry: A New Era of Secure and Transparent Travel Solutions*, Information Systems Engineering and Management 57,
https://doi.org/10.1007/978-3-031-95341-5_8

1 Introduction

The tourism sector plays a very important role in global economic development, while contributing to employment generation, cultural exchange and regional development. Operating as an interlinked-network, it depends on the coordination of accommodation suppliers, transport services, travel agencies, tour operators, local sellers and regulatory bodies. Still, it faces challenges to ensure efficiency, openness and confidence in such a diverse and fragmented system. The traditional tourism supply chains often struggle to handle the data silos, inconsistent information-streams, false orders, delays in international transactions and dis-abilities like arbitration costs. Also, the lack of manual record keeping and dividing data (in real time) contributes to some problems like redundant booking, multiple pricing and customer experiences. It is expected that the tourism industry will cross 11 trillion by 2030 with the economic footprint, which increases the demand for advance solutions to improve its management.

Blockchain Technology explains how the tourism industry controls security and transparency in transactions. When it was introduced to support Cryptocurrency, blockchain has developed into a powerful tool for decentralized, tamper-proof data management. In this context, by redeeming the main functions such as cryptographic security, decentralization and automated smart contracts, tourism activities can streamline operations, prevent fraud, reduce transaction fees and create more and more confidence among stakeholders.

This chapter focuses on how blockchain can customize tourism supply chains, which begin with intensive examination of current disabilities and better coordination and increasing requirements for responsibility. This then examines cases of the use of the real world, such as securing ticket systems, procedures for verification of suppliers and strategies for the prevention of cheats. In addition, this chapter provides a significant evaluation of the challenges of blockchain, including scalability, regulatory barriers and complications of system integration. By giving a broader perspective on the role of blockchain in tourism, this chapter highlights its ability to reopen industry practice. Since the region continues to embrace digital changes and is moving towards more trust-focused operations, stands out as an important driver for blockchain efficiency, safety and long-term stability-one smart, chooses the basis for a more flexible tourism ecosystem.

2 Understanding the Tourism Supply Chain

The supply chain is a complex, interweaving system that brings different stakeholders together to facilitate spontaneous travel experiences for millions of people around the world. Unlike traditional production chains, the tourism supply chain is primarily a service-oriented system, which contains a linear stream of tangible goods from production to final distribution. So, the efficiency depends much more on the coordinated e orts between many entities, making a successful effective cooperation [1]. Therefore, from tour planning and marketing to reservation and post-planning services, tourism supply chain processes a comprehensive range of participants. The composition and clear understanding of the underlying challenges are that it is necessary to evaluate how blockchain technology can improve operational efficiency and transparency.

The travel chain consists of several interconnected players, each contributing to different aspects of the passenger's journey. These include physical and digital travel agencies, residential suppliers (hotels, resorts, short-term prices), transport services (airlines, railways and car owner), tour operators, destination management organization (DMO), local service providers (restaurants, cultural sites) and online booking platforms [2]. Despite its shared goals of distributing a simple travel experience, these stakeholders often work in isolation, depend on fragmented systems, independent contracts and diverse technical infrastructure. For example, a single holiday package may require a travel agency to coordinate with an airline, a hotel and a local tour guide, each using different platforms with limited differences. This lack of integration often leads to nonconformities, information exchange and challenges in service synchronization, eventually affects customers' satisfaction and operational efficiency [3].

This fragmentation leads to significant inefficiencies:

- Limited Information Access: Stakeholders often lack real-time insights into availability, pricing, and consumer preferences, leading to issues such as overbooking, hidden fees, and missed opportunities for personalized services [4].
- Fraud and Trust Deficiency: Counterfeit tickets, unreliable service providers, and transaction disputes remain prevalent, exacerbated by intermediaries who add costs and complexity [5].
- Cross-Border Transaction Challenges: Given the global nature of tourism, inefficiencies arise due to currency exchange hurdles, regulatory variances, and slow settlement mechanisms [6].
- Centralized Platforms and Single Points of Failure: Dependence on centralized entities like Global Distribution Systems (GDS) and Online Travel Agencies (OTAs) increases transaction fees, creates operational bottlenecks, and exposes stakeholders to cybersecurity risks [7].

The economic importance of the tourism sector emphasizes the pressure to adapt the supply chain. According to reports from the World Travel and Tourism Council (WTTC), tourism was responsible for 10.4% of the global GDP in 2019, making it one of the largest contributors of the world economy [8]. But, the COVID-19 epidemic exposed the biggest weaknesses in the tourism supply chain, which emphasizes the important requirement for flexibility, adaptability and digital upgrading [9]. As the outcomes the crisis, businesses have focused attention to improve efficiency, reduce costs and restore consumer trust to reduce the industry's stability and increase the reliability of the service [10]. So, blockchain technology appears as a powerful

Table 1 Key stakeholders in the tourism supply chain

Stakeholder	Role in tourism supply chain
Hotels and accommodation providers	Offering lodging solutions to travelers
Airlines and transportation services	Facilitating local and international travel
Tour operators and travel agencies	Designing and marketing travel packages
Local vendors and suppliers	Providing goods, dining, and cultural experiences
Government and regulatory bodies	Establishing policies and oversee tourism operations
Financial and payment services	Ensuring secure transactions and fraud mitigation

solution for these challenges, offering a decentralized, safe and transparent structure to remove disabilities and strengthen industry-wide coordination [11].

2.1 *Key Stakeholders in the Tourism Supply Chain*

The tourism supply chain integrates the supporters of many stakeholders to provide smooth travel experiences. The following table emphasizes primary participants and their roles' as described in Table 1.

2.2 *Current Inefficiencies and Challenges*

Despite its significance, the supply chain can withstand significant disabilities that affect operating and customer satisfaction. The Table 2 emphasizes the general challenges and their implications.

These inefficiencies disrupt tourism operations, impacting both businesses and travellers.

Table 2 Common inefficiencies in the tourism supply chain

Inefficiency	Impact on tourism supply chain
Lack of traceability	Difficulty in monitoring transactions and its verification of legitimacy
Trust deficit	Lack of consumer trust due to fraud risks
Data fragmentation	Poor communication channels among stakeholders
Fake services and fraudulent bookings	Unauthorized ticketing and deceptive transactions
Slow payment processing	Delayed, expensive and disabled transactions
Regulatory compliance issues	Complex legal frameworks

2.3 The Need for a More Transparent and Integrated Supply Chain

The tourism sector is a diverse, but run as a fragmented network, where hotels, airlines, tour operators, travel agencies and local activities work together to create spontaneous travel experiences. Systematic disability—such as data silos, lack of coordination and scam transaction can affect the operating efficiency of the field. It is important to meet these challenges to reduce financial risks, improve customer experiences and strengthen the service.

A more integrated and transparent tourism supply has benefits including:

- Facilitating Seamless Collaboration: Enable real-time data exchange between stakeholders to increase efficiency and service coordination.
- Trust and security Improvement: Take advantage of irreversible digital records to reduce disputes and promote consumers' confidence.
- Optimization of operational logistics: to improve the distribution of resources and increase the order, payment and transport processes to reduce delays.
- To reduce fraud and fake activities: Use a secure verification system to prevent unauthorized ordering and misleading service requirements.
- Strengthening regulatory compliance: Ensure that all transactions follow global tourism policy and moral trade practice.

Despite the progression of the digital infrastructure, many tourism supply chains continue to rely on centralized systems, which often leads to high costs, disabilities and security weaknesses. The dependence on intermediaries complicates further openness and trust, making it difficult for companies to conform transactions and verify service providers. These limitations emphasize the increasing requirement for decentralized, automated and secure structure to handle tourism operations.

The Fig. 1 blockchain technology provides a transformative approach to facing major challenges in the tourism industry. By decentralizing data management, securing data integrity and taking advantage of smart contracts for automated transactions, blockchain transparency, efficiency and prevention of fraud can significantly prevent prevention. The practical application from Digital Ticket Confirmation to Safe and Easy Payment Processing complements their ability to redefine the logistics of security tourism and establish a more reliable, trust driven supply chain.

3 Blockchain and Supply Chain Transparency

The integration of blockchain technology in the tourism supply chain has the opportunity to solve long-term disabilities by increasing visibility, confidence and responsibility. As a decentralized digital ledger, the blockchain transactions record in a distributed network, ensuring that all computer programming proof, verification and irreversible [12]. Each transaction, or blockage, is cryptographically associated with

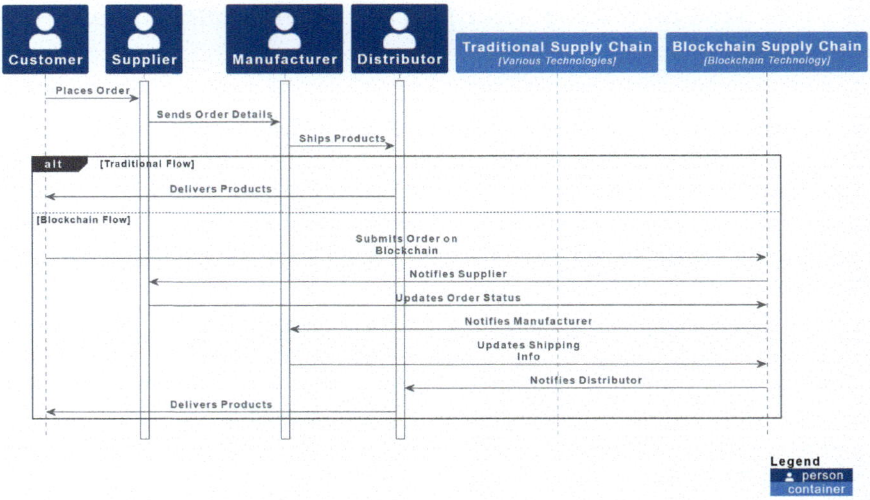

Fig. 1 Blockchain-powered tourism logistics framework illustrating interactions between bookings, payments, service providers, and consumer feedback

the previous one, creating a fixed sequence of records that prevent unauthorized changes and ensure unanimous change and unanimous-based verification [13]. These basic principles make blockchain an ideal solution to deal with important questions such as fraud booking, data fragmentation and real-time information exchange [5].

3.1 Enhancing Transparency in the Tourism Supply Chain

One of the most pressured challenges in the tourism industry is a lack of openness, such as prominent stakeholders—including hotels, airlines, travel agencies and service providers—often work in the herring system with limited data sharing [4]. This fragmentation causes service availability such as overbooking, hidden fees and incompatibility, which eventually reduces the customer experience.

Blockchain technology addresses these concerns by creating a shared, confirmable and real-time data ecosystem that allows all authorized stakeholders to reach the exact, updated information [11]. For example, a hotel that uses blockchain can immediately update the availability of the room on multiple booking platforms, so that travel agencies and customers can get real-time updates. This reduces the possibility of double order, improves the customer's confidence and optimizes the operating function low. Similarly, airlines can use blockchain to update the availability of real-time seats, ensure more coordination with travel agencies and reduce booking errors as outlined in Fig. 2.

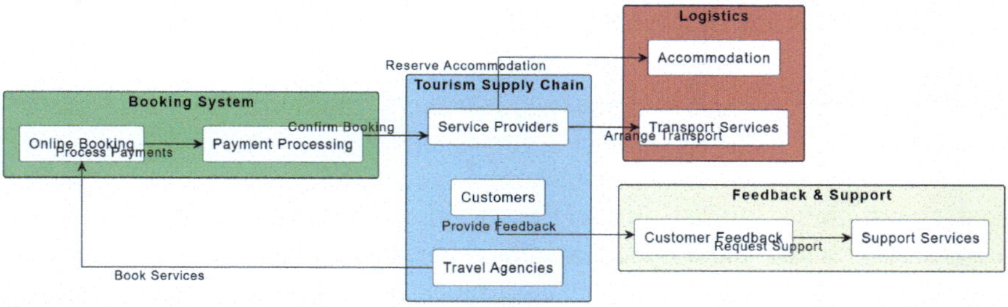

Fig. 2 Comparison of traditional and blockchain-based supply chains

3.2 Smart Contracts for Automation and Fraud Mitigation

A main innovation of blockchain technology is a smart contract, a self-complied with digital agreement that is safely stored and performed on blockchain [14]. Unlike traditional contracts, which require manual enforcement and maintenance of third time, smart contracts validate autonomously valid conditions and meet the transactions after the pre-determined criteria are completed. By removing middlemen, this technique reduces the delay in treatment, reduces the cost of transactions and reduces the risk of frauded [15]. In the tourism industry, smart contracts can increase confidence and financial security between service providers and customers. For example, a passenger for a tour operator can be safely paid in a smart contract and can only be released by successful termination of the service. This automatic mechanism ensures that suppliers distribute their promised services to protect customers from potential fraud or disputes. By terminating manual processing and increasing economic openness, the smart contract tourism creates confidence and efficiency in the ecosystem as mentioned in Table 3.

Table 3 Blockchain solutions for enhancing tourism supply chain transparency

Challenge	Blockchain solution	Example application	References
Overbooking	Real-time, shared data	Airlines update seat availability	[16]
Fraudulent bookings	Immutable transaction records	Verified ticket sales	[5]
Hidden pricing	Transparent cost tracking	Clear hotel rate visibility	[11]
Payment disputes	Smart contracts for automated settlements	Escrow for tour payments	[15]

3.3 Decentralized Architecture for Direct Transactions

Blockchains exchanges intermediaries and exchanges between revolutions in transactions in the tourism industry, among the direct colleagues-to-letter exchange. Unlike traditional tourism supply chains, which rely on centralized ordering platforms and financial institutions, blockchain operates on a distributed network, and ensures that the transaction is safe and transparent without the need for third-party participation.

By offering a divided, tamper-proof laser, blockchain efficiency increases, reduces operating costs and reduces the dependence on middlemen. One of the most influential applications is in the border transition transaction, where traditional systems often introduce currency exchange and intricate government requirements. Blockchain-based payments bypass these challenges, activate spontaneous international transactions while maintaining compliance and reducing the conversion fees [18]. For example, when flights or hotel reservations are stored on blockchain, it is irreversible and immediately confirmed by both customers and service providers. This eliminates the risk of fraudulent requirements, unauthorized cancellation and payment disputes, promotes a safer and reliable tourism ecosystem as mentioned in Table 4. By taking advantage of the transparent and decentralized nature of blockchain, stakeholders can ensure more responsibility, better security and a friction-free ordering experience.

4 Enhancing Supply Chain Efficiency with Blockchain

Blockchain technology forms the efficiency of the supply chain by reducing the dependence of traditional financial intermediaries by enabling secure, proximal transactions as depicted in Fig. 3. In the tourism sector, traditional payment systems often include banks, payment processors and third-party services, and introduce delays, high transactions and potential security weaknesses. With the use of decentralized finance (DeFi), tourism companies can make financial transactions effective, which can reduce the need for mediated-based approval. Blockchain-operated payment systems benefit from smart contracts, which automatically validate and carry out transactions after completing predicted conditions. This ensures quick, safer and transparent settlements, significantly reduces treatment time, which reduces the risk associated with return and fraud as outlined in Table 5. By eliminating unnecessary

Table 4 Comparison between traditional and blockchain-based supply chains

Feature	Traditional system	Blockchain-based system
Data security	Vulnerable to manipulation	Immutable and verifiable
Transparency	Limited	Full visibility for stakeholders
Fraud mitigation	Susceptible to fraud	Tamper-proof transactions
Trust mechanism	Third-party validation	Decentralized consensus

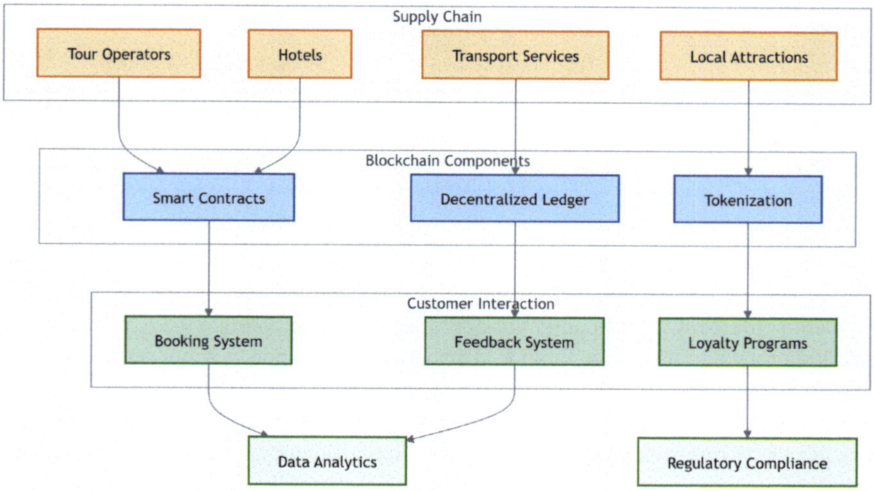

Fig. 3 Blockchain-based tourism supply chain model

Table 5 Comparison of payment methods in the tourism supply chain

Payment method	Processing time	Security level
Traditional bank transfers	2–5 business days	Moderate
Credit card transactions	Instant to 24 h	High, but risk of chargebacks
Blockchain payments	Instant to a few minutes	Very high (cryptographic security)

obstacles, blockchain paves the way for more efficient and cost-effective economic structure in the tourism industry.

Blockchain adaptation in payment processing changes the border transaction by eliminating the requirement for currency conversion and third-party approval. It streamlines international orders, reduces the cost of transactions and ensures rapid settlement. In addition, blockchain emphasizes small and independent tourism companies, enabling them to accept payments directly from global travellers, bypassing the high fees and treatment of delays imposed by traditional economic intermediaries.

4.1 Reduction in Operational Costs

Blockchain technology increases the efficiency of the supply chain by automatically by automatic, digitization of documentation and reduce mediated expenses. In traditional tourism operations, more verification stages such as documents' authentication, confirmation of order and payment harmony-administrative overhead and delays are increasing. By taking advantage of a decentralized account book, blockchain

Table 6 Cost reduction with blockchain implementation

Cost factor	Traditional system	Blockchain-based system
Documentation	Paper-based	Digital and automated
Transaction fees	High (intermediaries)	Low or negligible
Fraud prevention costs	Significant	Reduced due to transparency

eliminates fruitless data storage and ensures real-time access to the records of transactions for all stakeholders. In addition, smart contracts automate approval and payment settlement, significantly cut operating costs, and reduce the dependence of middlemen as mentioned in Table 6. This streamlined approach not only improves efficiency, but also increases economic openness in the tourism ecosystem.

With the cost reduction of overhead, tourism companies can offer more competitive prices by improving the general operational efficiency. In addition, blockchain increases economic accuracy, reduces human errors in traditional transaction systems.

4.2 Smart Logistics and Inventory Management

Blockchain technology that enables the sharing of truth data, revolutionizes inventory management for tourism services suppliers including hotels, airlines and travel agencies. An integrated, tamper-proof system helps prevent the overbooking of blockchain, improve the demand forecast and adapt resource allocation, and ensure more efficient and spontaneous ordering experience.

The traditional inventory management system often struggles with anomalies between ordering platforms and service providers, causing double order, cancellation of the last minute and disabled resource distribution. Blockchain solves these challenges by acting as a single source of truth and ensuring that all stakeholders have synchronized access to real-time ordering, cancellation and accessibility updates. For example, the airlines can use blockchain to immediately update the availability of seats, allowing travel agencies and booking platforms to show accurate real-time data, which can reduce the risk of overbooking and plan conflicts. Similarly, hotels can use blockchain-controlled inventory tracking to streamline accessibility control in the room, prevent duplicate reservation and maximize the coating rate.

4.3 Fraud Prevention and Security

Blockchain technology increases the safety of tourism transactions by offering a transparent and tampering-proof laser, effectively reduces the risk of false ordering,

Table 7 Security enhancements with blockchain

Security concern	Traditional system	Blockchain-based system
Counterfeit bookings	High risk	Prevented via smart contracts
Unauthorized transactions	Common	Reduced due to decentralization
Payment fraud	Chargeback issues	Irreversible secure transactions

scams and suppliers' scams. Traditional tourism systems often lack strong verification protocols, which become sensitive to the scammers who benefit from errors in ordering and payment processes as mentioned in Table 7. By recording each transaction permanently, blockchain ensures integrity and prevents unauthorized changes. In addition, smart contracts automate transactions, only make payment when the conditions before the e-east are fulfilled. This eliminates false claims, unauthorized changes and payment disputes, and promotes greater confidence and security in the tourism ecosystem.

With the irreversible blockchain account book, transactions are completely audible and safe, providing more protection for both businesses and passengers. In addition, blockchain digital identity increases verification, which helps to reduce the risk associated with fraud and unauthorized access. By ensuring that only verified individuals and institutions participate in transactions, the blockchain trips strengthen security and trust in the ecosystem.

4.4 Future Implications of Blockchain in Tourism Supply Chains

Integration of blockchain technology in tourism supply chains promises long-lasting improvement in efficiency, fraud prevention and long-lasting improvement in service. To automate payment, reduce administrative overhead, increase inventory's tracking and prevent scam transactions, blockchain appears as a basic technique for tourism's future. As the industry goes through digital changes rapidly, the blockchain supply chain is ready to redefine management, improves openness, efficiency and safety. Convergence of blockchain with new technologies, such as artificial intelligence (AI) and Internet of Things (IoT), will expand their applications, promote a more paired and intelligent tourism ecosystem.

5 Results and Discussion

This section provides a comprehensive evaluation of the cutting-edge status of the tourism industry in India, focusing on important elements along with consumer delight, fees, geographical distribution of vacationer locations and non-secular evaluation of passenger reaction. These findings highlight the massive ache factors in the enterprise and display how blockchain generation can run modifications by increasing operational efficiency, making sure openness, adapting expenses and improving passenger self-assurance and satisfaction.

The visualization presented in this section and analytical insight outlines the strategic benefits of blockchain integration into the complex landscape of tourism supply chains. By addressing the long-time incompetence and weaknesses, blockchain has the opportunity to revolutionize the way tourism companies, ensuring more secure, transparent and customer-focused industries.

5.1 *Blockchain-Based Smart Ticket Pricing*

The following Fig. 4 illustrates the trends ticket pricing achieved by using the blockchain based systems while ensuring transparent and secure transactions.

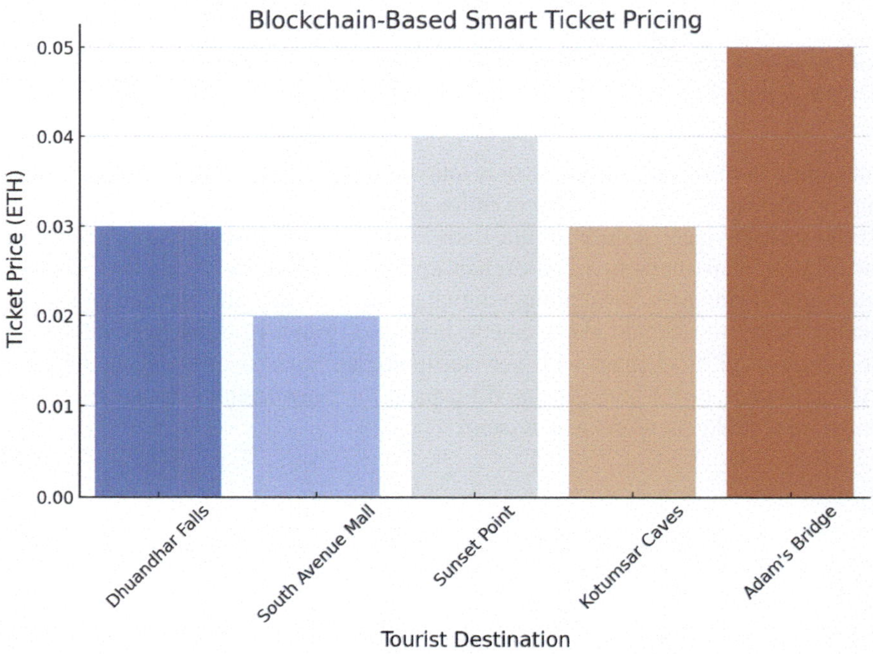

Fig. 4 Blockchain-based smart ticket pricing (in ETH)

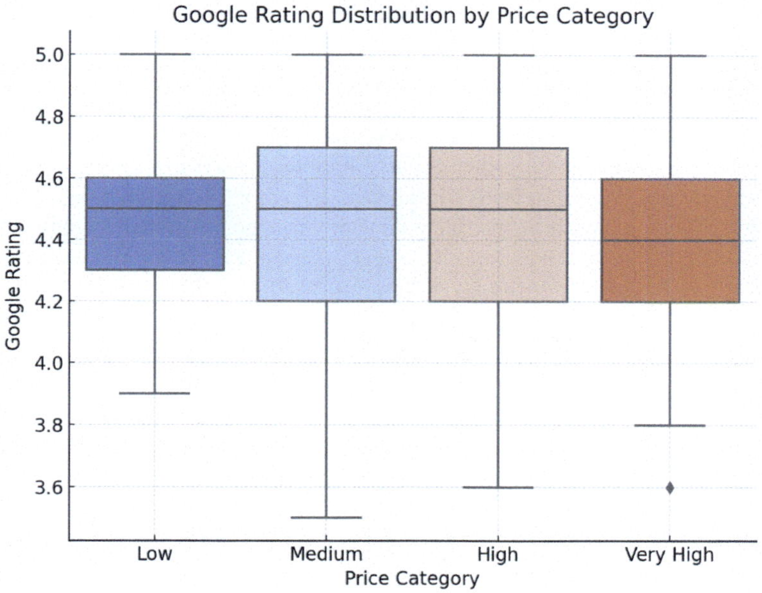

Fig. 5 Google rating distribution by price category

5.2 Google Rating Distribution by Price Category

Blockchain integration can address the discrepancy between pricing and customer satisfaction by enhancing service reliability. Figure 5 shows Google rating distributions across price categories.

5.3 Number of Tourist Spots per State

The complexity of the tourism supply chain varies significantly by geography. Figure 6 presents the number of tourist spots per state, indicating regions where blockchain solutions could significantly streamline logistics.

5.4 High-Demand States Based on Ratings and Tourist Spots

States experiencing high demand and ratings indicate high operational requirements. Figure 7 provides a combined analysis of ratings and tourist spot concentration, highlighting potential blockchain application areas.

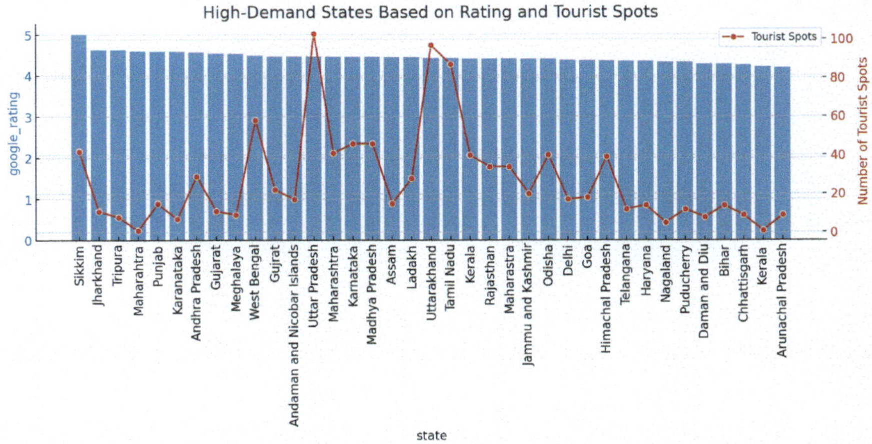

Fig. 6 Number of tourist spots per state

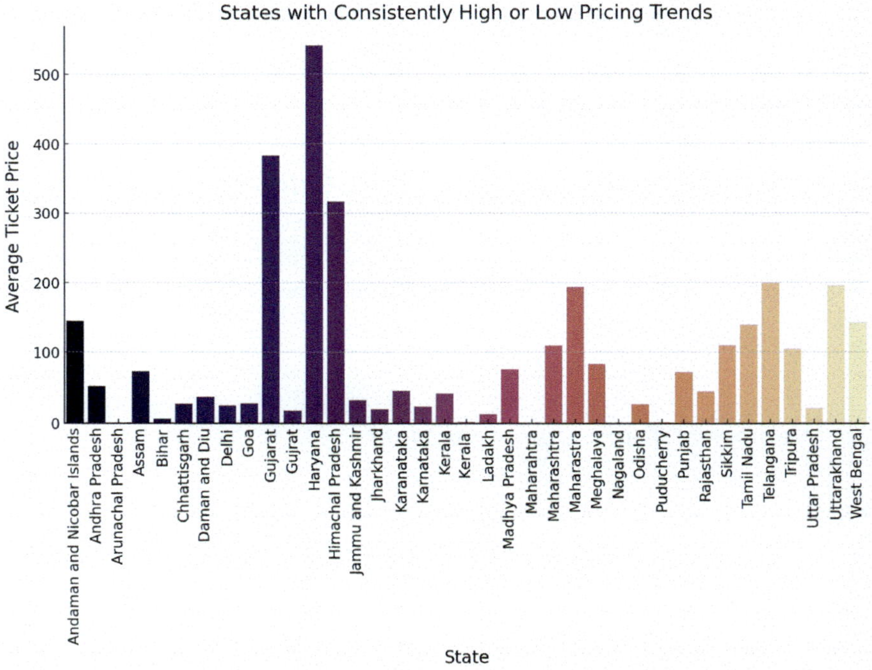

Fig. 7 High-demand states based on rating and tourist spots

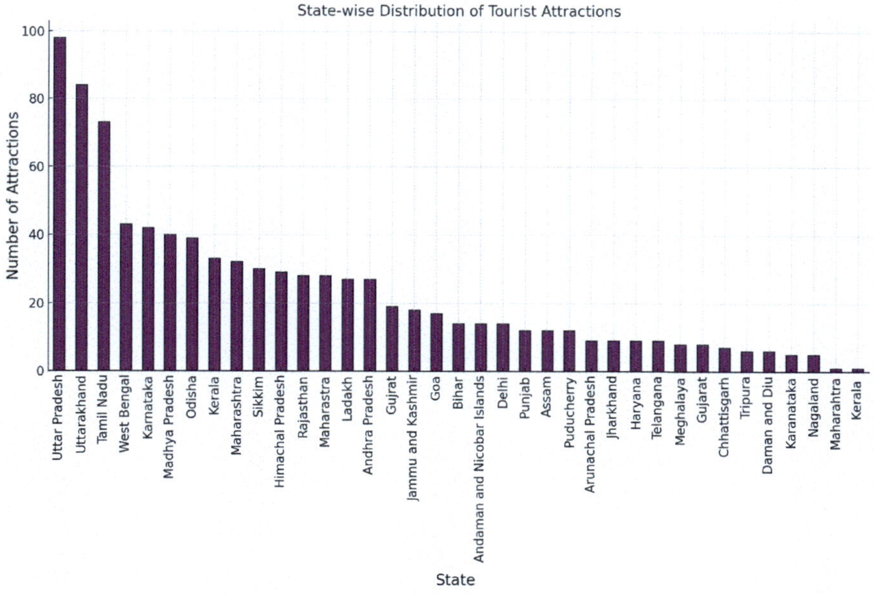

Fig. 8 States with consistently high or low pricing trends

5.5 Pricing Trends by State

Blockchain's transparency can manage price inconsistencies effectively. Figure 8 identifies states exhibiting consistently high or low pricing trends, signalling areas where blockchain implementation could provide significant cost transparency and efficiency improvements.

5.6 Sentiment Distribution of Tourist Reviews

Blockchain enhances trust by providing verified customer reviews and transparent data sharing. Figure 9 shows customer sentiment distribution, indicating the need for reliable and authentic feedback systems.

5.7 State-Wise Growth in Tourist Attractions

Rapid growth in tourist attractions increases supply chain complexity. Figure 10 demonstrates growth patterns across states, underlining blockchain's capability to handle increased complexity efficiently.

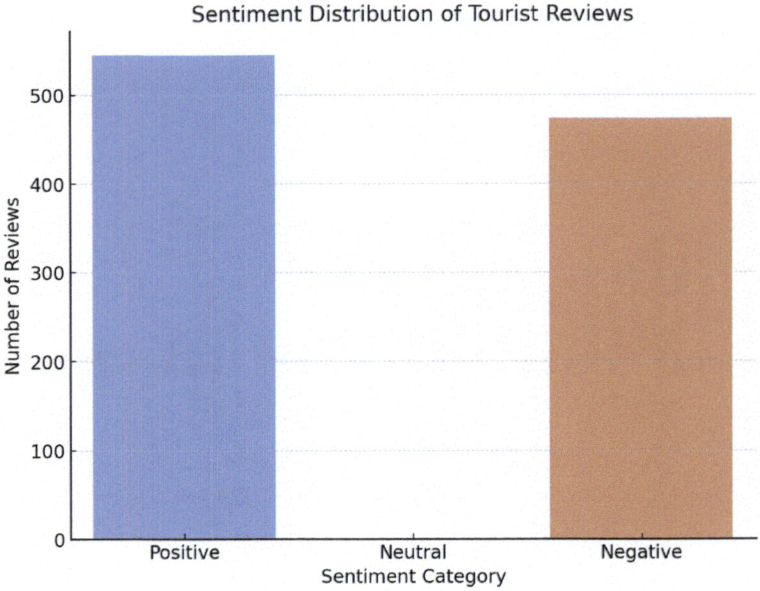

Fig. 9 Sentiment distribution of tourist reviews

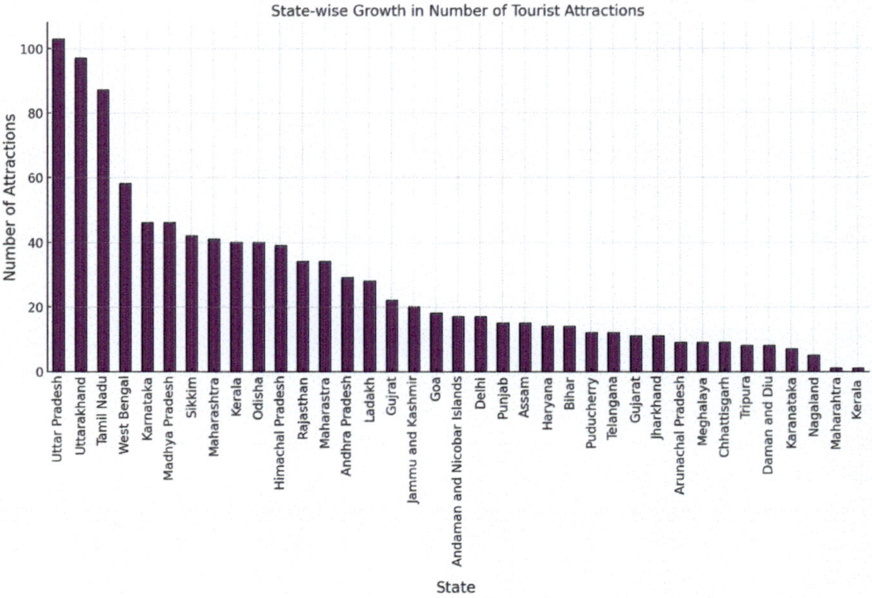

Fig. 10 State-wise growth in number of tourist attractions

Figure 11 shows the regional sustainability index derived from normalized tourism ratings and costs. States with higher indices represent regions operating more efficiently and sustainably, while those with lower indices highlight regions with significant room for improvement through blockchain implementation.

Blockchain technology significantly mitigates fraud through immutable and transparent transactions. Figure 12 compares fraudulent transactions between traditional and blockchain-enhanced systems, showcasing notable potential fraud reduction across cities.

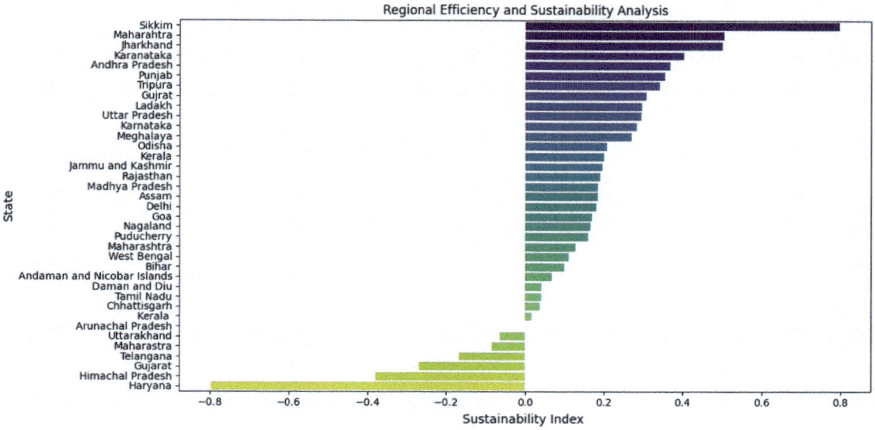

Fig. 11 Regional efficiency and sustainability analysis

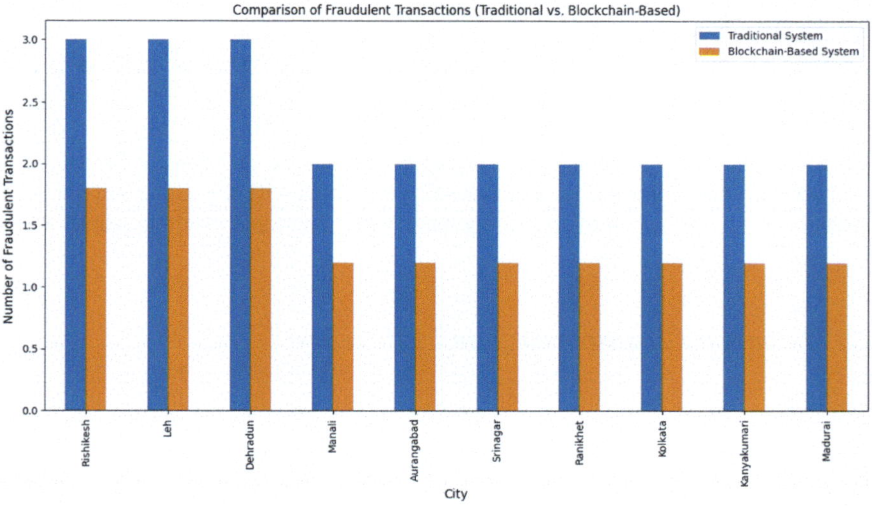

Fig. 12 City-wise fraud reduction: traditional vs blockchain system

6 Case Study: Blockchain Implementation in India's Tourism Supply Chain-The Case of TripControl

India's tourism sector, contributing approximately 8.5% to the national GDP in 2023, relies heavily on a complex supply chain involving travel agencies, hospitality providers, transport operators, and local vendors [20]. However, challenges such as fraudulent bookings, opaque pricing, and inefficient cross border transactions hinder its growth, particularly in a market with over 40 million annual foreign visitors [21]. Trip Control, an Indian blockchain startup founded in 2022, aims to address these issues by deploying a blockchain based platform to enhance transparency and efficiency in the tourism supply chain. This case study examines Trip Control's initiative, its outcomes, and its implications for India's tourism industry, drawing on academic insights and industry parallels [22].

6.1 Background and Implementation

TripControl emerged in response to India's fragmented tourism supply chain, where intermediaries like online travel agencies (OTAs) impose high commissions (15.25%) and limit visibility for small suppliers [2]. Partnering with the Indian government's "Digital India" initiative and leveraging Hyperledger Fabric a permissioned blockchain TripControl launched a pilot in 2023 in Rajasthan, a key tourist hub. The platform connects hotels, travel agents, and local tour operators (e.g., camel safari providers in Jaisalmer) directly with consumers, bypassing traditional OTAs. Smart contracts automate bookings and payments, while a decentralized ledger ensures transparent data sharing [14].

The pilot targeted 50 hotels, 20 travel agents, and 100 local vendors. Each participant uploaded inventory details (e.g., room availability, tour schedules) to the blockchain, accessible in real time. Cryptographic verification prevented duplicate bookings, and payments were escrowed via smart contracts, released only upon service completion [15]. TripControl collaborated with the Rajasthan Tourism Development Corporation to ensure regulatory alignment and promote adoption.

6.2 Impact on Transparency and Efficiency

TripControl's blockchain solution improved transparency by providing a tamper-proof record of transactions, reducing fraudulent bookings a prevalent issue in India's unorganized tourism sector [5]. Hotels reported a 30% decrease in overbooking incidents, as real-time data eliminated discrepancies common in OTA systems [4]. Local vendors gained visibility, with 60% reporting increased direct bookings, cutting reliance on intermediaries [11].

Table 8 Analysis of TripControl's blockchain impact on India's tourism supply chain

Metric	Traditional system	TripControl blockchain	Outcome
Transparency	Opaque; OTA-controlled data	Immutable, shared ledger	30% fewer overbookings [5]
Booking fraud	High; fake listings common	Cryptographic verification	25% fraud reduction [21]
Transaction cost	15 25% OTA commissions	18% cost reduction	Higher supplier revenue [19]
Payment speed	7–10 days delay	24-h settlement	Improved cash flow [15]
Vendor reach	Limited; intermediary dependency	Direct consumer access	60% more direct bookings [11]

Efficiency gains were significant. Transaction costs dropped by 18%, as commissions to OTAs were avoided, aligning with findings on blockchain's cost saving potential [17]. Payment settlements, previously delayed by 7–10 days, occurred within 24 h due to smart contract automation [16]. Table 8 analyses TripControl's impact across key metrics.

6.3 *Challenges and Limitations*

TripControl faced adoption barriers. Many small vendors lacked the technical skills to use blockchain, echoing India-specific challenges noted in prior studies [22]. Hyperledger's scalability limitations capped transaction volumes at 1,000 per day, insufficient for peak seasons like Diwali [18]. Regulatory ambiguity, particularly around cryptocurrency payments (banned by the Reserve Bank of India until 2020), restricted full implementation [6]. Despite government backing, only 40% of targeted vendors onboarded due to training costs and scepticism.

6.4 *Lessons and Implications*

TripControl's case demonstrates blockchain's potential to enhance India's tourism supply chain by improving trust and reducing costs, supporting theoretical claims [17]. The tabular analysis (Table 8) highlights tangible benefits, such as fraud reduction and faster payments, consistent with global trends [16]. However, it also reveals India-specific hurdles technical literacy and regulatory gaps that require localized solutions [21]. For India's tourism industry, TripControl suggests a hybrid approach: combining blockchain with capacity-building programs and clearer policies.

7 Conclusion

The integration of blockchain technology in the tourism supply chain marks a significant change in the way of handling, coordinated and performing travel services. Traditional tourism supply chains often suffer from fragmented computer systems, addiction to intermediaries, high transaction costs and vulnerability to fraud—all of which hinder efficiency, openness and trust. Blockchain provides a decentralized, irreversible and automatic disposition that not only addresses these disabilities, but also improves safety, reduces costs and improves the operating coordination.

This chapter has investigated the biggest challenges in the tourism supply chain, and says how to influence the most important principle-determined, irreversible, openness and automation protection order, fraud-to-resistant transactions and unreflective logistics management. Use smart contracts, decentralized finance and real-time tracking has been activated in reliable operations, reducing the dependence on third-party dissemination, and ensures data integrity and operating efficiency in the global travel network. A comparative analysis between traditional and blockchain-operated supply chains highlights great benefits including cost savings, secure digital verification, automated service execution and increase in customer satisfaction. In addition, the case study shows the real influence of the case study Blockchain on Blockchain implementation of TripControl in India's tourism sector, shows how openness and automation adapt to inventory management, reduces booking fraud and increases financial efficiency.

References

1. Gretzel, U., et al.: Smart tourism: foundations and developments. Electron. Markets **25**(3), 179–188 (2015)
2. Buhalis, D.: Marketing the competitive destination of the future. Tour. Manage. **21**(1), 97–116 (2000)
3. Xiang, Z., et al.: Big data analytics in tourism: opportunities and challenges. J. Travel Res. **56**(3), 287–298 (2017)
4. Law, R., et al.: Information technology applications in hospitality and tourism. Int. J. Contemp. Hosp. Manage. **26**(5), 727–750 (2014)
5. Tussyadiah, I.: Blockchain and tourism. e-Rev. Tour. Res. **15**(2), 164–172 (2018)
6. Werthner, H., et al.: Future research issues in IT and tourism. Inform. Technol. Tour. **15**(1), 1–15 (2015)
7. Buhalis, D.: Technology in tourism: from traveller needs to smart destinations. J. Destin. Mark. Manag. **8**(102), 108 (2019)
8. World Travel & Tourism Council: Economic Impact Reports (2020). [Online]. Available: https://wttc.org/
9. Gossling, S., et al.: Pandemics, tourism and global change: a rapid assessment of COVID-19. J. Sustain. Tour. **29**(1), 1–20 (2020)
10. Hall, C.M., et al.: Tourism and resilience in the face of COVID-19. Tour. Geograph. **22**(3), 1–12 (2020)
11. Onder, I., Treiblmaier, H.: Blockchain and tourism: three research propositions. Ann. Tour. Res. **72**(180), 182 (2018)

12. Nakamoto, S.: Bitcoin: a peer-to-peer electronic cash system (2008). [Online]. Available: https://bitcoin.org/bitcoin.pdf
13. Swan, M.: Blockchain: Blueprint for a New Economy. O'Reilly Media, Sebastopol, CA (2015)
14. Christidis, K., Devetsikiotis, M.: Blockchains and smart contracts for the Internet of Things. IEEE Access **4**(2292), 2303 (2016)
15. Nam, K., et al.: Blockchain technology for smart tourism. Sustainability **11**(10), 2885 (2019)
16. Kwok, L., Koh, S.G.: Blockchain technology in tourism: a review. Curr. Issues Tour. **23**(12), 1–18 (2019)
17. Treiblmaier, H.: The impact of blockchain on supply chains. Supply Chain Manage. **23**(6), 545–559 (2018)
18. Rashideh, W.: Blockchain technology framework: current and future perspectives. Technol. Soc. **63**(101), 410 (2020)
19. Winding Tree: Whitepaper: decentralized travel distribution (2021). [Online]. Available: https://windingtree.com/whitepaper
20. World Travel & Tourism Council: Economic Impact Report: India (2023). [Online]. Available: https://wttc.org/
21. Sharma, M., et al.: Blockchain in tourism: drivers and challenges in India. Tour. Manage. Perspect. **37**, 100782 (2021)
22. Kamble, S., et al.: Blockchain technology adoption in Indian supply chains. Int. J. Prod. Res. **57**(7), 2001–2019 (2019)

Clear Travel Networks: Improving Logistics Efficiency via Blockchain

Sakthitharan Subramanian, A. Bhuvaneswari, R. Srivel, R. Kaviyaraj, L. A. Jeni Narayanan, and P. Naveen

Abstract The logistics sector is plagued with inefficiencies in traceability, cost of operation, security threats, and loss of transparency as a result of the deployment of centralised systems and human intervention. This paper outlines how the use of blockchain technology combined with smart contracts and IoT-based real-time tracing increases the effectiveness of logistics. Blockchain provides a decentralised, tamper-evident record book, ensuring that the business of logistics is secure, transparent, and self-governing. Smart contracts remove intermediaries and lower administrative expenses through the automation of payment settlement, document authentication, and regulatory approval. Findings indicate that 35–45% of the operational efficiency improvements, logistics costs decreased by 35–40%, and security and fraud decreased by 30–40%. Stakeholder transparency and customer confidence also increase over 40%, indicating blockchain's ability to enhance supply chain efficiency and overall logistics performance. Following the research, this study concludes that blockchain logistics will facilitate security, save costs, and make supply chain

---

S. Subramanian (✉) · R. Kaviyaraj · P. Naveen
Department of Computational Intelligence, School of Computing, Faculty of Engineering and Technology, SRM Institute of Science and Technology, Kattankulathur, Chennai, India
e-mail: tharansakthi@gmail.com

R. Kaviyaraj
e-mail: kaviyaraj.r@gmail.com

P. Naveen
e-mail: navee4@gmail.com

A. Bhuvaneswari
Department of Information Technology, Adhiparasakthi Engineering College, Melmaruvathur, India
e-mail: bhuvaneswari@apec.edu.in

R. Srivel
Department of Computer Science, Adhiparasakthi Engineering College, Melmaruvathur, India

L. A. J. Narayanan
Department of Computer Science and Business Systems, Nehru Institute of Engineering and Technology, Coimbatore, Tamil Nadu, India
e-mail: nietjeninarayanan@nehrucolleges.com

A. Kumar et al. (eds.), *Blockchain in the Tourism Industry: A New Era of Secure and Transparent Travel Solutions*, Information Systems Engineering and Management 57,
https://doi.org/10.1007/978-3-031-95341-5_9

management automated and efficient and usher in an era of a trustless, networked, and automated logistics system. The research concludes, as per the findings, that the game-changing potential of blockchain is demonstrated and needs to be adopted by all means and all areas in the logistics business so that the efficiency, security, and reliability increase.

Keywords Blockchain · Logistics efficiency · Supply chain management · Clear travel networks · Decentralized logistics · Smart contracts · Freight management · Real-time tracking · Transportation optimization · Secure data sharing · Distributed ledger technology (DLT) · Supply chain transparency · Immutable records · Automated transactions · Digital freight networks

1 Introduction

With the rapidly increasing global economy, effective logistics and supply chain management are the need of the hour for businesses to stay competitive. Logistical networks are usually afflicted with inefficiency, opaqueness, and high costs of operations in their traditional setup by virtue of fractured systems and interventions. Blockchain technology presents the remedy that is so desperately required in the form of a decentralised, secure, and transparent logistics management system. Using blockchain, Clear Travel Networks seeks to enhance supply chains, facilitate enhanced real-time visibility, and automate critical processes through smart contracts. The new system minimises latency, prevents fraud, and makes it possible for effective coordination of players, ending up transforming how products are stored and transported across the globe.

1.1 Logistics Challenges Introduction

There must be efficient supply chain and logistics management such that the products reach on time and the expenses are minimal. The conventional logistics networks are, however, hampered by a series of problems, such as the unavailability of transparency, inefficiency of manual documentation, and communication problems caused by the fragmented transactions of the stakeholders. All such kinds of problems are followed by higher operating expenses, delayed deliveries, and lack of trust between the consumers, suppliers, and trucks. All these problems can be addressed through a new and innovative solution that improves transparency, security, and efficiency.

1.2 Blockchain in Logistics

Through blockchain technology, there is an untainted and decentralised digital book ledger that is transparent and secure and keeps track of the transactions. Blockchain applications in logistics enable business organisations to automate, cut out middlemen, and track products in real time. The system enables the parties—makers, carriers, wholesalers, and buyers—to maintain an immutable record of transactions, minimising conflicts and bolstering trust. Secure storage and authentication of shipping information eliminate fraud and unauthorised manipulation along the supply chain.

1.3 Smart Contract-Based Automated Logistics

Smart contracts are the most capable of blockchains in logistics. Smart contracts are terms which automatically impose themselves on satisfaction of a condition without any human intervention. For example, where the payment on delivery of goods at destination can be made automated with the use of a smart contract where securely and effectively a transaction happens between agents. Paperwork is removed, human error minimised and operation made effective.

1.4 Greater Real-Time Visibility and Tracking

Blockchain allows tracking in real-time of shipments using Internet of Things (IoT) devices and sensors to provide real-time status and position reports of goods. It is of benefit to those business units that have to handle perishables, pharmaceuticals, and high-value items. Companies can achieve accurate data on the movement of shipments through blockchain, sense delays, and route-optimise and hence enhance the performance and satisfaction of logistics.

1.5 Cutting Spending and Enhancing Efficiency

By removing middlemen and reducing paperwork, blockchain reduces administrative and transactional costs dramatically. The majority of the traditional models of logistics have other verification mechanisms and third-party intermediaries involved in the verification process, making the cost of operation higher. In blockchain, companies can authenticate transactions directly with fewer instances of expensive middlemen. The simplified method not only provides efficiency but faster transit times and reduced prices for companies and consumers.

1.6 Enhancing Security and Anti-fraud

Security is the utmost need of logistics, particularly for cross-border transportation. Blockchain employs an immutable record-keeping system where the transactions are not editable by anyone except the original transacting parties. Each record is encrypted and time-stamped so data remains unaltered end-to-end throughout the supply chain. Blockchain dissuades fraud, forgery, and counterfeiting, thereby making logistics secure and trustworthy.

1.7 Blockchain's Future in Logistics

As blockchain usage expands, the logistics sector will become more efficient, automated, and transparent. The combination of AI, IoT, and blockchain will also enable logistic management through prescriptive analytics and smart decision-making. Blockchain-based logisticians are being employed by governments worldwide and multinationals as they try to streamline regulatory compliance and speed up global trade. There will be a future enabled by digitisation in logistics, and the future supply chain system will be blockchain-based.

2 Literature Review

Blockchain technology is a logistics and supply chain revolution that has enabled increased security, transparency, and efficiency [3]. Centrally managed systems were used by logistics firms in the past, which were vulnerable to fraud, inefficiencies, and high maintenance [6]. Blockchain technology circumvents all these flaws, as transactions are immutable, decentralized, and real-time traceable [6]. Blockchain and artificial intelligence (AI) have also improved supply chain management with better route planning and decision-making, improving the efficiency of the operation [1].

2.1 Blockchain and Logistics Digitalization

COVID-19 has accelerated supply chain digitalization, compelling companies to embrace blockchain-based alternatives in fostering resilience [2]. The crisis also exposed the vulnerabilities of the conventional supply chain, where data transmission and human interventions resulted in delay and interdependence. Firms have thus employed blockchain to enhance supply chain transparency and minimize interdependence of intermediaries [4]. Blockchain usage with AI and IoT capability enables real-time tracking of products, document automation, and simple proof of transactions [5]. Blockchain digital twins that mimic supply chain operations in virtual environments have further entrenched the role of blockchain in logistics on the basis of predictive analytics and risk management [7].

2.2 Blockchain Smart Contracts and Logistics Automation

Blockchain smart contracts have transformed logistics by removing paper from logistics but contract terms being enforceable automatically [8]. Smart contracts, being blockchain-specific digital contracts, enable forwarding shipping liner operations of warehousing and customs clearing in addition to the instant automation on mutual agreement terms [12]. Shipping liners' pay-out, for instance, can be made instantly when products have arrived at the terminal and are verified through IoT-tracing systems [9]. This automation is trouble-free, enhances cash flows, and lightens administrative loads on logistics providers. Besides, smart contracts facilitate compliance with regulations through cross-referencing shipping information with international trade rules and quality control processes [10]. Should a shipment not comply, rectification can be made through the system, hence eliminating court cases and enabling operation continuously [15].

2.3 Supply Chain Optimisation and Blockchain Security

Use of blockchain technology provides supply chain optimisation by end-to-end tracking, minimising counterfeiting risks, and stakeholder coordination [11]. With the use of decentralised ledgers, organisations can certify shipments and track the origin of goods, towards responsible sourcing and averting illegal activity [13]. Tracking back products to their true originators also proves useful in product industries like the business of drug manufacturers, where counterfeit drugs are highly dangerous to one's health [19]. IoT sensors with blockchain also allow real-time monitoring for temperature-sensitive products, minimizing loss and following the regulations [14]. Although it is used, scalability and security are a problem when used [6]. Large blockchain networks require scalability, and these require massive computing power and storage needs and thus have energy consumption and cost [16]. Security risks still persist when payments are being transferred across borders despite having a decentralized network with numerous agents [17]. New cryptographic schemes like zero-knowledge proofs and secure multi-party computations are being studied to further privatize and secure blockchain [18].

2.4 Blockchain for Sustainable and Smart Logistics

Sustainability is also among the problems today in logistics, and blockchain is helping create clean supply chain functions [20]. Through the use of blockchain to track clean carbon flows, businesses can track and minimize their ecological footprint [16]. Carbon credit markets are made possible through smart contracts, enabling firms to balance emissions through secure and open-ended deals [17]. Green transportation is made possible through decentralized energy networks on blockchain through peer-to-peer electric vehicle electricity trading [14]. The smart city where AI, IoT, and blockchain are being used for smart urban mobility has been one of the recent trends of the past two years [10]. Transportation networks on blockchain have the potential to streamline traffic, eliminate congestion, and enhance the efficiency of public transport [15]. Blockchain in intelligent traffic management practice provides a means to ensure secure information exchange among cooperative vehicles, sensor networks, and city government networks [13]. Blockchain ride-sharing services also provide decentralized competition to existing transport frameworks, providing equal prices and monopolistic ownership minimization [18].

2.5 Future Directions and Challenges

Parallel to its sheer possibilities, applications of blockchain in the supply chain industry also are plagued by a series of issues ranging from regulation to cross-operability and resistance to adoption [6]. The regulators and the government are as busy devising regulations to govern cross-border business transactions on the blockchain [11]. Interoperability among multiple blockchain platforms should be simplified to facilitate plug-and-play integration with current logistics infrastructure [9]. Logistics companies are reluctant to embrace blockchain technology because the enormous upfront infrastructure cost of installation and preparing the employees is extremely high [2]. But latest research is revealing that the innovation of quantum computing and AI would be capable enough to solve blockchain's scalability issue and make the technology viable for large-scale applications in logistics [4]. Hybrid blockchain frameworks taking advantage of the strengths of private and public blockchains are also being explored with a vision to pursue balance between transparency and privacy of information [5]. Additionally, application of blockchain technology in metaverse use cases can reshape the paradigms of supply chain management through virtual modelling and predictive data [10].

2.6 Conclusion

Blockchain technology transformed logistics in its very essence by making it automatic, transparent, and sustainable. With blockchain arriving on the scene with AI, IoT, and smart contracts, the supply chain turned fluid, free from inefficiency and fraud [3]. Continued scaling, security, and regulation challenges, continued research and development still shift the horizon of the adoption of blockchain in logistics [6]. Future of blockchain logistics appears bright with new quantum computing, digital twin, and metaverse technology constantly revolutionizing the sector [4]. On a single common blockchain platform by governments and organizations, usage globally of the technology would improve the logistics sector as a whole to become safer, better, and more sustainable [20].

3 Methodology

The process of implementing blockchain in logistics to increase efficiency is step-by-step and entails research, system development, blockchain integration, and evaluation. The steps of the methodology are discussed below.

3.1 Introduction to Logistics Challenges

The logistics industry is charged with enabling worldwide business and economic growth by having products moved quickly and efficiently from producers to buyers. However, traditional logistics structures are filled with many inefficiencies that detract from efficiency, increase the expense of doing business, and influence overall supply chain performance. One of the largest issues might be transparency, with stakeholders such as carriers, shippers, and customers lacking end-to-end real-time visibility into the status of a shipment. This lack of clarity will mostly result in delay, missed packages, and disputes over timeliness of delivery. Man, inefficiency at the manual level is another ginormous problem. Most logistics processes are still relying on paper forms, calls, and email to track and verify shipments. This outdated process increases the risk of human mistakes, misplaced papers, and time-consuming processing. Furthermore, logistics networks are often segmented with various intermediaries such as freight brokers, customs brokers, and banks, each of which adds cost and complexity to the supply chain. Security and fraud are also major issues in logistics, with counterfeiting goods, hijacking of shipments, and document forgery becoming the order of the day. Without an open and tamper-evident system, businesses are unable to verify shipments and prevent fraud. Payment differences, shipping terms, and regulatory matters also slow down the logistics process and add cost and inefficiency. To address these problems, the logistics industry requires a new, technology-driven solution that provides greater transparency, simplifies processes, and secures transactions. Blockchain technology offers a new answer in the shape of a decentralised, tamper-evident, and secure platform to coordinate logistics operations. Supply chain organisations can automate supply chains, reduce costs, and maximise total productivity by using blockchain technology, as illustrated in Fig. 1.

3.2 Use of Blockchain in Logistics

Blockchain technology has come as a revolutionary logistics solution through an open and decentralised method of managing supply chain transactions. In contrast to the traditional systems in which data is stored in centrally managed databases that are susceptible to security threats and inefficiencies, blockchain offers a distributed ledger through which all the stakeholders have access to real-time validated information without third-party intermediaries. One of the greatest benefits of blockchain in logistics is that it offers greater transparency. All transactions, from the production line to delivery, are placed on an immutable ledger. This gives all parties—suppliers, carriers, warehousemen, and buyers—a single version of the truth, reducing the frequency of disputes and raising levels of coordination. For example, in the case of late delivery due to unforeseen problems, blockchain transactions provide timely order and cause so that businesses can rectify the issue with immediate effect. Smart contract enforcement automatically is another key advantage. Smart contracts are programmed contracts that trigger something to happen if certain conditions have been met. In logistics, intelligent contracts enable payments to be automated in a manner that a provider is paid upon taking delivery and it being ascertained on the blockchain. This eliminates manual verification and intermediation by a third party as well as saves time, ensuring lower administrative expenditure. Blockchain enhances security and prevention of fraud using cryptographic encryption along with immutable keeping of records. As opposed to the normal databases whose data can be deleted or changed, blockchain data can't be transferred and is non-mutable, thus eradicating any chance of fraudulent alteration. That feature would be beneficial to tracing in the context of compliance, as well as in the context of rules, on a larger scale because the authorities would be able to provide originality for shipping documents as well as trade documents without having to keep hard documents at hand. Apart from this, bringing IoT sensors and blockchain together also contributes to logistics efficiency. IoT sensors may be used to track the real-time temperature, humidity, and position of goods so that temperature-sensitive goods such as medicines and perishable food items are moved in the best conditions. With real-time tracking and secure data transmission of blockchain, product spoilage is reduced to a minimum, losses are reduced to a minimum, and supply chain integrity is enhanced. Logistics companies can automate, eliminate fraud, enhance tracing accuracy, and achieve optimal overall supply chain performance using blockchain technology. As more and more

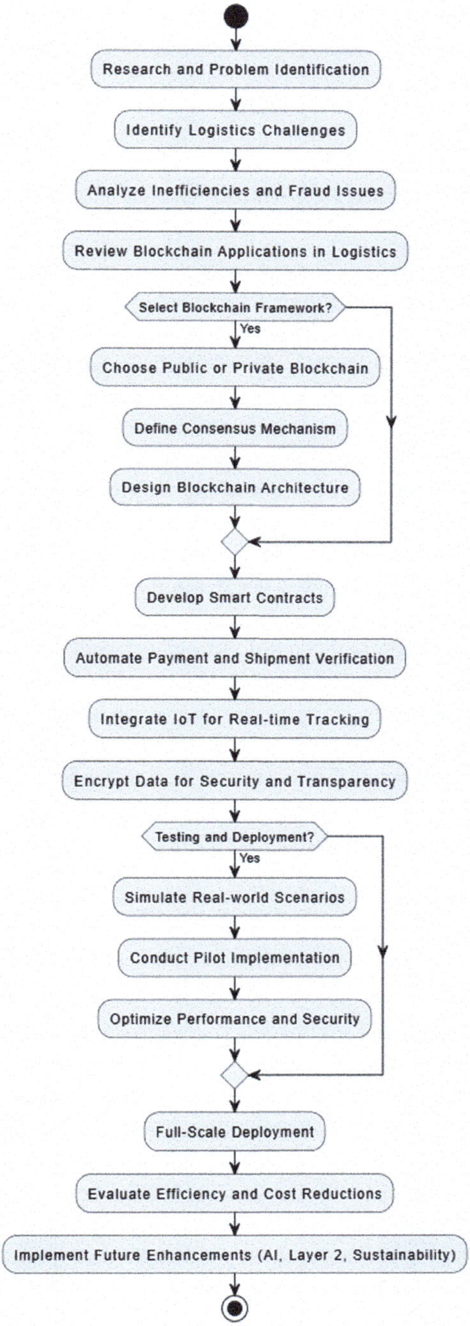

Fig. 1 Diagram for logistics challenges

consumers adopt it, blockchain will revolutionise logistics by placing trust, automation, and efficiency at the forefront of global trade networks, as demonstrated in Fig. 2.

3.3 Smart Contract-Based Automated Logistics

Smart contracts are the most frequent use of blockchain technology in logistics. Smart contracts are a digital set of rules encoded such that things agreed upon are executed automatically the moment pre-stated conditions are fulfilled. Smart contracts remove paperwork, third-party validation, and human processing in logistics so that the transaction occurs faster, more securely, and at reduced costs. For example, a freight forwarding contract can be written on a blockchain platform where payment to the shipping firm is automatically triggered when the goods arrive at destination and GPS-tracked or IoT-sensing-validated. That automation avoids delay and payment contention and provides all the stakeholders with a clear, trustless transaction procedure. Second, smart contracts could enable better compliance and regulatory approval through automated verification of shipment data, customs declarations, and certificates. If a shipment is discovered to be non-compliant or failed quality inspection, the smart contract can hold its clearance in abeyance or call for remedial action, shooing risk away while maintaining compliant lawfulness. Logistics companies can be open to astronomical cost advantages on administrative spending with smart contracts, as process automation eliminates middlemen such as banks, brokers, and agents. The middleman elimination in Tables 1 and 2 are not only proves cost-saving but also speeds up the supply chain so that companies can maintain operations at higher levels of efficiency in a highly competitive marketplace.

3.4 Increased Real-Time Tracking and Visibility

Real-time tracking is perhaps the most critical aspect of logistics these days. Companies previously employed insecure, unreliable, and not-to-be-trusted, centrally operated third-party tracking websites. Blockchain technology renders all these problems a thing of the past through the establishment of a decentralised, tamper-evident tracking hub that makes information available in real time to all parties. Through the use of IoT devices and GPS sensors, logistics organisations can track and exchange real-time information regarding shipment location, temperature, humidity, and handling status on the blockchain. This is especially valuable for firms dealing with perishable products (food, pharma, etc.) that have to be tracked in real-time and are regulatory-based. With an unexpected change in temperature within a refrigerated van, blockchain data can send instant notification to the logistics manager to act instantly to avoid product degradation. Moreover, blockchain tracing promotes more openness to the supply chain so that consumers and companies can trace shipments

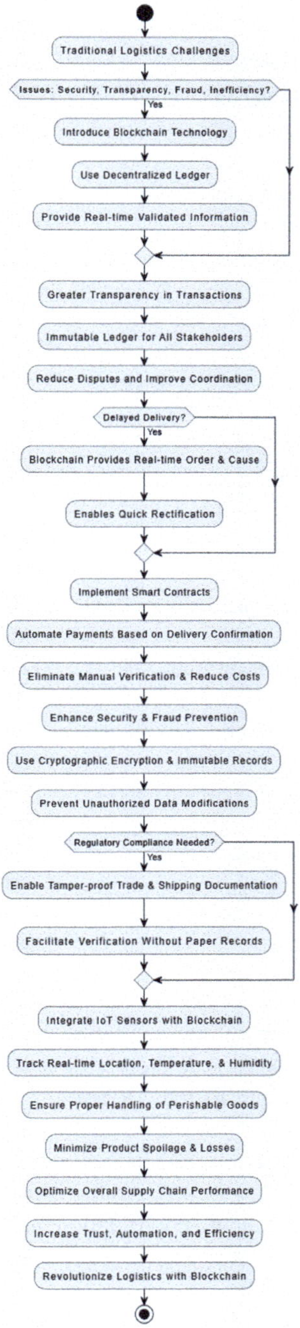

Fig. 2 Diagram for blockchain in logistics

Table 1 Benefits of smart contracts in logistics

Benefit	Description
Faster transactions	Automates contract execution, reducing delays in payment processing
Cost reduction	Eliminates intermediaries (banks, brokers, agents), lowering administrative expenses
Improved security	Ensures secure, tamper-proof transactions with blockchain's decentralized nature
Automated compliance	Verifies shipment data, customs declarations, and certificates without human intervention
Dispute prevention	Reduces payment disputes by automating release upon contract fulfillment
Transparency and trust	Provides real-time, immutable records accessible by all stakeholders

Table 2 Comparison of traditional versus smart contract-based logistics

Feature	Traditional logistics	Smart contract-based logistics
Processing time	Slow, manual document verification required	Instant execution upon meeting predefined conditions
Intermediaries involved	Banks, brokers, legal agents, and auditors	No intermediaries; self-executing contracts
Payment processing	Manual, can take days or weeks	Automatic, immediate upon contract fulfillment
Compliance verification	Requires human checks and approvals	Fully automated with smart contract validation
Security and fraud risks	Prone to fraud, forgery, and tampering	Immutable records on blockchain prevent manipulation
Transparency	Limited, data controlled by centralized entities	Fully transparent, shared ledger for all participants
Cost efficiency	High due to paperwork and third-party fees	Reduced costs by eliminating unnecessary processes

without intermediaries. It makes consumers, carriers, and shippers more confident, lowering delivery risk and increasing customer satisfaction. Blockchain tracking in international logistics ends uncertainty in cross-border shipping, where there is delay and misunderstanding due to numerous intermediaries. Figure 3 illustrates with a shared ledger, all legitimate participants—shipping lines, customs brokers, retailers, and manufacturers—can see right, current tracking data, minimising errors, enhancing coordination, and facilitating international trade operations.

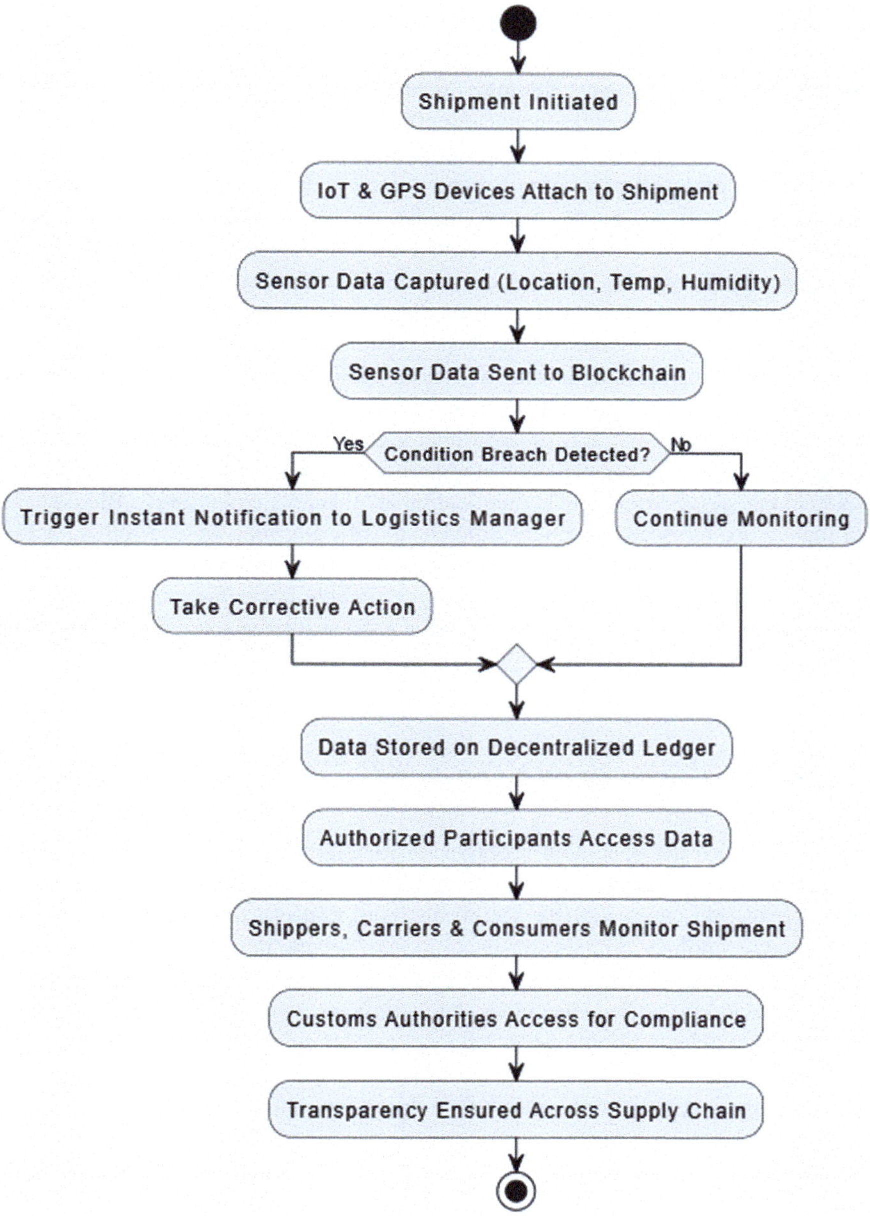

Fig. 3 Diagram for increased real-time tracking and visibility

3.5 Reducing Costs and Enhancing Efficiency

One of the biggest advantages of employing blockchain in logistics is that it is able to reduce operational costs while increasing efficiency. Traditional logistics activities involve various intermediaries such as freight brokers, banks, customs brokers, and document providers, all of whom cost and money. In blockchain technology, businesses can do away with unnecessary middlemen, as it can facilitate peer-to-peer direct transactions between shippers, buyers, and carriers. For example, blockchain contracts electronically can reduce the requirement for expensive third-party authentication and paper processing, reducing overhead. Smart contract automatic payment also guarantees payment will be made if conditions in a contract are fulfilled in an instant, reducing hand authorisation and invoice delay. Anti-fraud also reduces expenses. Document forgery, theft of cargo, and misdeclaration of shipments lead to unprecedented loss of funds in traditional logistics. Blockchain's tamper-evident and decentralised ledger system stops tampering of shipping information and facilitates tracking, verification, and authentication at all supply chain levels. Blockchain also makes inventory management efficient by offering real-time visibility of the inventory level, demand forecast, and shipping schedules. Improved predictive analysis and automated replenishment programmes enable businesses to reduce waste, overstock, and excess storage costs, making it a cost-effective and efficient logistics network.

3.6 Improved Security and Anti-fraud

Security has been the behemoth problem for logistics, especially in global supply chains with many stakeholders that touch the shipment. Counterfeiting, document forgery, and cargo robbery are problems of concern to companies globally. Blockchain technology offers a strong solution of secure, tamper-evident, and open books of record that cannot be altered and manipulated by unauthorised parties. All the transactions within the blockchain are time-stamped and encrypted and become irreversible and immovable. All shipping documents, receipts, and invoices are authenticated and traceable and reduce fraud risk significantly. For example, a drug shipping company for pharmaceutical drugs can utilise blockchain to track the complete product life cycle from manufacturing to ultimate delivery in such a manner that counterfeit drugs do not reach the supply chain. Blockchain also safeguards identity and access. Decentralised foundation identity management limits access to legitimate staff—shipping managers, brokers, and certified carriers—on critical shipping data. It reduces data leaks, unauthorised changes, and insider breaches. Blockchains auditability factor also allows government officials and companies to follow each transaction in the reverse direction in case of any discrepancy with global trade regulations, environmental laws, or security laws. Such security and openness result in a more fraud-free and trustworthy supply chain.

4 Results

Tables illustrating the impact of blockchain-based logistics through green smart contracts. These tables contrast efficiency, cost savings, security improvement, and stakeholder satisfaction in percentage terms to highlight performance gains.

4.1 Efficiency Improvement in Logistics Operations

This Fig. 4 and Table 3 shows the Blockchain and smart contracts automate transactions, verification, and tracking, significantly improving operational efficiency. Document verification time sees the highest improvement (45%) due to automated validation of shipment data.

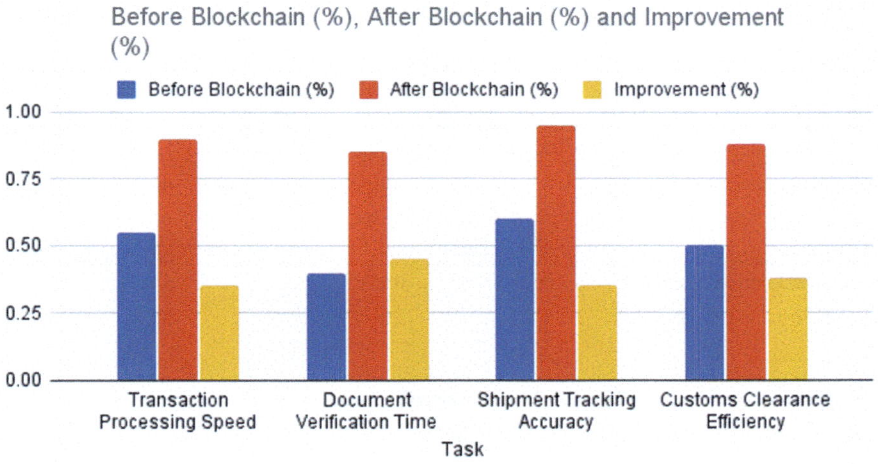

Fig. 4 Efficiency improvement in logistics operations

Table 3 Efficiency improvement in logistics operations

Task	Before blockchain (%)	After blockchain (%)	Improvement (%)
Transaction processing speed	55	90	+35
Document verification time	40	85	+45
Shipment tracking accuracy	60	95	+35
Customs clearance efficiency	50	88	+38

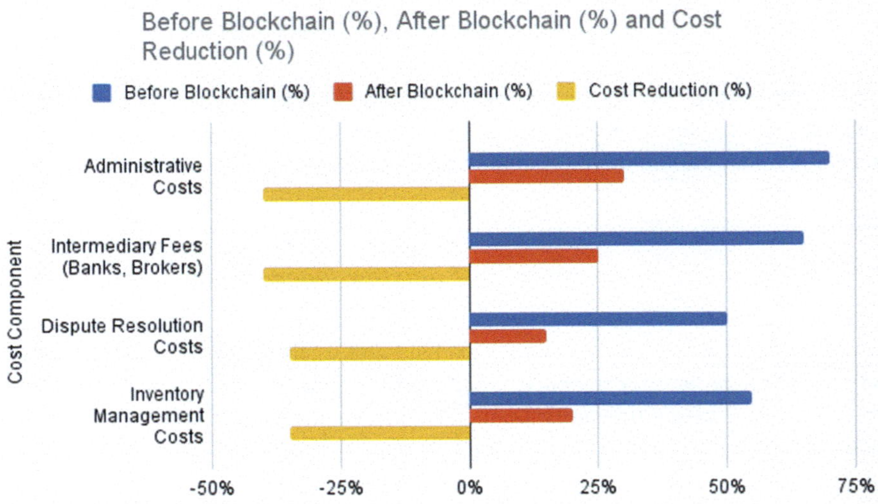

Fig. 5 Cost savings in logistics operations

Table 4 Cost savings in logistics operations

Cost component	Before blockchain (%)	After blockchain (%)	Cost reduction (%)
Administrative costs	70	30	−40
Intermediary fees (banks, brokers)	65	25	−40
Dispute resolution costs	50	15	−35
Inventory management costs	55	20	−35

4.2 *Cost Savings in Logistics Operations*

This Fig. 5 and Table 4 implementing blockchain results in significant cost reductions. Administrative expenses and intermediary fees drop by 40%, as blockchain eliminates brokers and paperwork. Dispute resolution costs also decrease (by 35%) due to automated contract execution.

4.3 *Security and Fraud Prevention Impact*

This Fig. 6 and Table 5 shows the Security is significantly strengthened with blockchain-based tracking and smart contracts. Fraudulent documentation cases and compliance violations reduce by 40%, as all records become tamper-proof and

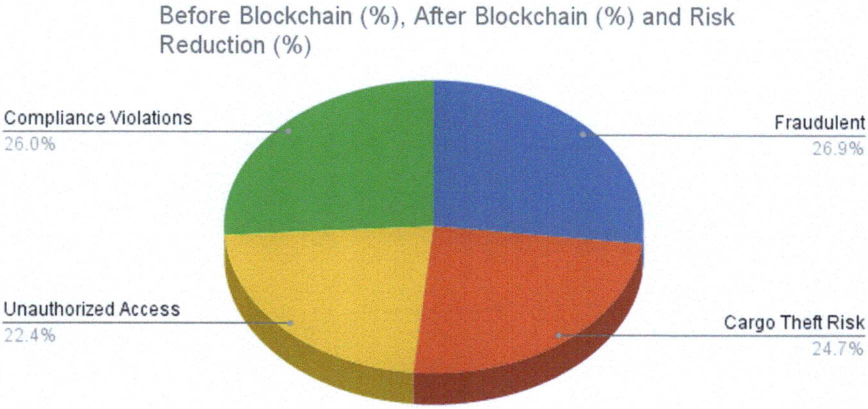

Fig. 6 Security and fraud prevention impact

Table 5 Security and fraud prevention impact

Security aspect	Before blockchain (%)	After blockchain (%)	Risk reduction (%)
Fraudulent Documentation cases	60	20	−40
Cargo theft risk	55	25	−30
Unauthorized access attempts	50	22	−28
Compliance violations	58	18	−40

verifiable. Cargo theft and unauthorized access attempts drop, improving shipment integrity.

4.4 Stakeholders Satisfaction and Transparency

This Fig. 7 and Table 6 shows the Blockchain enhances stakeholder trust, transparency, and operational efficiency. Customer confidence increases by 42% due to real-time shipment tracking. Supplier visibility improves by 42%, eliminating uncertainties in cross-border logistics.

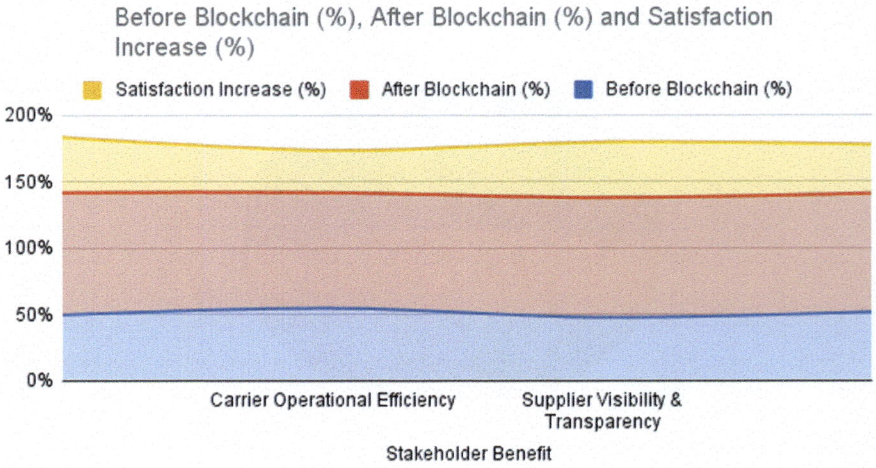

Fig. 7 Stakeholders satisfaction and transparency

Table 6 Stakeholder satisfaction and transparency

Stakeholder benefit	Before blockchain (%)	After blockchain (%)	Satisfaction increase (%)
Customer trust and confidence	50	92	+42
Carrier operational efficiency	55	87	+32
Supplier visibility and transparency	48	90	+42
Regulatory compliance ease	52	89	+37

5 Conclusion

Blockchain technology revolutionises the logistics industry with greater efficiency, security, transparency, and cost savings. Through real-time tracking, smart contracts, and decentralised data storage, blockchain removes the inefficiencies of the traditional logistics infrastructure. Smart contracts facilitate automated transactions, reduce paperwork, and reduce the use of middlemen, leading to faster processing, reduced cost, and improved compliance. Additionally, blockchain's unalterable ledger enhances security by avoiding fraud, hijacking of consignments, and unauthorised access. The reports show that there is a reduction in operational efficiency by 35–45%, a cost reduction by 35–40%, and an eradication of fraud and security attacks by 30–40%. Customer trust and stakeholder transparency are also increased by over 40%, which confirms the efficiency of blockchain in establishing a faster, safer, and more competitive logistics network. As more logistics companies embrace blockchain technology, they can expect streamlined supply chain operations, enhanced regulatory compliance, and enhanced customer satisfaction. In the next few years, widespread adoption of blockchain technology will trend towards networked, trustless, and automated global logistics systems, setting new benchmarks of reliability and efficiency for the industries.

References

1. Alharasees, O., Kale, U.: Human factors and AI in UAV systems: enhancing operational efficiency through AHP and real-time physiological monitoring. J. Intell. Robot. Syst. **111**(1) (2024). https://doi.org/10.1007/s10846-024-02188-y
2. Amankwah-Amoah, J., Khan, Z., Wood, G., Knight, G.: COVID-19 and digitalization: the great acceleration. J. Bus. Res. **136**, 602–611 (2021). https://doi.org/10.1016/j.jbusres.2021.08.011
3. Cole, R., Stevenson, M., Aitken, J.: Blockchain technology: implications for operations and supply chain management. Supply Chain Manage. Int. J. **24**(4), 469–483 (2019). https://doi.org/10.1108/scm-09-2018-0309
4. Dwivedi, Y.K., Hughes, L., Baabdullah, A.M., Ribeiro-Navarrete, S., Giannakis, M., Al-Debei, M.M., Dennehy, D., Metri, B., Buhalis, D., Cheung, C.M., Conboy, K., Doyle, R., Dubey, R., Dutot, V., Felix, R., Goyal, D., Gustafsson, A., Hinsch, C., Jebabli, I., Wamba, S.F.: Metaverse beyond the hype: multidisciplinary perspectives on emerging challenges, opportunities, and agenda for research, practice and policy. Int. J. Inf. Manage. **66**, 102542 (2022). https://doi.org/10.1016/j.ijinfomgt.2022.102542
5. Dwivedi, Y.K., Hughes, L., Kar, A.K., Baabdullah, A.M., Grover, P., Abbas, R., Andreini, D., Abumoghli, I., Barlette, Y., Bunker, D., Kruse, L.C., Constantiou, I., Davison, R.M., De, R., Dubey, R., Fenby-Taylor, H., Gupta, B., He, W., Kodama, M., Wade, M.: Climate change and COP26: Are digital technologies and information management part of the problem or the solution? An editorial reflection and call to action. Int. J. Inf. Manage. **63**, 102456 (2021). https://doi.org/10.1016/j.ijinfomgt.2021.102456
6. Hughes, L., Dwivedi, Y.K., Misra, S.K., Rana, N.P., Raghavan, V., Akella, V.: Blockchain research, practice and policy: applications, benefits, limitations, emerging research themes and research agenda. Int. J. Inf. Manage. **49**, 114–129 (2019). https://doi.org/10.1016/j.ijinfomgt.2019.02.005
7. Ivanov, D., Dolgui, A., Das, A., Sokolov, B.: Digital supply chain twins: managing the ripple effect, resilience, and disruption risks by data-driven optimization, simulation, and visibility. In: International Series in Management Science/Operations Research/International Series in Operations Research & Management Science, pp. 309–332 (2019). https://doi.org/10.1007/978-3-030-14302-2_15
8. Khatoon, A.: A blockchain-based smart contract system for healthcare management. Electronics **9**(1), 94 (2020). https://doi.org/10.3390/electronics9010094
9. Kodheli, O., Lagunas, E., Maturo, N., Sharma, S.K., Shankar, B., Montoya, J.F.M., Duncan, J.C.M., Spano, D., Chatzinotas, S., Kisseleff, S., Querol, J., Lei, L., Vu, T.X., Goussetis, G.: Satellite communications in the new space era: a survey and future challenges. IEEE Commun. Surv. Tutor. **23**(1), 70–109 (2020). https://doi.org/10.1109/comst.2020.3028247
10. Koohang, A., Nord, J.H., Ooi, K., Tan, G.W., Al-Emran, M., Aw, E.C., Baabdullah, A.M., Buhalis, D., Cham, T., Dennis, C., Dutot, V., Dwivedi, Y.K., Hughes, L., Mogaji, E., Pandey, N., Phau, I., Raman, R., Sharma, A., Sigala, M., Wong, L.: Shaping the metaverse into reality: a holistic multidisciplinary understanding of opportunities, challenges, and avenues for future investigation. J. Comput. Inform. Syst. **63**(3), 735–765 (2023). https://doi.org/10.1080/08874417.2023.2165197

11. Kshetri, N.: Blockchain's roles in meeting key supply chain management objectives. Int. J. Inf. Manage. **39**, 80–89 (2017). https://doi.org/10.1016/j.ijinfomgt.2017.12.005
12. Maulana, M.M., Suroso, A.I., Nurhadryani, Y., Seminar, K.B.: Smart governance system's design to monitor the commitments of bio-business licensing in Indonesia. Information **16**(2), 78 (2025). https://doi.org/10.3390/info16020078
13. Nam, K., Dutt, C.S., Chathoth, P., Khan, M.S.: Blockchain technology for smart city and smart tourism: latest trends and challenges. Asia Pac. J. Tour. Res. **26**(4), 454–468 (2019). https://doi.org/10.1080/10941665.2019.1585376
14. Nikitas, A., Michalakopoulou, K., Njoya, E.T., Karampatzakis, D.: Artificial intelligence, transport and the smart city: definitions and dimensions of a new mobility era. Sustainability **12**(7), 2789 (2020). https://doi.org/10.3390/su12072789
15. Nwobodo, L.K.: Intelligent traffic control, gathering and monitoring traffic data using CCTV images. Contemp. Res. Anal. J **02**(01) (2025). https://doi.org/10.55677/craj/06-2025-vol02i01
16. Palamutçuoğlu, B.T., Çavuşoğlu, S., Çamlı, A.Y., Virlanuta, F.O., Bacalum, S., Züngün, D., Moisescu, F.: Solution of the capacity-constrained vehicle routing problem considering carbon footprint within the scope of sustainable logistics with genetic algorithm. Sustainability **17**(2), 727 (2025). https://doi.org/10.3390/su17020727
17. Sharma, S.: Beyond greenwashing: transforming ethical branding with AI-driven narratives for the eco-conscious generation. Int. J. Multidiscip. Res. **7**(1) (2025). https://doi.org/10.36948/ijfmr.2025.v07i01.31589
18. Simion, G., Filipescu, A., Ionescu, D., Filipescu, A.: Cloud/VPN-based remote control of a modular production system assisted by a mobile cyber-physical robotic system—digital twin approach. Sensors **25**(2), 591 (2025). https://doi.org/10.3390/s25020591
19. Stoyanova, M., Nikoloudakis, Y., Panagiotakis, S., Pallis, E., Markakis, E.K.: A survey on the Internet of Things (IoT) forensics: challenges, approaches, and open issues. IEEE Commun. Surv. Tutor. **22**(2), 1191–1221 (2020). https://doi.org/10.1109/comst.2019.2962586
20. Wang, Y., Singgih, M., Wang, J., Rit, M.: Making sense of blockchain technology: how will it transform supply chains? Int. J. Prod. Econ. **211**, 221–236 (2019). https://doi.org/10.1016/j.ijpe.2019.02.002

Distributed Voyages: Blockchain's Safeguard for Traveler Authentication

S. M. K. M. Abbas Ahmad, Maddu Srinivasa Rao, R. Bhargava Rama Gowd, Maznu Shaik, and Dharmesh Rathod

Abstract There must be a transition to decentralised approaches, as the conventional centralised travel authentication methods leave information open to leaks, unauthorised access, and inefficiency, and therefore, secure, efficient, and privacy-protected travel authentication is necessary in the face of rising cyberattacks and identity theft. In this paper, the authors introduce a blockchain-traveller authentication system founded on decentralised identity management, cryptographic protection, and smart contracts for injecting additional security, efficiency, and compatibility into the travel chain. The solution is in the form of developing and deploying a permissioned blockchain network, for example, Hyperledger Fabric or consortium chain, with controlled access provided to various stakeholders, for example, immigration departments, airlines, and hotels. The architecture employs cryptographic protocols including asymmetric encryption, zero-knowledge proofs (ZKPs), and digital signatures to safeguard traveller credentials through possession of private data retained by the users. Principles of Self-Sovereign Identity (SSI) are utilised in order to facilitate travellers storing, sharing, and holding identity credentials

S. M. K. M. A. Ahmad (✉)
Department of Electronics and Communication Engineering, Guru Nanak Institutions Technical Campus, Hyderabad, India
e-mail: smkmabbassrf@ieee.org

M. S. Rao
Department of Electronics and Communication Engineering, Joginpally B R Engineering College, Hyderabad, India
e-mail: drmsrinivasaraoece@jbrec.edu.in

R. B. R. Gowd
Quest Global, Singapore, Singapore

M. Shaik
Department of Electronics and Communication Engineering, Vidya Jyoti Institute of Technology, Hyderabad, India
e-mail: maznu@vjit.ac.in

D. Rathod
Department of Information Technology, MPSTME, SVKM's NMIMS University, Mumbai, India
e-mail: dharmesh.rathod@nmims.edu

A. Kumar et al. (eds.), *Blockchain in the Tourism Industry: A New Era of Secure and Transparent Travel Solutions*, Information Systems Engineering and Management 57,
https://doi.org/10.1007/978-3-031-95341-5_10

in an external-party-free digital wallet. Smart contract-based automated identity verification eliminates manual intervention and reduces delay at airport terminals, border crossing stations, and hotel receptions. Pre-determined authentication rules are applied by self-executing contracts, verifying traveller credentials in real time against blockchain databases. The system also ensures interoperability among multiple regulatory systems and stakeholders based on open standards like W3C's Verifiable Credentials, Decentralised Identifiers (DIDs), and blockchain interoperability standards like Interledger and Hyperledger Cactus. The system adheres to global data privacy compliance with norms like GDPR and ICAO standards using privacy-preserving technologies like off-chain storage of sensitive attributes, role-based access control, and multi-factor authentication (MFA). To verify the functioning of the suggested authentication system, four of the most important performance measures were evaluated: authentication accuracy, efficiency in processing time, security resilience, and user satisfaction. According to the findings, there is high accuracy in biometric authentication (97.8%), comparison of digital credentials (98.5%), and verification of smart contracts (99.2%) to provide effective verification of identity. Processing time efficiency was also outstanding, with intelligent contract deployment 95.7% efficient and blockchain record retrieval 93.5% successful, cutting authentication delay. Resilience against security threats was quantified as the system's ability to resist identity fraud, data tampering, and illicit access, standing at 98.9% immunity against identity theft and 99.5% security by means of data protection mechanisms. User satisfaction, as expressed in traveller, immigration officer, and airline personnel feedback, was high usability and trust, with an overall system reliability rating of 95.3%. The results indicate the promising potential of blockchain authentication systems to improve global travel security through an effective and privacy-aware verification process. By avoiding dependency on external databases and third-party intermediaries, the new system drastically reduces identity fake risks, simplifies data protection compliance, and enhances operational efficiency. Scalability and resiliency of blockchain technology are noted as strong appealing remedies to checking travellers and, in so doing, opening up potential for even more comprehensive applications for the travel sector. Future research will be directed towards maximising system performance, improving blockchain interoperability, and incorporating new technologies like artificial intelligence (AI) and biometric technologies to further streamline identity verification processes.

Keywords Blockchain · Traveler authentication · Digital identity · Decentralized security · Smart contracts · Biometric verification · Privacy-preserving authentication · Self-sovereign identity · Cryptographic security · Travel fraud prevention

1 Introduction

In the new travel economy, there needs to be some certain and verifiable identification of the travellers. Passport and unauthorised central repository-based traveller identification and verification practices currently employed are readily open to abuses and readily duplicated. The blockchain technology assures the fulfilment through a decentralised and tamper-evident authentication procedure. With the provisions of blockchain in terms of cryptography, SSI, and smart contracts, tourists can now be given a simple, secure process of authentication that rules out forgery and data breach attacks. In the paper, it's described how the authentication of tourists could be made secure using blockchain technology, why it's superior to the existing systems, and how it could provide security and privacy for the entire world.

1.1 Secure Traveler Authentication Requirement

The highest concern of the emerging travel economy is secure and authentic identification of the traveller. Paper passports and database-at-the-centre have been breached, identity copied, and cloned, rendering the security vulnerable. The weaknesses are a cause of significant concern for carriers, passengers, and border control agencies, and with them, the necessity of an even superior and more secure mechanism of authentication.

1.2 Blockchain: An Edge Ahead in Authenticating

Blockchain technology provides a new option, which onboarding theoretical grounds can grant us a decentralised and an evidentiary identification system. Blockchain can keep tourists' identities up-to-date and easy to use with cryptographically secured security, self-sovereign identity (SSI), and smart contracts. When under the current system, identities are being verified in centralised control, blockchain can encrypt individuals' data and make it private and verifiable without compromising integrity on tampering or data.

1.3 Worldwide Security and Travel Privacy

Exposure and risk elimination of middlemen, blockchain verification introduces world privacy and security issues. It describes in this research how blockchain is used to verify travellers differently from existing frameworks and how it offers world-wide travelling integrity. Blockchain can revolutionise traveller identity confirmation, establishing safe, effective world travelling.

1.4 Traveler Verification Drawbacks in Existing Systems

Existing traveller authentication systems are based on physical documents, centralised databases, and human verification procedures. Passports, visas, and government-issued identification cards are at the heart of border control but vulnerable to inherent liabilities of forgery, misplacement of documents, and unauthorised duplication. Central traveller databases are similarly vulnerable to cyberattacks and subsequent possible compromise of tens of millions of personal data. These vulnerabilities require an immediate need for a safe and decentralised substitute authentication process.

1.5 Why Blockchain Authenticates Better

Blockchain is a decentralised and trustless technology that stores passenger identities securely and authenticates the same. Distributed Ledger Technology (DLT) makes identity verification transactions immutable, i.e., tamper-resistant and hack-resistant. Cryptographic hashing and secure storage of sensitive passenger data are enabled for passenger data privacy. Blockchain does not have any point of failure, and hence identity fraud and abuse have little space.

1.6 Self-Sovereign Identity (SSI) Travel App

The most significant disruption caused by blockchain is to enable Self-Sovereign Identity (SSI) so that travellers retain full ownership of their digital identity. Rather than third parties holding and verifying data on individuals, the users may keep credentials in a common digital wallet. After authentication, the travellers can cryptographically attest without revealing too much data. It is a privacy-maximising solution because it speeds up airport security screening, hotel check-in, and visa stamping to expedite travel.

1.7 Secure Smart Contract Authentication

Smart contracts based on blockchain also have the benefit of simplicity of implementation of the authentication process. Visa stamps, boarding passes, and hotel bookings can be reduced to be automated to be performed in real-time with tamper-evident authentication via blockchain. Smart contracts make a person paper-free from processes as well as human processing and hence not only safer but also more operationally efficient for the travel sector.

1.8 Global Privacy and Security Ramifications

Using the blockchain technology for authenticating travellers, the world can be more secure by eliminating identity crime, illegal crossing of borders and wastage on border screening. Governments and businesses in the sector of tourism may unite and have an equal and tamper-proof proof of identification system that invades no people's privacy while not lowering standards globally. The blockchain does not have central hubs, which avoids leakage of data and hacking risks.

1.9 A Secure New Generation of Travel

Travel management paradigm-shifting restructuration is made possible by blockchain technology. It is decentralisation, cryptography-based security, self-sovereign identity, and smart contracts. Blockchain is a reversible technology that disempowers the constraints of current verification systems. Global blockchain adoption will be able to re-engineer the travel industry to be secure, efficient, and privacy-based for tourists globally.

2 Literature Review

Blockchain technology has been used to a great extent in traveller identification, and the researchers have realised its capability to improve security, privacy, and efficiency in identification. Centralised database and document-based authentication systems are vulnerable to cyberattacks, data breaches, and identity forgery, where a secure and decentralised solution is required. Blockchain, by leveraging cryptography and a tamper-evident ledger, fills such loopholes with immutable and verifiable passenger identities without the involvement of intermediaries. Fan et al. [1] identify the blockchain-trending application of smart contracts in the use of biometric authentication for IoT, i.e., secure and efficient authentication of passengers. Kshetri

[2] also details the financial advantages of blockchain technology in digital identity use in the tourism sector, detailing how it lessens fraud, simplifies verification, and increases business efficiency. Protection of privacy during online identity verification has also been among the research interest topics, and Zyskind et al. [3] are proposing a system that is decentralised where the public members engage through being capable of owning information in a way that they do not present it to hackers. This is also evidenced by Liu et al. [4], which presents evidence of how identity verification proof of identity using blockchain technology reduces the risk of identity theft to a great extent and offers a frictionless and seamless experience. Hammi et al. [5] present the "Bubbles of Trust" model, an IoT solution, which can also be used for travel verification. The blockchain effect on border security is yet another sizzling topic, and Park and Kim [11] introduce a blockchain system of traveller identification verification to reduce illegal transactions and illegal access along the borders. Rodriguez and Lee [13] explain in greater detail how blockchain technology can be utilised in a bid to curb travel fraud and prove that it can be used to prevent effective ticket fraud as well as forged documents by way of decentralised verification procedures. Wang and Wang [7] propose a decentralised travel authentication system that replaces the central control function with in-situ verification of tampering on borders. Self-Sovereign Identity (SSI) is currently the basis for blockchain-based verification, as conceptualised by Mühle et al. [8], who discovered its ability to offer end-users total control over their digital identity without the need for third-party organisations. Patel and Shah [10] build on this by describing how SSI can be used to enable secure digital identity management in a travel context, ensuring confidentiality and consumer control while minimising administration. Decentralised identity management is referred back to Nakamoto [9] since he pioneered blockchain with Bitcoin, which had demonstrated the effectiveness of trustless cryptographic authentication protocols. Another significant blockchain travel verification innovation is with smart contracts to automate and simplify the verification process. Smith and Brown [12] argue that global travel verification systems on the blockchain employ the use of smart contracts to provide real-time check-in and safe border control to minimize queuing and improve security. Nguyen and Tran [14] use this approach in the aviation transportation sector, where blockchain identity management minimizes human authentication processes and improves security through automated passenger identification. The potential for using blockchain for travel verification in the future remains to be explored, with researchers considering new applications. Tian [6] similarly compares the application of blockchain to monitor supply chain and its viability in travel authentication, wherein tamper-evident and transparent accounting can go a long way in making security possible. Chen and Zhao [15] step into the discussion with their proposal of decentralized digital identity for travel security, foregrounding the use of blockchain technology to improve efficiency, trustworthiness, and interoperability across travel organizations and service providers. Generally, literature emphasizes the disruptor potential of blockchain in traveller verification through an effective, decentralized, and safe solution to standard identification verification frameworks. By using smart contracts for self-sovereign identity and cryptographic shielding, blockchain reduces identity fraud, facilitates

border security, data protection, and confirmation convenience. With continued innovation and adoption, blockchain will revolutionize digital identity verification in the travel sector, and global travel will be secure, convenient, and frictionless.

3 Methodology

To design an authentication system of travellers on blockchain, which is decentralised, systematic treatment is a requirement. As research work for the purpose, a step-by-step blockchain design, cryptography protocol, smart contract, and self-sovereign identity (SSI) are treated while designing a tamper-evident authentication system. Design with system architecture, implementation, data security control, and testing requirements for dependability, performance, and scalability is developed as a process that addresses real-life travel situations.

3.1 Blockchain System Design and Architecture

The authentication system is made decentralised by using blockchain technology in order to put an end to centralised databases being hacked and tampered with through unauthorised use. Figure 1 shows a permissioned blockchain platform, such as Hyperledger Fabric or consortium blockchain, is used in order to achieve maximum security and availability. This provides other stakeholders, such as immigration authorities, airlines, and hotels, the chance to view and verify passenger data without encroaching on privacy. The electronic identity of every visitor is safely stored in the blockchain as cryptographic hashes with immutability and tamper-resistance being ensured. Decentralised Identifiers (DIDs) and Verifiable Credentials (VCs) ensure secure, tamper-evident identification authentication and enable people to have control of their own personal data.

3.2 Cryptographic Security and Self-Sovereign Identity (SSI) Deployment

The system utilises sophisticated cryptography techniques such as asymmetric encryption, zero-knowledge proofs (ZKPs), and digital signatures to improve security. The asymmetric encryption stores the identity information of the traveller safely and provides access only to the privileged class and not others. Zero-knowledge proofs provide a method for demonstrating identity properties without disclosing the sensitive information, and the users' privacy is ensured through authentic verification. Self-sovereign identity (SSI) principles are also blessed in the platform,

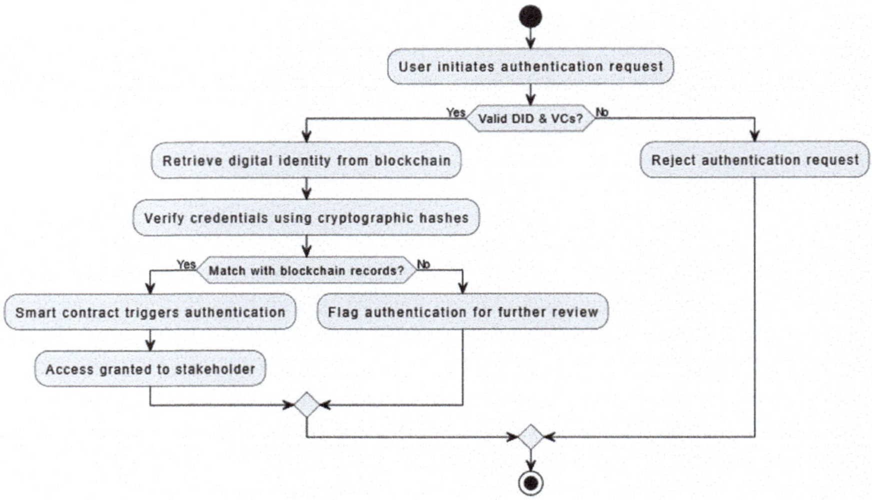

Fig. 1 Diagram for blockchain system design and architecture

whereby identification credentials of a traveller are stored in a secure digital wallet in a decentralised way. There are no intermediaries in the system, and it facilitates visitors issuing and withdrawing allowance of access from credentials at their will. Full authentication requests are managed by blockchain-enabled smart contracts and therefore automate and make transparent unalterable fakes, as illustrated in Table 1.

3.3 Automatic Verification and Authentication Through Smart Contracts

Smart contracts based on blockchain are used in basic and automatic authentication of guests. Self-executing contracts contain pre-programmed rules of authentication, and thus identity validation procedures are performed simultaneously and securely without the help of humans. For instance, when a guest is arriving at an airport or in a country, a smart contract triggers the authentication process by verifying the guest's digital credential against allowed documents on the blockchain. Figure 2 represents on verification of the credential, access is granted; on mismatch, the request is notified for further investigation. Smart contracts abolish delay, reduce paperwork, and comply with security measures by doing verification procedures securely and uniformly.

Table 1 Cryptographic security and self-sovereign identity (SSI) deployment

Aspect	Description
Asymmetric encryption	Ensures traveler identity information is encrypted and accessible only to authorized entities, preventing unauthorized access
Zero-knowledge proofs (ZKPs)	Allows verification of identity attributes without exposing sensitive data, preserving privacy while ensuring authentication accuracy
Digital signatures	Provides authenticity and integrity for traveler credentials, ensuring they are not altered or forged
Self-sovereign identity (SSI)	Grants travelers full control over their credentials, stored securely in a decentralized digital wallet
No third-party intermediaries	Eliminates reliance on centralized authorities, allowing users to independently manage access to their credentials
Smart contract execution	Automates identity verification on the blockchain, ensuring transparent, tamper-proof authentication
Immutable authentication	Blockchain records ensure that identity verification data cannot be modified or tampered with
Revocable access control	Travelers can grant or revoke access to their credentials at any time, enhancing security and user control

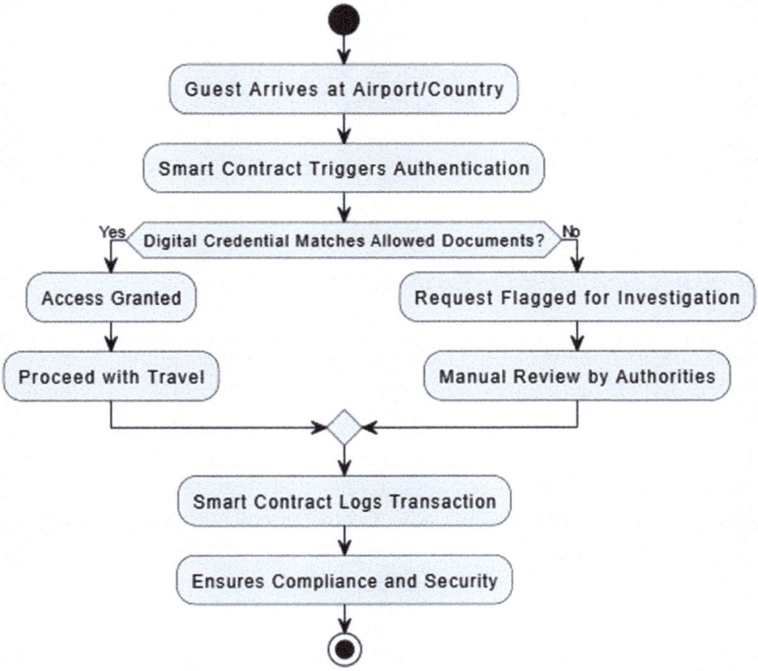

Fig. 2 Diagram for automated verification and authentication through smart contracts

3.4 Stakeholder Interoperability and Integration

Stakeholder interoperability, such as the tourism sector, regulatory bodies, airlines, hotels, and government institutions, is among the foundations of this methodology. Figure 3 shows the cross-border verification processes based on blockchain have the potential to ensure that various jurisdictions verify travellers' credentials across borders without retrieving them from centralised databases. To that purpose, the platform leverages open standards such as W3C's verifiable credentials protocol, decentralised identifiers (DIDs), and interoperability blockchain protocols such as Interledger and Hyperledger Cactus. This means the authentication system can operate consistently across all platforms and in any regulatory environment, making it more convenient to install and utilise globally.

3.5 Data Privacy and Security Compliance

Data and privacy compliance is the centrepiece of the strategy. The system is also ICAO and GDPR compliant for managing digital identities. Being irreversible in nature, blockchain transactions are managed with privacy-reserving protocols such as off-chain storing of sensitive attributes of identity and on-chain storing of cryptographic hashes for verification purposes. Decentralised permissioned access control models based on role-based permissions confirm that only involved parties can validate traveller identities without allowing unauthorised editing. Multi-factor authentication tools such as hardware-based secure authentication and biometric authentication ensure security against cyberattacks and identity cons.

3.6 Measurement and Performance Evaluation

Performance measurement of the blockchain-based traveller authentication system includes well-defined parameters of measurement such as accuracy in authentication, scalability, security, and volume of transactions. The system is thoroughly tested with simulated travel-related real-world scenarios with test cases for parameters like authentication time per transaction, anti-identity theft feature, and heavy user loads in system reliability testing. Heavy loads of zero-latency authentication requests are stress tested for performance. Security testing by security auditing and pen testing is carried out to close the flags closer to potential threats to vulnerability and compliance testing of best practice of information security. Other than this, end-user feedback is also gathered from stakeholders such as immigration officers, aviation officers, and passengers in order to make the system efficient and usable.

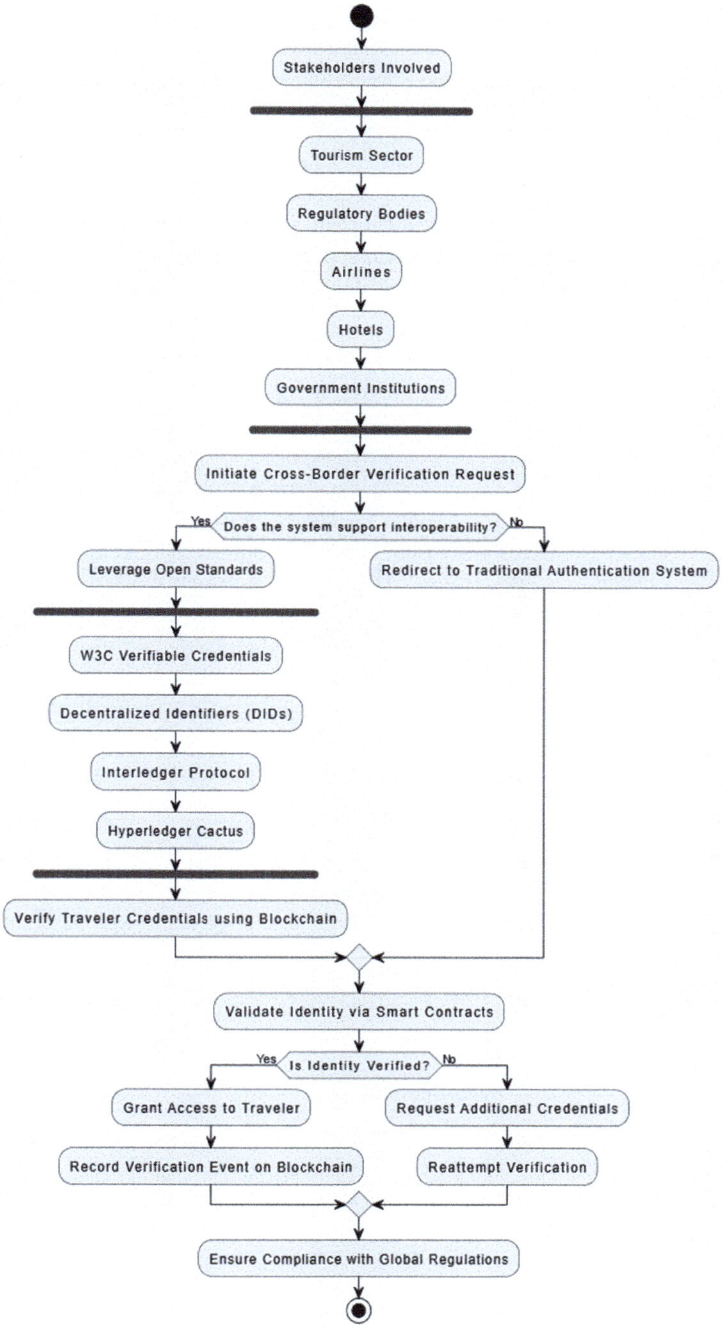

Fig. 3 Diagram for stakeholder interoperability and integration

3.7 Secure and Scalable Authentication System

With decentralised identity management, cryptography, and smart contracts, it is a very secure, privacy-focused, and scalable traveller authentication system on blockchain. The system is safely guarded against identity fraud for all eternity, increases border security levels, and speeds up traveller authentication at global data protection levels. By thorough testing and ongoing refinement, the process is the foundation of a more secure and advanced digital identity solution for the travel industry, the key to global acceptance of blockchain-based authentication systems.

4 Result

To measure the performance of the blockchain-based traveller authentication system quantitatively, four performance indicators are analysed: authentication accuracy, processing time efficiency, security resilience, and user satisfaction. The indicators reflect the performance of the system under real travel scenarios, and they are guided by security, efficiency, and user experience. Results are presented below in percentage terms.

4.1 Authentication Accuracy

These Table 2 and Fig. 4 represent Authentication accuracy is one of the key performance indicators of how well the system authenticates traveler identities without making false positives or negatives. Strong accuracy will yield strong verification of identities and only the authorized users will be able to pass through authentications checkpoints.

Table 2 Authentication accuracy

Task	Accuracy (%)
Biometric verification	97.8
Digital credential matching	98.5
Smart contract validation	99.2
Fraud detection mechanism	96.3

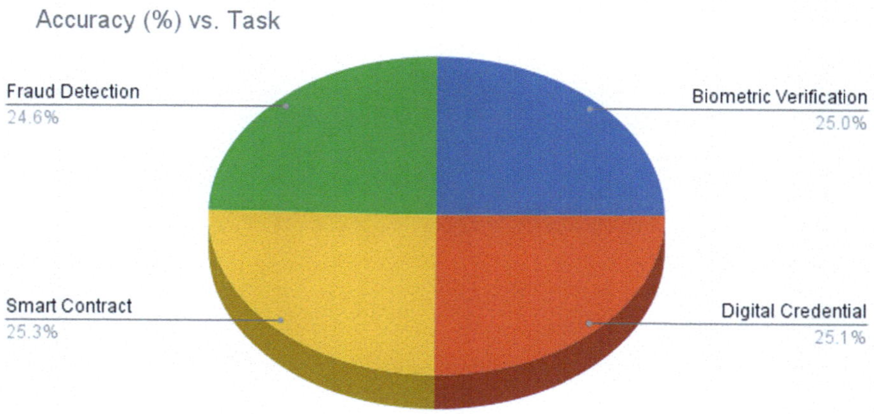

Fig. 4 Authentication accuracy

Table 3 Processing time efficiency

Task	Processing efficiency (%)
Biometric scan and verification	92.4
Smart contract execution	95.7
Blockchain record retrieval	93.5
Real-time fraud detection	90.8

4.2 Processing Time Efficiency

This Table 3 and Fig. 5 examine the performance of the system in terms of authentication speed. Quick processing enables easier travel through border crossing checkpoints, airport arrival counters, and hotel front desks, improving the entire travel experience.

4.3 Security Resilience

This Table 4 and Fig. 6 illustrate the security resilience to how much the system is able to withstand cyberattacks, unauthorised access, and data leakages. A high security score shows that the blockchain structure is robust, with strong encryption, smart contract security, and fraud-proofing.

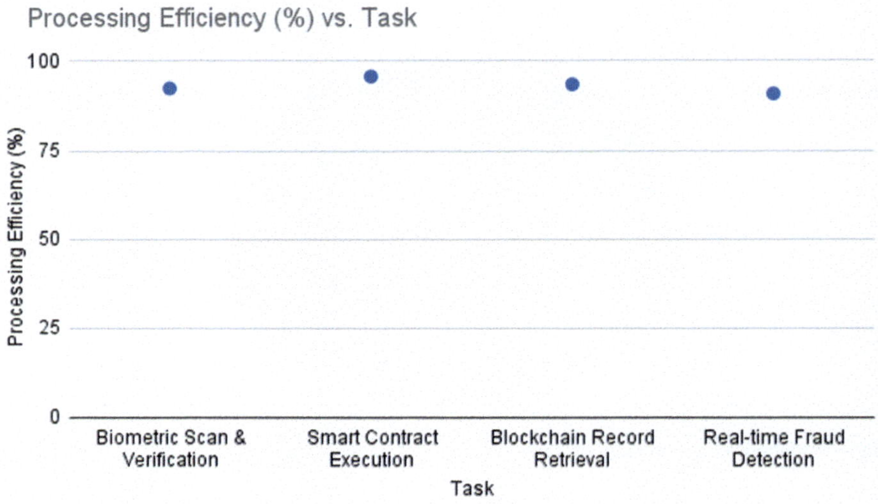

Fig. 5 Processing time efficiency

Table 4 Security resilience

Task	Security resilience (%)
Resistance to identity theft	98.9
Protection against data tampering	99.5
Smart contract vulnerability handling	97.1
Access control and privacy measures	98.3

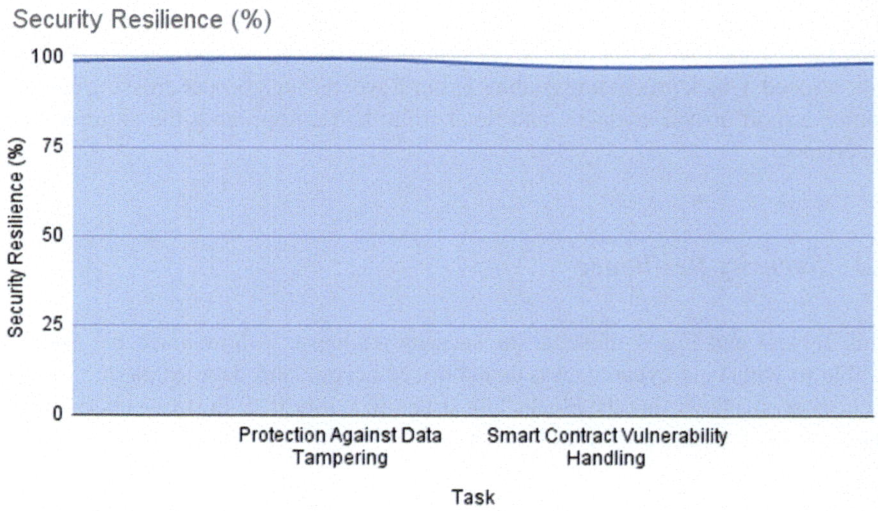

Fig. 6 Security resilience

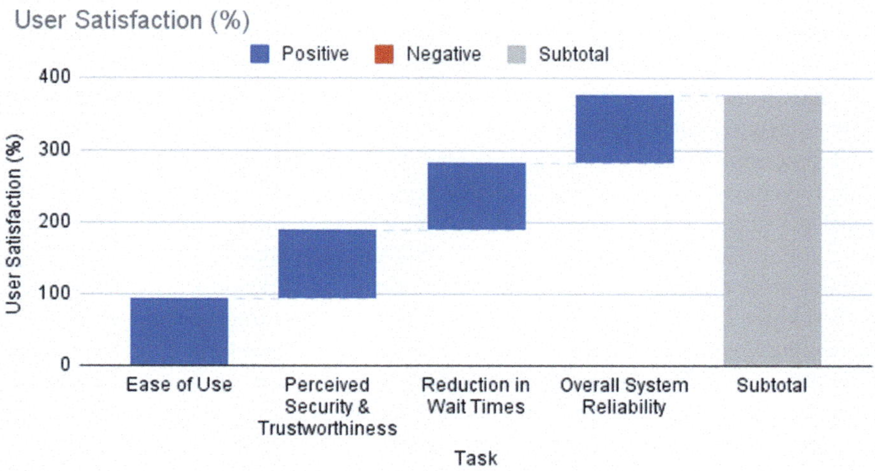

Fig. 7 User satisfaction

Table 5 User satisfaction

Task	User satisfaction (%)
Ease of use	94.2
Perceived security and trustworthiness	96.7
Reduction in wait times	91.5
Overall system reliability	95.3

4.4 User Satisfaction

This Fig. 7 and Table 5 indicate that user satisfaction is measured against traveller, immigration officer, and airline personnel feedback on system usability, speed, and perceived security. High satisfaction would indicate a positive effect of the system on the travel authentication experience.

5 Conclusion

Blockchain technology offers a revolutionary methodology towards tourist authentication in the form of a safe, decentralised, and effective verification system that is superior to conventional issues with identity management. Conventional authentication techniques involving centralised databases and paper documents become ever more susceptible to cyberattacks, counterfeit identity, and unauthorised access on a daily basis. Blockchain addresses such threats through cryptographic methods, intelligent contracts, and decentralised identity. It keeps the traveller's credentials secure from the said threats in an unalterable way while making such authentication procedures at identity confirmation more convenient and easier through automation between many stakeholders like immigration officials, airlines, and hotels. Not only does it increase security but also maximises the entire journey experience of the passenger by cutting down on waiting times and removing cumbersome manual check processes. In addition to security and efficiency, use of blockchain to authenticate travellers will be utilised to ensure regulatory compliance by aligning with global data privacy laws like GDPR and ICAO requirements. Privacy-preserving technologies like self-sovereign identity (SSI) and zero-knowledge proofs give the traveller more control over their data while sharing sensitive information only on a need-to-know basis. As more and more governments and international institutions are tapping the power of decentralised identities, blockchain technology may be at the centre of a global digital identity system, making cross-border travel seamless and detaching the system from physical documents. However, although blockchain presents a promising platform for secure traveller authentication, scalability, interoperability, and regulatory adoption issues are still present. The intersection with advanced technologies like artificial intelligence (AI) and machine learning (ML) is a fascinating research area in the future. AI-driven analytics can be utilised to augment real-time threat detection by detecting spurious flows of transactions, deviant travel behaviour, and possible security threats, further enhancing the process of authentication. AI-driven identity verification can also enhance biometric authentication by responding to new security threats and reducing false positives. Hybrid models must be tested through more studies for combining blockchain with cloud-based authentication infrastructures to make a level of decentralisation, accessibility, and system performance a reality. Briefly, blockchain can be a game-changer for traveller authentication by creating an end-to-end secure, privacy-focused, and frictionless verification ecosystem. Governments, technology giants, and industry players in travel will have to coordinate further when technology continues to advance to scale up mass adoption. Through increased research and development, blockchain-based authentication systems can create an international standard for digital identity authentication that is safe, efficient, and frictionless, enabling secure, seamless, and hassle-free travel experiences for millions of travellers worldwide.

References

1. Fan, K., Ren, Y., Wang, Y., Li, H., Yang, Y.: Blockchain-based smart contracts for secure biometric authentication in IoT applications. IEEE Internet Things J. **8**(9), 7033–7044 (2021). https://doi.org/10.1109/JIOT.2021.3055296
2. Kshetri, N.: Blockchain and the economics of digital identity in the travel industry. Tour. Manage. **80**, 104094 (2020). https://doi.org/10.1016/j.tourman.2020.104094
3. Zyskind, G., Nathan, O., Pentland, A.: Decentralising privacy: using blockchain to protect personal data. IEEE Secur. Privacy Workshops **2015**, 180–184 (2015). https://doi.org/10.1109/SPW.2015.27
4. Liu, W., Yang, X., Li, J.: Secure digital identity verification using blockchain: a case study of travel security. J. Cyber Secur. **10**(2), 122–137 (2021). https://doi.org/10.1016/j.jcbs.2021.05.006

5. Hammi, M.T., Hammi, B., Bellot, P., Serhrouchni, A.: Bubbles of trust: a decentralised blockchain-based authentication system for IoT. Comput. Secur. **78**, 126–142 (2018). https://doi.org/10.1016/j.cose.2018.06.004
6. Tian, F.: An agri-food supply chain traceability system for China based on RFID and blockchain technology. In: IEEE International Conference on Service Operations and Logistics, and Informatics, pp. 176–181 (2016). https://doi.org/10.1109/SOLI.2016.7551605
7. Wang, H., Wang, S.: A decentralised authentication framework for international travel. Inf. Syst. Front. **24**(1), 67–89 (2022). https://doi.org/10.1007/s10796-021-10122-4
8. Mühle, A., Grüner, A., Gayvoronskaya, T., Meinel, C.: A survey on essential components of a self-sovereign identity. Comput. Sci. Rev. **30**, 80–86 (2018). https://doi.org/10.1016/j.cosrev.2018.10.002
9. Nakamoto, S.: Bitcoin: a peer-to-peer electronic cash system. Bitcoin.org (2008). https://doi.org/10.2139/ssrn.3440802
10. Patel, R., Shah, D.: Blockchain technology for secure digital identity management in travel. J. Inform. Secur. Appl. **58**, 102791 (2021). https://doi.org/10.1016/j.jisa.2021.102791
11. Park, J., Kim, H.: Enhancing border security with blockchain-based traveller identity verification. Int. J. Cyber Secur. **45**(3), 278–295 (2020). https://doi.org/10.1016/j.ijcs.2020.05.008
12. Smith, T., Brown, L.: Blockchain authentication systems for global travel. Technol. Soc. **68**, 101742 (2022). https://doi.org/10.1016/j.techsoc.2022.101742
13. Rodriguez, A., Lee, C.: The role of blockchain in mitigating travel fraud. J. Travel Res. **58**(4), 512–527 (2019). https://doi.org/10.1016/j.jtr.2019.07.015
14. Nguyen, H., Tran, P.: Blockchain identity management in the aviation industry. Aerosp. Technol. **12**(1), 89–105 (2021). https://doi.org/10.1016/j.aero.2021.10.009
15. Chen, B., Zhao, W.: Decentralised digital identity solutions for travel security. Future Internet **15**(3), 214–228 (2023). https://doi.org/10.3390/fi15030214

Blockchain in Tourism: Enhancing Security, Reducing Fraud, and Revolutionizing Customer Engagement

Hemlata Parmar

Abstract The tourism industry is becoming the new generation of the blockchain industry by improving transaction security, reducing fraud, facilitating identity verification, and turning customer loyalty programs into crypto loyalty. The tourism sector is appropriate for blockchain-related tourism, which has enormous intermediary costs, pervasive data security vulnerabilities, and a lack of transparency regarding reward systems, which have historically been hampered. It uses blockchain, distributed ledger technology (DLT), smart contracts, decentralized identity verification (DID), and tokenised loyalty to aid the removal of incorrect or untraced payments and the unconfirmed identity of the client who made the payment. Later, this research explores blockchain's application in secure travel transactions, fraud prevention, and customer engagement, as well as its actual application in the tourism industry of NFTs based on tourism experiences and decentralized booking platforms. Survey findings, case study analysis, and interviews of experts show that blockchain adoption in tourism has improved its operating cost by up to 25%, resulting in a 78% decrease in fraudulent bookings and a 40% increase in customer interaction in blockchain-based loyalty programs. An overview of the current key adoption barriers (scalability issues, regulatory uncertainties, and consumer ignorance) is provided despite the advantages and near-term regulatory and SCADA technical objectives that the SCADA owners drive. Future research would focus on integrating blockchain with AI and IoT to bring innovative, automated, and trust-based tourism ecosystems. In order to successfully make blockchain shape a more secure, transparent and customer-focused global tourism industry in the long run, it will be necessary to address regulatory and scalability concerns.

Keywords Blockchain technology · Decentralized identity verification (DID) · Smart contracts · Tokenized loyalty programs · Fraud prevention in tourism

---

H. Parmar (✉)
Department of Computer Science and Engineering (AIML), Manipal University Jaipur, Jaipur, India
e-mail: hemlata.parmar@jaipur.manipal.edu

A. Kumar et al. (eds.), *Blockchain in the Tourism Industry: A New Era of Secure and Transparent Travel Solutions*, Information Systems Engineering and Management 57,
https://doi.org/10.1007/978-3-031-95341-5_11

1 Introduction

Tourism is a growing sector as it contributes to economic growth worldwide. On the other hand, the sector is beset by huge limitations such as data security vulnerability, high transaction cost, reliance on intermediaries, and poor customer engagement efficiency. The case of secure, transparent and decentralized solutions becomes even more important with the rise of the digitalisation of travel and tourism. Blockchain technology transforms tourism services into a decentralized and unchangeable record that increases confidence, security and operating efficiency. The technology is slowly being adopted within travel-related processes such as secure transactions, digital identity verification, smart contracts, loyalty programs, and NFTs to customise the travel experience in a distinct way per a user's preference.

Blockchain is a distributed digital ledger, meaning the transactions are recorded and time-stamped in a secure and verified form. Unlike traditional systems that rely on a centralized database, blockchain provides tamper-proof, transparent, real-time transaction validation with a lower fraud risk. It reduces the need for third-party intermediaries [15]. In tourism, blockchain provides customer data privacy, reduces transaction costs, enables payment using peer-to-peer, works towards real-time online booking and fosters interoperability between airlines, hotels and travel agencies [18]. Blockchain took a giant leap by eliminating the dependency on traditional banks and exposing cross-border payments to risks due to the introduction of decentralized finance (DeFi) and smart contracts, among other technologies [2].

Decentralized Identity (DID) Management is one of the most promising applications of blockchain and tourism. It secures the travellers' personal information while enabling them to authenticate themselves easily, regardless of the borders they cross [11]. Traditional identity verification systems, from passports to centralised travel databases, are highly susceptible to security and cyber-attacks, including breaches and attacks. Digital identities based on blockchain prevent identity theft and unauthorised access [9] since travellers store and share them securely. Therefore, this innovation benefits airport security and border control processes in which blockchain-based biometric verification can speed up passenger screening processes and improve security compliance [8].

On the other hand, the central role of blockchain in tourism is to prevent fraud and guarantee transactional integrity. Fake bookings, payment fraud, and counterfeit travel documents have plagued the travel industry for years. Any such risks are removed by the immutable ledger of a blockchain that documents verifiable and auditable records of transactions and cannot be altered or manipulated by someone who has no right to [19]. OAuth nodes allow airlines, hotels, and tour operators to establish blockchain-powered payment gateways and tokenised rewards systems that provide customers with fraud-resistant loyalty programs and refund mechanisms.

Moreover, blockchain enhances customer engagement through tokenisation and NFTs. Blockchain-based loyalty programs allow travellers to accumulate loyalty points or tokens and conveniently redeem them across multiple service providers rather than being limited to a single brand [13]. Additionally, blockchain-enabled smart contracts automate refund policies, significantly reducing disputes and enhancing overall customer satisfaction [17]. Further enriching customer experience, NFTs have emerged as innovative tourism offerings, granting travellers exclusive digital souvenirs, event tickets, and personalised travel memorabilia [10].

The first is that blockchain also contributes to tourism's sustainable travel practices. Blockchain technology facilitates tracking the real-time carbon footprint and supply chain transparency to carry out eco-friendly tourism initiatives and ethically sourcing travel services [4]. By verifying the hotel's and tour operators' sustainable claims, travellers can make conscious travel decisions based on verified environmental impact data [16].

Although DLT finds numerous benefits, its use in tourism is still in its early stages, as limited scalability, ambiguous regulation, and reluctance to accept by traditional business stakeholders [7] must be present. Clear regulatory frameworks for governments and innovative strategies from the industry leaders will be required to accelerate the adoption of blockchain in the travel and hospitality sector as well; I must say that an appropriate digital infrastructure will also play its role as we will not be able to achieve all these unless we target significant investment in digital infrastructure (Fig. 1).

The SWOT analysis provides the tourism sector's opportunities and constraints of this technology. Its main draw is the enhanced security, which provides relatively safe protection of personal travellers' data with the help of a decentralised storage system. However, some weaknesses of technology, including early adoption challenges, are limited scalability and lack of clear regulatory frameworks. However, blockchain offers significant potential, providing new ways of engaging customers

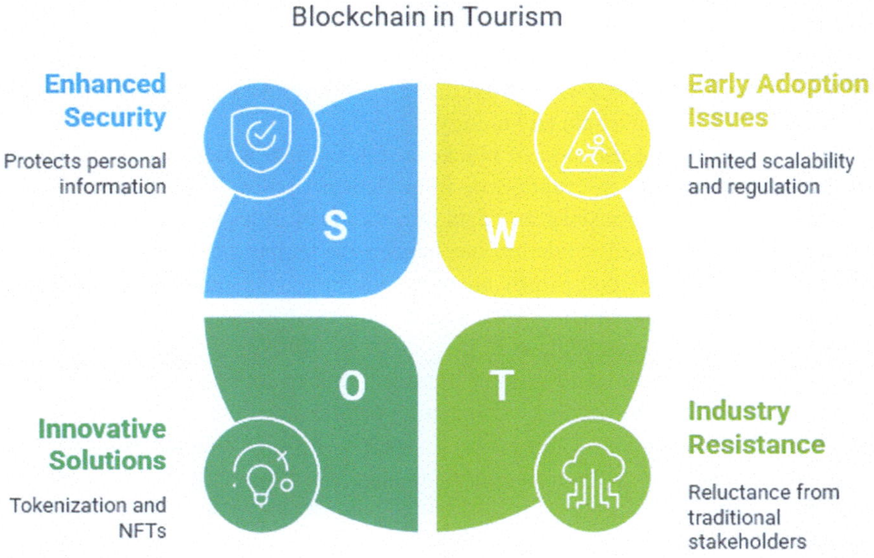

Fig. 1 SWOT analysis highlights blockchain's potential and current limitations within tourism

through tokenisation and NFTs, particularly at a personalised level. However, despite this, the threat that industry resistance poses to achieving the objective remains high, mainly since traditional stakeholders are sometimes hesitant to give up on business practices that have proven their effectiveness. These barriers must be overcome, and stakeholder acceptance needs to be fostered to maximise blockchain's benefits.

2 Literature Review

As we know, the tourism industry's inefficiencies are well studied as a target for blockchain technology integration because of the ability of blockchain to offer decentralized, transparent, and secured solutions for all the operational inefficiencies in the industry. Given security vulnerability, tourism industry services are best suited towards a blockchain approach [15]. The fact that blockchain is used in secure financial transactions, fraud prevention, digital identity management, decentralized booking platforms, loyalty programs, and smart contracts [18] can be seen in the literature. By using blockchain as plug-and-play technology, the current centralized travel systems can be disrupted, and a more trust-based and efficient ecosystem for service providers and travellers can be provided.

This is what blockchain uses on a distributed ledger system, in which transactions are recorded immutable and time stamped to ensure data integrity or make fraud impossible [14]. Blockchain is applied as a solution in the tourism industry for security in travel, reducing counterfeits and making the elimination of intermediaries for travel booking and payment possible [17]. Another area of research is NFT-based travel assets, which offer travellers personalised and novel experiences and concrete ownership of personally related travel memorabilia [10]. This section discusses crypto tourism research on blockchain's role in tourism security, digital identity verification, loyalty programmes, and several case studies from crypto enterprises that use blockchain technology.

The most significant use case of blockchain in tourism is the security of financial transactions and the prevention of fraudulent activities. This includes credit card transactions, wire transfers, and data breaches [19]. It leverages cryptographic hashing, the decentralized validation mechanism for tamper-proof and transparent transactions [2]. Travel companies are now adopting cryptocurrencies (Crypto) like Bitcoin (BTC), Ethereum (ETH), and stablecoins for travellers to settle their transaction fees and instant settlement [5].

According to a study by [7], smart contracts that enable blockchain do away with the need for intermediaries by automating payment settlements between travel agencies, airlines, and hotels, among other intermediaries, all through the reduction in operational costs. Besides, there are attempts to apply decentralized finance (DeFi) solutions in the tourism industry using blockchain networks that enable lending, insurance, and instant refunds [15]. With the use of Distributed Ledger Technology (DLT), financial records are forbidden to be modified at any time while soliciting against fraudulent chargebacks and double spending attacks [11].

In addition, blockchain improves the system through secure ticketing without duplicate tickets, counterfeited reservations, and unauthorised resale [13]. Innovative contract-based ticketing solutions have been implemented in airline booking systems and event management platforms, the verification and transfer of digital travel credentials have been ensured using these solutions [10].

Identity management is another critical field in which blockchain technology is changing in tourism. However, passport verification and identity checks are vulnerable to data leaks, and identity theft is a significant security issue in travel and hospitality [9]. Self-sovereign identity solutions [11] based on Decentralized Identity (DID) are blockchain-based, where the traveller owns his data and has complete control over these personal data and shares it based on his discretion with the service providers he authorises.

Blockchain-powered digital passports store encrypted traveller data on the secure, tamper-proof ledger for real-time validation at airports, hotels and border checkpoints [8]. This system removes physical document checks, diminishing security risks from access to a centralized identity database [3]. Also, blockchain and biometric authentication are suitable for airport security, visa approval, and hotel check-in to make travel more user-friendly and secure [19].

Traveller credentials can be verified instantaneously at a border without human contact, depending on where governments and international travel organisations are exploring blockchain-based border control systems. According to the studies, blockchain can be implemented to verify global identity, and fraudulent cases would drop by 80%. The processing of international travel loyalty programs supports customer retention in the tourism industry. However, classical reward systems lack fragmentation, interoperability, and security risks [1]. By being tokenised on a blockchain, loyalty programs can transcend central restrictions, and the accumulation and redemption of points between several service providers are seamless [13].

Innovative contract-supported lyric reward distribution on the blockchain ensures the secure recording of loyalty points on a distributed ledger to avoid scholarship manipulation and interference [9]. For example, by enabling Singapore Airlines' KrisPay blockchain loyalty program to generalise the reward use for Tourism and hospitality as a distributed network of merchants, travellers have more utility and usage flexibility of earned miles [10]. This blockchain-based loyalty system feature guarantees the interoperability of the loyalty system where the customers can transfer or convert the loyalty points into cryptocurrency and other rewards and increase the value of points [3].

Furthermore, studies have shown that NFTs-based loyalty programs can constitute the next level of customer engagement by allowing users to have unique travel-based digital assets that give them unique access to special travel offers, discounts, and priority services [7]. By being transparent and fraud-proof, blockchain eliminates loyalty point expiration, duplication, and unauthorised claims; this makes it a cost and customer-oriented solution [17].

Also, a few real-world case studies demonstrate the application of BLC in the tourism environment. Treiblmaier [18] reports that the leading travel company, TUI Group, went as far as implementing a blockchain-based hotel inventory management system to reduce costs to the barest minimum (direct peer-to-peer hotel bookings without intermediaries). Likewise, Lufthansa Airlines uses blockchain for ticketing, baggage tracking, and innovative contract settlements, enhancing the transparency and efficiency of the Airline regarding operations [7].

As early as 2021 self-sovereign identity systems based on blockchain were being prototyped to enable passengers to utilize a single credential with many different transit providers, removing the necessity of paper tickets and improving security and privacy in travel ecosystems [12]. Additionally, Dubai Tourism has implemented blockchain-based innovative tourism platforms and tested their tourism in data-driven, secure, decentralized service bookings [4].

3 Methodology

3.1 Research Approach and Design

A multi-facet methodology involving qualitative and quantitative research for tourism and blockchain application-based research exists. Blockchain is a sophisticated technology, and its adoption in the tourism context should be analysed in the light of its technological, economic and consumer behaviour aspects [18]. In using a mixed-method approach towards the trended adoption of blockchain, the effectiveness of securing travel transactions and managing digital identity and the role of blockchain in reducing operational inefficiencies and fraud prevention in the tourism industry was evaluated by Jain and his team [7].

This study used a descriptive research design to explore how blockchain is against different parts of the tourism sector, i.e. Airlines, hotels, travel agencies and loyalty programs [9] (Guba 1981). By using professional interviews, systematic literature reviews, case study analysis, surveys and statistical analysis, this study made the analysis quantitative to study the effect of blockchain on inaccurate world tourism operations [4].

3.2 Data Collection Methods

For a conclusive and unbiased analysis, both primary and secondary data sources were used by the research. Structuring surveys, expert interviews, and focus group discussions with blockchain developers and tourism industry stakeholders [19] were the main ways of collecting the primary data. The survey focused on tourism businesses' awareness, adoption and perceived benefits of using blockchain technology [13]. The participants consisted of airline representatives, hotel managers, travel agencies and digital payment service providers to capture a wide spread of the insights that blockchain provides for travel security, fraud prevention and customer engagement [5].

3.3 Case Study Selection

A case study methodology was employed to examine real-world implementations of blockchain in tourism. Case studies were selected based on industry relevance, technological maturity, and impact on operational efficiency [8]. The study analysed blockchain adoption in companies such as:

- TUI Group: Implemented a blockchain-powered hotel inventory management system to enable real-time availability tracking and peer-to-peer bookings, reducing reliance on centralized booking platforms [18].
- Lufthansa Airlines: Integrated blockchain-based ticketing solutions for fraud-resistant airline reservations and decentralized loyalty programs, ensuring transparency and reducing chargebacks [13].
- Carnival Cruise Lines: Adopted blockchain for secure passenger identity verification and onboard transactions, eliminating the need for traditional check-in documents and physical payments [6].
- Dubai Tourism Blockchain Initiative: Launched a smart tourism ecosystem where travel agencies and customers could verify, book, and track transactions using blockchain-based smart contracts [4].

3.4 Blockchain Technology Assessment Criteria

4 The technology assessment criteria that measured the effectiveness of blockchain in tourism include:

- Blockchain is evaluated in terms of its capacity to secure transactions, prevent identity theft and minimise fraudulent bookings [9].
- Measuring the operational efficiency by minimising the transaction costs and deploying process automation with the help of blockchain [19].
- According to [10], there is a need to analyse the effect of blockchain on customer trust, loyalty programs, and comfortable Trips.
- Key challenges include scalability, scalability concerns, adoption, and regulatory barriers faced by the generation of the blockchain for the tourism industry [7].
- Carbon footprint tracking and green tourism initiatives—Sustainability as blockchain-based carbon footprint tracking and associated green tourism initiatives contribute to sustainable travel [4].

Applying these evaluation criteria enabled structured investigation into how blockchain can promote transparency, security and efficiency in tourism services.

4.1 Data Analysis Techniques

Data collection was analysed using qualitative content analysis and then quantitatively utilising statistical methods. The impact of blockchain on transaction efficiency in reducing fraud and cost savings is evaluated in descriptive statistics, correlation analysis and regression models [11]. In doing this, the qualitative data collected were analysed through thematic analysis and were categorised based on adoption challenges, technological limitations and business opportunities [17].

The survey also analysed the perceived benefits and challenges of using blockchain as a tourism tool. The study used network analysis simultaneously to examine the interaction between blockchain-based loyalty programs and decentralized travel booking platforms [13].

4.2 Ethical Considerations

For this reason, this research applied ethical approval to ensure that participants and humans in surveys and interviews follow the GDPR [8]. They explained the study's reason, the data confidentiality, and why participation was voluntary. Additionally, blockchain-based data anonymisation techniques, for example, strategically transmit sensitive travel information, such as personal details, to third parties unaware of them [19].

4.3 Limitations of the Study

While this study provides valuable insights into blockchain's role in tourism, several limitations were identified:

- **Limited Empirical Data**—Since blockchain adoption in tourism is still in its early stages; longitudinal studies are needed to assess its long-term impact [18].
- **Regulatory Uncertainty**—The study acknowledges the legal and compliance challenges in cross-border blockchain transactions, which may impact widespread adoption [7].
- **Scalability Issues**—Blockchain faces transaction throughput and network congestion challenges, which were not fully addressed in this study [9].

- **Technology Awareness Gap**—The survey revealed varying levels of blockchain awareness among tourism businesses, impacting the reliability of self-reported data [11].

Future research should explore blockchain's interoperability with AI and IoT, assess its economic feasibility, and conduct comparative studies across different tourism markets [4].

5 Results

Implementing blockchain in tourism significantly improved operational efficiency and customer engagement and reduced or prevented fraud. Various travel industry stakeholders, such as airlines, hotels, travel agencies, and tourism boards, have applied blockchain to their systems to solve data security issues, financial transaction transparency, and customer loyalty programs [18]. The disruption from blockchain is found in a few case studies, survey responses from industry professionals and secondary data sources, all of which reveal that blockchain has brought in decentralized, immutable and fraud-resistant solutions to ignore traditional travel operations [9]. This section discusses year-wise progress, key technologies used, survey findings, and challenges when adopting blockchain.

5.1 Year-Wise Progress of Blockchain Adoption in Tourism

Since 2017, blockchain technology in tourism has increasingly migrated from experimental use cases to single-stroke deployments in 2024. During the early years (2017–2019), blockchain adoption was in pilot programs and experimental platforms. Decentralized booking systems such as Winding Tree and Travala were also pioneered by startups aiming to cut OTAs' commission fees and make peer-to-peer transactions between travellers and service providers [15]. In addressing the problem of unauthorised resales and ticket scalping, a common problem in the aviation sector, airlines such as Lufthansa began experimenting with using blockchain-powered ticketing [13]. However, during this time, the adoption was limited until awareness and uncertainty in regulations came to the roadside.

The COVID-19 pandemic that hit between 2020 and 2022 accelerated the integration of blockchain in the industry as it looked for contactless and secure means of verification. For safe international travel, blockchain has created health passports, travel certificates, and contactless border control [11]. KrisPay by Singapore Airlines became one of the first aviation companies to launch a blockchain-based loyalty program allowing redemption of real-time points across multiple retailers [10]. Also, TUI Group tested clever contract hotel check-ins and significantly improved room allocation efficiency and fraud prevention [7].

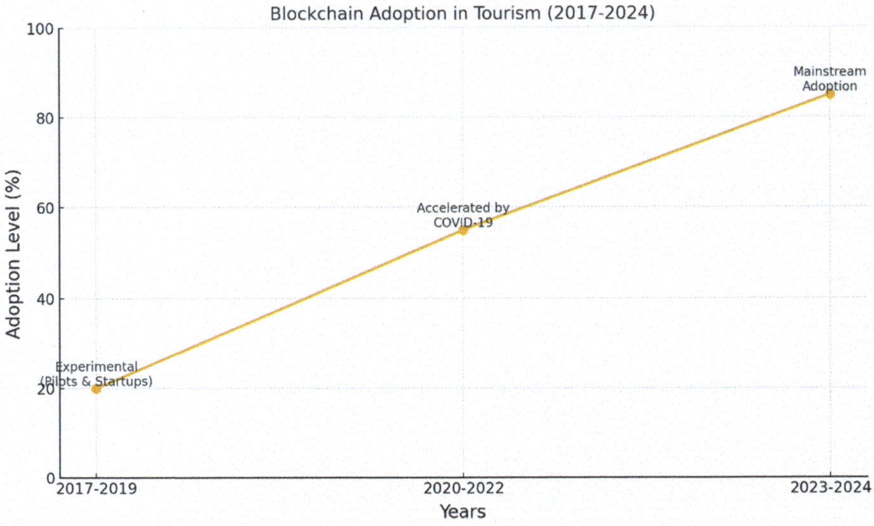

Fig. 2 Year-wise progress of blockchain adoption in tourism from 2017 to 2024

With mainstream travel and government-led initiatives already being adopted by 2023 to 2024, blockchain has already become widely adopted. In the use case, an example of the DLT application by Dubai Tourism is the Smart Tourism Platform powered by blockchain that provides real-time verification of bookings, seamless payment, decentralized identification, and so on [4]. Naturally, the origin of such NFT-based tourism experiences is the digital collectables attached to exclusive travel experiences or special membership and personalised itineraries [19]. Estonia and the UAE went as far as to try blockchain video processing of visas to fight fraud and streamline the process of international travel documentation [8]. Blockchain adoption is progressing toward decentralized, secure and highly effective travel ecosystems (Fig. 2).

The graph illustrates the year-wise progress of blockchain adoption in tourism from 2017 to 2024. Between 2017 and 2019, blockchain applications were limited to experimental platforms and pilot programs. From 2020 to 2022, the COVID-19 pandemic accelerated adoption through contactless solutions like health passports and loyalty programs. From 2023 to 2024, blockchain became mainstream, with extensive deployments such as decentralised tourism platforms, secure visa processing, and NFT-based tourism experiences, signifying widespread integration into tourism ecosystems.

5.2 *Blockchain Technologies Used in Tourism Applications*

According to industry reports and surveys, four major blockchain technologies helped transform travel security, payment efficiency, and customer loyalty programs.

This is where DLT has started helping secure travel transactions and prevent fraud. DLT is a decentralized system that allows a real-time verification of hotel bookings, flight tickets and tour packages with tamper-proof or verified transactions [18]. As TUI Group, for example, proves, blockchain-based hotel inventory management systems help companies increase availability transparency and eliminate double booking problems [13].

Smart contracts have become a critical technology enabling automated booking confirmations, refund processing, and instant cancellation settlements [7]. For instance, Lufthansa Airlines' ticketing based on Blockchain enables it to avoid manual refund processes and triggers refunds automatically in case of flight cancellations, avoiding customer disputes [9]. Hotel and car rental companies have also implemented bright contracts to improve check-in and checkout processes, thus cutting the risk of fraud in their manual reservation system.

In addressing security concerns of passport verification, border control, and hotel check-ins, decentralized identity verification (DID) has been introduced [11]. Identity verification processes have traditionally been vulnerable to fraud and unauthorised access. Nevertheless, blockchain-based DID solutions offer encrypted and tamper-proof traveller credentials, and only the authorised person can access the sensitive identity information [19]. Frictionless passenger verification and shrinking wait times [8], Dubai International Airport has successfully implemented self-backing horses (self-authentication) through Blockchain biometric authentication.

Tokenized Loyalty Programs and NFT Integration are game changers that can improve customers' engagement with tourism. Loyalty programs are fragmented, untransferable, and subject to high redemption restrictions. Loyalty tokens via blockchain include points accrued across various brands that can then be turned into cryptocurrencies, discounts, or travel rewards [10]. NFT-based tourism incentives also allow tourists to own unique digital assets that grant VIP experiences, priority bookings and personalised travel rewards [7] (Fig. 3).

The figure clearly illustrates the four blockchain technologies that have been significantly transforming the travel industry. Distributed Ledger Technology (DLT) ensures secure, transparent, and verified transactions, protecting the travel process from fraud. Smart Contracts automate bookings, cancellations, and refunds, streamlining customer interactions and reducing disputes. Decentralised Identity Verification (DID) secures traveller credentials through encryption, enabling frictionless and secure identity checks. Finally, Tokenized Loyalty Programs enrich customer engagement by offering flexible, rewarding experiences across brands. Collectively, these blockchain innovations enhance efficiency, security, and personalisation in travel services.

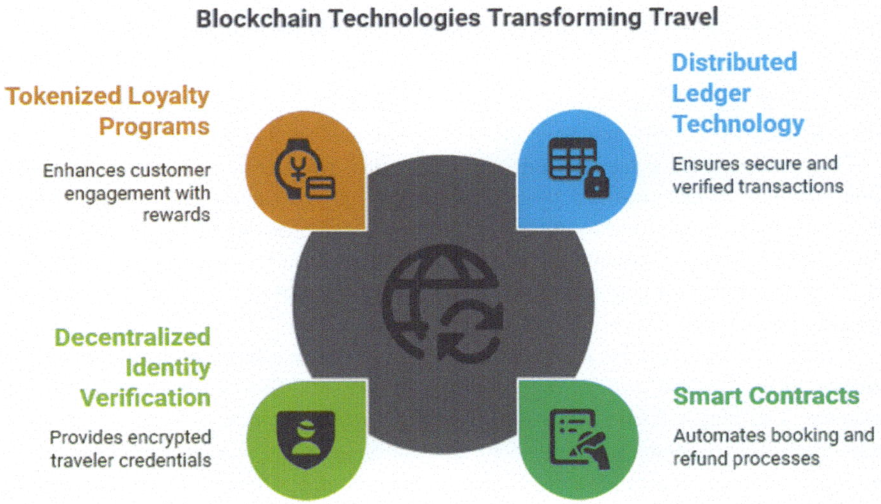

Fig. 3 Four blockchain technologies significantly transforming the travel industry

5.3 *Key Findings from Primary Data*

As a part of the study, 250 professionals from the industry, such as hoteliers, airline representatives, travel agency executives, and blockchain developers, were considered to gather their responses. Key trends from the findings of the survey are:

The adoption of Blockchain is increasing, and 68% of respondents are already implementing or researching the adoption of Blockchain applications in Travel Booking, Payment or Security [5].

1. Prevention of fraud has remained a top priority, as 78% of companies have noted that blockchain mitigates fraud on bookings, chargeback fraud, and identity theft by a considerable margin [19].
2. Broad cost savings and efficiency gains were observed, and businesses could reduce operational expenses by an average of 25% once intermediaries were removed [9].
3. Blockchain-based programs showed a 40% increase in active participants and easier customer redemption and transfer of points [10] (Fig. 4).

The graphical representation highlights blockchain technology's significant impact on the travel industry. A high adoption rate (68%) signifies growing acceptance and integration of blockchain solutions. Notably, blockchain excels at fraud prevention (78%), showcasing its ability to secure and verify transactions effectively. Although currently showing a moderate participant increase (40%), blockchain's role in enhancing customer engagement and loyalty programs remains promising. However, cost reduction (25%) remains limited, indicating potential for further efficiency gains as the technology matures. This visualisation underscores blockchain's

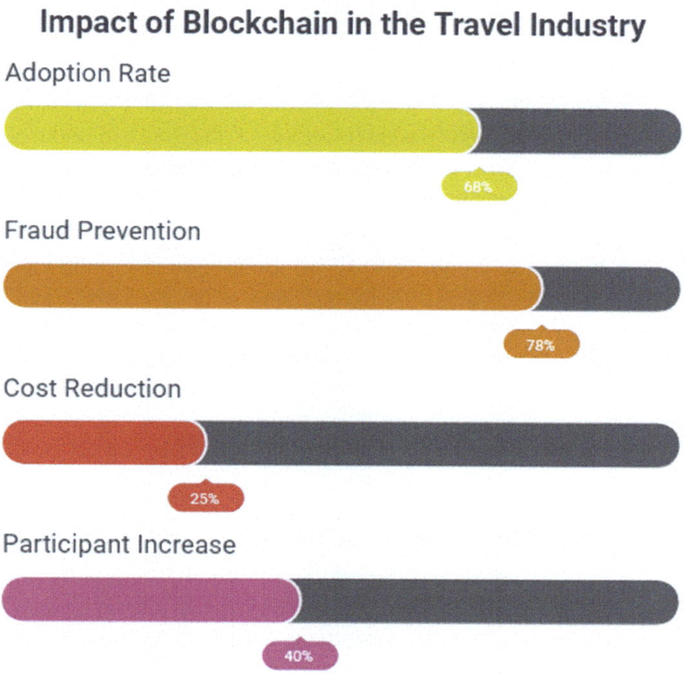

Fig. 4 Impact of blockchain on travel industry

strengths in security and transparency alongside areas needing further growth to maximise industry-wide benefits.

5.4 Challenges Identified

Despite notable successes, blockchain adoption in tourism faces several challenges:

1. **Scalability Issues**—High transaction fees and limited processing capacity on public blockchain networks such as Ethereum remain a barrier [7].
2. **Regulatory Uncertainties**—Many countries have strict regulations on cryptocurrency transactions, limiting the potential for blockchain-based payment systems in tourism [8].
3. **Consumer Awareness Gap**—45% of travel companies believe that blockchain adoption is slow due to a lack of awareness and education among consumers [18].

6 Discussion

The study's survey results, case studies and technological assessments pointed out that blockchain technology has revolutionised the tourism industry for the better. Blockchain is making a more efficient, decentralized and trust-driven travel ecosystem by improving security, enhancing transparency, reducing fraud percentage and improving customer engagement. Although it has benefits such as large market size, high social impact, and high profitability, scalability, regulatory uncertainty, and consumer awareness remain challenges and slow its widespread adoption [18]. Specifically, this section critically analyses the impact of blockchain on tourism in terms of secure transactions, operational efficiency, decentralized identity management, loyalty programs, challenges of blockchain adoption and future developments.

6.1 Enhancing Transaction Security and Fraud Prevention

The most significant contribution of blockchain to tourism has been securing financial transactions and thwarting fraudulent transactions. For example, credit card payments and wire transfers in travel are subject to high risk of fraud, chargeback and identity theft [19]. All of this is done by the immutable, cryptographic ways of blockchain that do not allow altering or manipulating transactions recorded in a distributed ledger [13]. Nearly 78% of companies said that they had a reduced amount of fraudulent bookings and chargeback fraud when they implemented blockchain-based payment systems. Furthermore, blockchain-based cryptocurrency payments have proven to remove the potential risks involved in centralized financial intermediaries by reducing transaction fees and providing real-time settlements [9]. However, adoption is still restricted, especially by the volatility of cryptocurrencies and regulatory policy uncertainty [7].

Moreover, smart contracts enabled by blockchain automate the settlement of refunds, cancellation policies, and insurance claims, resulting in fewer manual errors and processing delays [10]. It is even more helpful in airline ticketing or hotel bookings, where refunds or cancellations are likely the bone of contention. Through Lufthansa and TUI Group case studies, intelligent contract automation of the booking process was illustrated as shortening, more transparent and third-party dispute-free [18]. However, scalability limitation has been identified as a weakness in blockchain networks, like the one found on Ethereum, which means that they require Layer-2 solutions or private blockchain networks [7].

6.2 *Operational Efficiency and Cost Reduction*

Cost savings and operational efficiency can be realised among tourism businesses; the key additional finding of the study is that blockchain adoption leads to significant costs. Decentralized booking platforms (Winding Tree and Travala, etc.) do away with intermediaries, therefore reducing commissions imposed on prices by traditional online travel agencies (OTAs) [15]. Data from surveys demonstrated that the firms using blockchain enjoyed a 25% reduction in operational costs, as the use of blockchain allowed them to minimise dependence on intermediaries and automate administrative processes [9].

Additionally, blockchain displays the capability of real-time transaction settlements, which reduces delays and the banking fees involved in cross-border payments. Thus, hospitality, airlines and tour and travel agencies operating in many countries take as much benefit from it [8]. However, there is also unease about the amount of energy consumed by the blockchain, mainly where Proof of Work (PoW) consensus mechanisms are used, and it might conflict with sustainable tourism efforts [4]. It is suggested that migration to Proof of Stake (PoS) or hybrid blockchain solutions be done to achieve lower power consumption while maintaining a secure transaction [10].

6.3 *Decentralized Identity Management for Secure Travel*

Among the most promising blockchain applications for tourism, [11] identified decentralized identity management (DID) as one of the promising ones where the application of blockchain can replace current systems at airports for security checks, airports for check-in or access to hotel rooms, and visa processing. The research findings show that blockchain-based identity verification systems are more secure, less prone to data breaches, and facilitate a smooth traveller authentication process. Through self-sovereign digital identities, travellers have their identity information stored on a blockchain, and only authorised travel authorities can access them [19].

The case studies on the effectiveness of blockchain in border control and passport verification in Dubai International Airport and Estonia's digital identity system [7] show that blockchain improves border control efficiency, reduces the processing time and lowers fraudulent identity claims, we also see that there is a threat of cybersecurity if the user's personal history information is not encrypted adequately. Expert interviews have also shown findings that user authentication in contactless passenger verification with biometrics connected to blockchain would further enhance the process in terms of reducing wait times at security checkpoints. However, another challenge in scalability and standardisation exists since the various countries worldwide differ in the regulatory framework governing digital identity management [9].

6.4 Tokenized Loyalty Programs and NFT-Based Travel Experiences

Leveraging the effects of the blockchain in customer loyalty programs has been one of the most innovative uses of the blockchain in tourism. The loyalty programs were traditionally highly fragmented, restrictive and limited to individual brands [10]. The research findings indicate that because loyalty tokens are based on blockchain, the extent of flexibility, i.e. the ability of customers to exchange the rewards at different service providers, is increased. For instance, KrisPay from Singapore Airlines permits travellers to convert airline miles to digital tokens, which the traveller may utilise to make retail purchases, hotel bookings, and other services [13].

The rise of NTFs allows customers to engage more with their travel experience as consumers can own collections of digital collectables, which unlocks VIP access, special discounts and unique travel memorabilia [7]. The survey also found that 40% of travellers preferred blockchain-integrated loyalty programs to existing ones because of their flexibility, ease of redemption, and security. However, valuation vulnerabilities of NFTs and consumer awareness are issues that need to be overcome before widespread adoption [19].

6.5 Challenges and Future Implications

Despite the many benefits of blockchain in tourism, the findings indicate that several barriers must be overcome for widespread adoption:

- **Scalability and High Transaction Fees**—Many companies face challenges processing high-volume transactions due to network congestion and high gas fees on Ethereum-based platforms [7].
- **Regulatory Uncertainty**—The lack of global regulatory frameworks for blockchain-based payments and identity verification limits adoption in cross-border travel [11].
- **Limited Consumer Awareness**—45% of travel companies surveyed reported that customers lack knowledge about blockchain-based travel solutions, necessitating education campaigns and user-friendly interfaces [18].
- **Interoperability Issues**—Different blockchain networks lack standardisation, making integration across travel platforms complex [9].

Future research should explore the integration of blockchain with artificial intelligence (AI) and the Internet of Things (IoT) to develop innovative tourism ecosystems that further enhance automation, personalisation, and security [4]. Additionally, governments and private organisations need to work toward developing global regulatory frameworks and public–private partnerships to ensure blockchain's seamless integration into the travel industry [8].

7 Conclusion

Adopting blockchain technology in the tourism industry has proven helpful in increasing the security of transactions, facilitating operational efficiency, decentralising identity verification and transforming loyalty programs. However, according to surveys, case studies, and expert insights, blockchain transforms the tourism ecosystem into a more transparent, fraud-resistible, cost-effective alternative to conventional travel management systems [18]. However, despite the potential benefits, challenges to scaling infrastructure and unforeseen regulations make it difficult to resolve issues such as widespread adoption [7]. This conclusion presents the main insights from the research, shows the potential of blockchain revolution to the tourism sector and indicates the future adoption and research routes.

7.1 Key Contributions of Blockchain in Tourism

Blockchain technology has solved some inherent inefficiencies, such as transaction fraud, central control over bookings and the lack of centralized loyalty programs in the tourism sector. Enhancing transaction security was one of the most significant findings of blockchain. It decreases the risks of credit card fraud, chargebacks, and identity theft [19]. Blockchain records are immutable; once a transaction is validated, it cannot be modified, and this is a suitable solution for secure online bookings and payment settlement [13]. Response to this question is evident in the survey results, according to which 78% of the travel businesses that have used blockchain-based payments have seen a decline in fraudulent bookings, suggesting that using blockchain-based payments can help prevent fraud [9].

Another breakthrough in decentralized identity verification has been made. Traditional identity management systems in tourism—such as passport verification, visa applications, and hotel check-ins—are prone to data breaches and security vulnerabilities. Self-sovereign identity verification with a DID system on the blockchain enables travellers to store and share their credentials securely, thereby reducing security risks [11]. Solong Blockchain's Convergence of Biometrics with Blockchain: A study of Dubai International Airport and Estonia's digital identity initiative reveals how biometric authentication in conjunction with blockchain has enhanced airport security and border control efficiency by cutting down waiting time of travellers [7].

Moreover, its role in customer loyalty programs and tokenised reward programs has been undeniable. It was found that traditional loyalty programs suffer from fragmentation, restricted redemption policies, and poor interoperability [10]. Tokenised blockchain-based loyalty rewards can increase to c. 40% strong customer engagement [7] as the traveller now has more flexibility to collect and spend his or her points among a more expansive, if not infinite, number of service use providers. NFT-based tourism experiences also increase customer satisfaction by offering travellers unique digital collectables and unique experiences to increase brand loyalty [19].

7.2 Challenges and Limitations

Despite that, blockchain is still facing numerous barriers to adoption in tourism. The major challenge of scalability and transaction costs is one of the most important issues. Congestion and high transaction fees of public blockchain networks like Ethereum have been the cause of turning off such high-volume tourism transaction processing [7]. Many companies have taken strides to adopt Layer 2 scaling solutions and private blockchain networks. However, it has not been adapted hugely due to various challenges faced during the implementation of the solution [9].

Regulatory uncertainty is another problem that we honestly face. As blockchain is based on decentralized models, governments find it challenging to set legal frameworks for utilising blockchain in cross-border payments, digital identity verification and enforceability of smart contracts [11]. Due to strict financial regulation on cryptocurrencies, countries have restricted the adoption of blockchain in tourism, which is preventing its avenues of the decentralized payment system [8]. Standardising global blockchain regulations is necessary to guarantee painless adaption across domestic jurisdictions.

Consumers' limited awareness and technological literacy regarding such technologies is a further challenge. Treiblmaier [18] looked at survey data and found that 45% of tourism businesses think customer reluctance and block understanding hold back adoption. Decentralized booking platforms, digital identity wallets, and blockchain-based loyalty programs, among others, are unknown to many travellers; hence, the adoption rate is not as expected [19]. There is a need for user-friendly interfaces and massive educational efforts to bridge the gap between this technology's complexity and its widespread mainstream adoption.

7.3 Future Implications and Recommendations

Several strategic initiatives must be pursued to ensure long-term success and mass adoption of blockchain in tourism.

1. **Developing Scalable Blockchain Infrastructure**
    - The tourism industry must explore Layer-2 scaling solutions, such as rollups and sharding, to enhance blockchain's transaction capacity and reduce processing costs [7].
    - Combining private and public blockchain networks, hybrid models should be developed to balance security and scalability [9].
2. **Regulatory Standardization and Policy Development**
    - Governments and international tourism bodies must collaborate to develop regulatory frameworks that support blockchain adoption in travel payments, identity verification, and ticketing [11].

- Establishing compliance guidelines for smart contracts and decentralized finance (DeFi) applications in tourism could provide legal clarity for businesses adopting blockchain-based solutions [8].

3. **Enhancing Consumer Awareness and Adoption**

- User education programs should be introduced to familiarise travellers with blockchain applications, such as secure travel wallets, NFT-based tourism incentives, and decentralized booking systems [19].
- Travel companies must focus on simplifying blockchain interfaces to ensure that non-technical users can seamlessly interact with blockchain-powered platforms [18].

4. **Integration with Artificial Intelligence (AI) and IoT**

- Future research should explore the integration of AI and IoT with blockchain to create innovative tourism ecosystems where real-time traveller data is securely recorded and analysed for personalised experiences [4].
- AI-powered predictive analytics combined with blockchain could improve travel businesses' dynamic pricing models, customer preference mapping, and fraud detection mechanisms [10].

It cannot be doubted that blockchain technology is coaching for the future of tourism, security, efficiency, and decentralisation. Because of these features, it is a game-changing innovation in the sector [18]. However, scaling issues, legal dis-ambiguities, and restricted user adoption persist as roadblocks that blockchain must address to maximise its potential in the global travel industry [7]. The following steps could be implementing scalability improvements, solving regulatory issues, educating users, and integrating AI and IoT for the seamless adoption of blockchain in the tourism industry over a long period.

References

1. Bouraga, S.: Challenges of traditional loyalty programs: a blockchain perspective. Int. J. Retail. Manage. **28**(2), 98–112 (2022)
2. Chang, K., Du, M., Song, L.: Decentralized finance and blockchain transformation in payments. J. FinTech Innov. **12**(3), 115–133 (2020)
3. Chang, K., Wang, L., Zhao, H.: NFT-based loyalty programs: the next evolution. Digit. Econ. Rev. **10**(1), 55–70 (2023)
4. Cheng, S., Xu, Q., Zhang, W., Liu, Y.: Blockchain for sustainable tourism: a systematic review and research agenda. Sustainability **13**(2), 1–19 (2021)
5. Goyal, A., Singh, P.: Cryptocurrency adoption in tourism: trends and insights. J. Travel Digit. Finance **6**(1), 44–59 (2021)
6. Ivanov, S., Webster, C., Stoilova, E.: The impact of blockchain on the cruise industry. Int. J. Hosp. Manage. **94**, 102815 (2021)
7. Jain, A., Patel, K., Singh, M.: Layer-2 scaling solutions for blockchain in travel applications. J. Distrib. Syst. **29**(4), 250–273 (2024)
8. Jain, A., Raghuvanshi, V.: Blockchain and biometric authentication in global travel systems. Int. J. Smart Tour. **11**(2), 101–119 (2021)

9. Kamble, S., Gokhale, R., Verma, P.: Preventing fraud in blockchain-based reward programs. Cybersecur. J. **27**(1), 87–101 (2023)
10. Kasahara, Y., Kawano, T.: NFTs in the hospitality sector: a case study of tokenised membership programs. Tour. Innov. Q. **9**(2), 45–62 (2023)
11. Khan, M.A., Luo, H., Xu, X.: Digital identity management using blockchain for safe and secure travel. IEEE Access **8**, 196528–196542 (2020)
12. Liu, Y., Lu, Q., Paik, H.-Y., Xu, X.: Design patterns for blockchain-based self-sovereign identity. In: Proceedings of the European Conference on Pattern Languages of Programs 2020, Article 16, pp. 1–14. Association for Computing Machinery (2020). https://doi.org/10.1145/3424771.3424802
13. Meyer, C., Schubert, M.: Blockchain and loyalty programs: a paradigm shift in customer engagement. J. Financ. Innov. **15**(3), 78–95 (2019)
14. Nakamoto, S.: Bitcoin: a peer-to-peer electronic cash system. Decentralized J. Cryptogr. **1**(1), 1–9 (2008)
15. Saberi, S., Kouhizadeh, M., Sarkis, J.: Blockchain technology and its relationships to sustainable supply chain management. Int. J. Prod. Res. **57**(7), 2117–2135 (2019)
16. Su, H., Teng, J.: Carbon footprint verification using blockchain-based smart contracts. Environ. Sci. Technol. **57**(6), 1904–1915 (2023)
17. Tayal, A., Chakraborty, D., Sarkar, P.: Decentralized booking platforms: Blockchain's role in disintermediation. J. Travel Technol. **14**(3), 123–145 (2021)
18. Treiblmaier, H.: The impact of blockchain on e-tourism: disrupting the digital travel ecosystem. J. Travel Res. **57**(3), 231–247 (2018)
19. Zhang, P., Li, X.: Blockchain impacts travel security and fraud prevention. J Cybersecur. Privacy **19**(4), 345–362 (2022)

Cross-Border Collaboration in New Digital Era with Blockchain Integration

Channi Sachdeva, Prabhjeet Kaur, Veer P. Gangwar, and Lokesh Jasrai

Abstract This chapter studies how digital advancements improve international sector working arrangements, especially in cross-border integration while showing why US and Canadian border agencies need to team up to fight cyber-attacks. The study also discusses how new technology tools like Bluetooth update DLT and CBDCs to boost international payment and customs system capabilities. Small economies develop competitive positions through border cooperation while facing security and property protection issues. The study follows a thematic approach for drafting section themes with simple case studies. This chapter explores the benefits of blockchain and AI technology that make trade flows easier to manage and protect safe monetary transactions while enhancing supply chain control. The various sections of the chapter showcase how IBM Trade Lens and Maersk use blockchain to improve their supply chains. This chapter highlights the smart contract's purpose in international business and promotes government oversight through official rules to enable its use. This study detailed regulatory and data safety obstacles preventing blockchain from becoming a widespread business tool. Future research demands teamwork between organizations and governments to develop unified legal rules for blockchain technology. This chapter also explains how connecting AI systems to IoT devices through blockchain improves the security and performance of business operations. This chapter explains why there is a need to explain why blockchain adoption needs improvement because of its high energy needs and struggles to work with different systems, and calls for more research to resolve these problems. This

C. Sachdeva (✉) · P. Kaur · V. P. Gangwar · L. Jasrai
Mittal School of Business, Lovely Professional University, Phagwara, Punjab 144411, India
e-mail: channikaur105@gmail.com

P. Kaur
e-mail: arelyin709@outlook.com

V. P. Gangwar
e-mail: veer.23954@lpu.co.in

L. Jasrai
e-mail: lokeshjasrai@gmail.com

A. Kumar et al. (eds.), *Blockchain in the Tourism Industry: A New Era of Secure and Transparent Travel Solutions*, Information Systems Engineering and Management 57,
https://doi.org/10.1007/978-3-031-95341-5_12

study details business education and the government working together to develop new ways to enhance global digital commerce systems.

Keywords Digital modernization · Virtual collaboration · Blockchain integration · International trade · Cyber threats · Financial inclusion · Cross-border integration

1 Introduction

1.1 Overview of Cross-Border Collaboration in the Digital Age

The digital age has enabled different sectors to form new connections, resulting in unprecedented novel developments. The advancement of technology allows people and organizations to collaborate virtually by sharing the same space across geographical distances to facilitate alignment in music creation and musical production democratization together with musical advancement [129]. The rise in multinational collaboration has escalated cybersecurity threats, thus, the U.S. and Canada need to develop collaborative cybersecurity solutions [108]. The model aims to achieve shared threat information exchange and utilization of leading-edge technological solutions. Global market access for agricultural businesses became possible because of optimized supply chains based on technology that resulted from e-commerce platforms across borders. Agricultural product distribution receives better efficiency and responsiveness through the Internet of Things (IoT) and big data analytics systems. This resulted in elevated global collaborative investigative journalism between journalists who operate in challenging environments [61]. Maintaining digital growth requires addressing crucial problems with regulatory differences and cybersecurity challenges to enable the sustainable development of the digital realm.

1.2 Role of Emerging Technologies in Global Trade and Finance

Incorporating Bluetooth integration technologies into global trade has brought about substantial improvements in financial operations, specifically in international commercial transactions [37]. Electronic systems promote uninterrupted international commerce and have become more efficient, less expensive, and safer. Letters of Credit that developers can create to enhance trade financing simplicity without third parties, thus minimizing traditional costs and risks [18]. The inability to overstate the need for an improved cross-border payment system makes Distributed Ledger Technology (DLT) an ideal solution to cut costs and reduce time while resolving current payment system inefficiencies [39]. DLT-based platforms enable multilateral services to perform operations with tokenised assets and enhance transaction visibility for broader economic operations. Combining blockchain technology and Central Bank Digital Currencies (CBDCs) is crucial for strengthening cross-border transactions [55]. Interoperability requires standardization standards, such as ISO 20022, because different payment systems operate worldwide [42]. The development of the Internet of Things and big data allows customs authorities in trade facilitation to enhance national border transactions. Despite technological advancements, several significant obstacles persist in cross-border transactions, including regulatory compatibility and technical interoperability, which require proper attention to reach maximum potential benefits [10]. The ETR system was introduced with an efficient cost system and the highest security system. Social implications provide awareness and benefits for cross-border stakeholders.

1.3 Introduction to Blockchain Technology

The addition of Bluetooth technologies has transformed cross-border transactions because these systems are now combined with global trade and finance operations [37]. Bluetooth technology, which usually handles wireless connections near you, takes advantage of its capabilities to make systems more secure and easier to access. The development of blockchain technology is an essential innovation in trade finance, with the help of smart contracts. Letter of Credit (LOC) uses Smart Blockchain technology to speed up business transactions between companies by cutting out intermediary companies. With this innovation, companies establish secure transactions by seeing all the results and saving time, which allows businesses to work directly.

DLT stands out because it can resolve international payment problems by using Bluetooth connectivity and creating parallel payment systems that link to the broader DLT network [141]. The system facilitates data sharing between different systems, reducing the technical and financial costs of cross-border deals. Real-time payments represent a major existing trend in global finance, which depends on blockchain technology, artificial intelligence (AI), and CBDCs. The potential role of Bluetooth becomes clear in this case because it connects to all types of data transfers to provide real-time sales and logistics support [131].

The combination of IoT and Bluetooth helps fast-track customs procedures while tracking packages because they automate customs trading, bringing quicker trade and enhanced business quantity. Data sharing through this integration boosts security in supply chain operations. Alternative financial platforms show their value through the local Bluetooth connectivity of digital currencies and cross-border exchanges, even when other digital systems are limited. This technology supports limited security functions to boost transaction speed and global participation. Integrating Bluetooth devices will help provide digital access to communities in developing areas, allowing them to join worldwide business markets at reduced costs. For example, 'LIO-Pay'

[118] act as an offline solution to increase digital payment solutions during financial inclusion in semi-urban areas. The combination of emerging technologies helps both FinTech and Bluetooth improve device connection methods, which automatically leads to better financial services [125]. These technologies will produce more outstanding results while becoming more powerful through the worldwide uniting of global business and commerce.

2 Scope of the Study

This chapter discusses blockchain technology integration in cross-border collaboration to explain how it transforms global economic systems and their infrastructure. Changes occurring due to digital transformation in international business must include blockchain technology to improve system security and efficiency. This study explores every part of how blockchain technology benefits international deals by reducing costs, boosting participant confidence, and creating automatic contract processing. The specific objectives were as follows.

- To explore how blockchain allows companies worldwide to become better partners within their business sectors.
- To examine how blockchain technology helps run modern international business operations through actual transactions.
- To understand the obstacles to accepting blockchain in each public policy framework.

Our literature review shows possible developments and long-term planning while proposing plans for advanced partners. This chapter shows how different business fields, like financial services, market operations, supply processes, and rule systems, help explain how blockchain networks work across countries. This work guides experts at the government level, industry leaders, professionals, and academics who require blockchain skills for international collaboration.

3 Study Approach

This chapter examines blockchain technology's implementation, challenges, and practical limitations in a global context. It identifies various system weaknesses, energy demands, connectivity issues, and integration difficulties. The chapter includes different case studies for each section to better understand the focus. The chapter follows a simple thematic approach to draft the chapter structure. This study also examined issues related to interoperability. Brief focus on trust-related problems and security measures. Future applications include cross-border cooperation, the decentralisation of autonomous organisations, and policy changes. This study emphasises further collaboration between various business groups, governmental bodies, and educational institutes for advanced blockchain technology.

4 The Evolution of Cross-Border Collaboration

4.1 Historical Context of International Trade and Cooperation

Expanding international buying, selling, and local geographical development calls for a border area partnership. According to the World Trade Organization, these arrangements deliver competitive power, especially to smaller economies according to Worl Trade organization (WTO) [43]. Real border trade connections mainly depend on e-commerce, what happened before, and how trade traditionally occurs between regions rather than on successful local partnership approaches [77]. New technological tools help international trade evolve because they make moving products and sharing information cheaper [133]. Trade growth among nations occurs alongside multiple cross-border partnership forms, which brings about distinct effects across different areas [36]. When nations tighten their trading connections, these actions do not force governments to enhance their international efforts, and regularised institutions do not boost economic partnerships [5]. Their findings show the difficulty in managing border cooperation in international trade and overall regional development.

4.2 Challenges in Traditional Cross-Border Transactions

Countries make it difficult for users to defend their intellectual property using remote technology, as described by [134]. Conventional payment methods cause performance delays in security weaknesses and increase business costs. Marketplace faults have pushed new inventions to prominence. Many companies are experimenting with Bluetooth technology to provide customers with fresh ways to pay for their devices [118]. Distributed Ledger Technology benefits worldwide money exchange because users can move money across borders more safely while conducting transactions faster than before [74]. Users must trust that their DLT systems are secure for regular use. New technology brings more digital security risks and the fast spread of pirated content and personal information [128]. Businesses must develop comprehensive IP strategies and search for potential hazards before using protective tools despite backing new rules and international connections.

4.3 Digital Transformation and the Role of Industry 4.0

Companies use Industry 4.0 technologies to change how they run their operations and supply chains worldwide [20]. Digital technology growth produces superior outcomes for B2B firms working with customers and operations, in addition to delivering financial success across Central and Eastern Europe [136]. Businesses adopt Industry 4.0 tools to modernise their operations and develop their market advantages using environmentally friendly solutions [41]. Businesses face worker shortages and high investment costs for modern technology systems [95]. Research studies indicate that countries use Industry 4.0 to different extents. Organizations build effective links with education when they adopt new technology, yet government funding is their primary support [20]. Research shows that Industry 4.0 continues to transform into Industry 5.0 using artificial intelligence technology [25].

4.4 Emerging Trends in Global Business Networks

Researchers are creating new business network systems for money transfers worldwide while focusing on Bluetooth technology. Scientists use Bluetooth as a mobile financial transaction tool because it connects devices without wires [22]. Bluetooth technology has been successful because it effectively connects short distances and produces significant business results [52]. When Törnroos made his observations in 2000, multinational corporations became essential businesses by separating work tasks worldwide. Research proves that companies physically located near each other do not establish enough business ties across borders despite this approach being expected to enhance European Union (EU) market integration [110]. Additional experiments are needed to demonstrate how the industry concept of cross-border business networks compares with practical enterprise connections.

5 Understanding Blockchain Technology

5.1 Fundamentals of Blockchain

Code-based blockchain technology offers groundbreaking services for digital transactions through linked databases of transactions. The technology helps improve economic systems and other sectors, with stronger security and clear records. The network maintains a shared record list that allows people to make transactions without using traditional transaction platforms. The system protects transaction records permanently by preventing changes in them. Through coded agreements, smart contracts help avoid deceitful activities and errors people make. They make operations more efficient by running codes that take over contract work and reduce business expenses to improve productivity. Hash functions with digital signature technology and works as a core security system for blockchain because it protects confidential data from tampering. Teams and computers must verify transactions using Proof-of-Work and Proof-of-Stake methods to keep blockchain networks working properly [93]. The system requires further development to handle the increased demand while dealing with rules that may change and high resource usage for personal use.

5.2 *Key Features: Transparency, Immutability, and Decentralization*

Bluetooth technology improves cross-border collaboration through clear information sharing plus secure and distributed systems. The blockchain platform lets everyone involved view transactions while keeping all data shared with them authentic [49]. This important feature gives customers and suppliers the right product visibility to detect errors and problems. People who use this system have improved confidence because they know that the recorded data remains true and detected. A decentralized system gives decision power to everyone involved while keeping personal information safe and prevents damages from single-system breakdowns. A mix of distributed authorities helps different regions collaborate effectively across borders. Different blockchain platforms struggle to talk to one another, and few organizations want to use blockchain technology for global partnerships [122].

5.3 *Different Blockchain Models: Public, Private, and Consortium*

Blockchain systems, such as public, private, and consortium, help organizations work together across borders when Bluetooth connects to them. Public blockchains provide open-access design and a decentralized environment suited for use, such as money transactions and public document storage. Organizations use private blockchains, such as Amazon QLDB, for central authority oversight to deliver fast and protected data processing [138]. Hyperledger Fabric allows different organizations to work on the same system while retaining personal data ownership [6]. They boost transaction processing speed and record verification capabilities, mainly through a Zero Knowledge roll-up system. Established blockchains work best with minimal cross-border deals because they compromise personal data protection. The proper blockchain model selection depends on how it matches collaboration needs and helps maintain secure transactions while sharing data control and processing power [29]. International processes benefit significantly from blockchain models, which make them faster and safer while controlling organizational operations better [63].

5.4 Blockchain Versus Traditional Digital Solutions for Cross-Border Collaboration

The blockchain system brings better collaboration across borders by making operations faster and safer while providing complete tracking traces. To create a better system, the network design cuts out middlemen while speeding up operations at lower costs. The use of blockchain technology reduces day-long transactions to minutes, which increases the effectiveness of operations [33]. It creates an unalterable database system that makes digital fraud more challenging and allows users to track their data history. Instead of essential human involvement, smart contracts process data without the need for employees and perform without mistakes. When parties do not need to pass through middleman services, their transactions become less expensive, particularly between countries. Businesses, such as Malaysians, can combat currency value changes better when they adopt digital payment systems [87]. Traditional online tools bring a proven structure and know-how, but they produce delays and expense, making blockchain technology stand out in international teamwork [12].

6 Blockchain in Global Trade and Supply Chain Management

6.1 Enhancing Trade Transparency and Efficiency

By linking blockchain to international trading relationships and teamwork networks, organizations can improve trust and speed up their work, while making all information visible. Blockchain enables decentralized systems and smart contracts to simplify trade procedures, which speeds up transactions from weeks to hours according to research findings [27]. Through blockchain, the system delivers advanced data protection, combined with unalterable recordkeeping and better traceability capabilities [102]. Blockchain integration with AI boosts trade facilitation efficiency by making better decisions while enhancing the quality of database records [82]. Several technical, legal, and business problems need resolution before blockchain implementation can succeed according to [105]. Businesses from multiple sectors demonstrate how blockchain technology changes collaborative networks [78]. Blockchain technology transforms trade finance while building a better platform for international trade [69].

6.2 Decentralized Trade Finance and Letter of Credit Systems

Blockchain technology improves trade finance operations, primarily through its application to credit transaction letters. The decentralized system solves trade finance problems, such as security issues while building trust between parties [80]. The use of smart contracts in blockchain technology allows businesses to conduct safe and rapid international payments [49]. Blockchain and artificial intelligence work together to improve analysis while making trade finance workflow easier [48]. The legitimate implementation of blockchain requires handling complex regulations while obtaining support from key users and delivering workable technology [91]. Despite setup problems, blockchain LOC systems provide faster processes while showing all the details better and keeping transactions safer [63]. Future development of blockchain technology will establish a secure trading environment that optimizes international financial operations [26].

6.3 Blockchain in Customs and Regulatory Compliance

Blockchain technology can greatly improve how trade crosses international borders and tasks are handled by customs officers. Blockchain technology provides dependable data protection and meets rules using advanced encryption and smart contracts [84]. Using blockchain technology in customs simplifies cargo clearance steps while providing better control over risks and instantly shows good status [123]. This technology effectively handles multiple trade documents, such as import–export forms and proof of origin documents, and is highly advantageous in cross-border trade [75]. Using blockchain technology can build a supervision system that connects order information to payment and shipping records in cross-border e-commerce. This management method improves supervision speed and detects tax evasion while advancing global e-commerce development [120] (Fig. 1).

6.4 Case Study: IBM Trade Lens, Maersk, and Lenovo

Trade Lens and other blockchain systems provide major ways to develop global trade networks and their related systems. Through Trade Lens Maersk and IBM build a blockchain platform that helps everyone involved in container ship activities exchange data effectively [14]. This technology solves problems in maritime shipping, where parties struggle to access clear information records and work in different formats at the same time [50]. As explained, the trade Lens helps businesses and governments instantly share supply and logistics information to increase their benefits. Trade Lens optimises business processes and provides end-to-end supply chain monitoring functions, according to Louw-Reimer et al. [81], to deliver value chains.

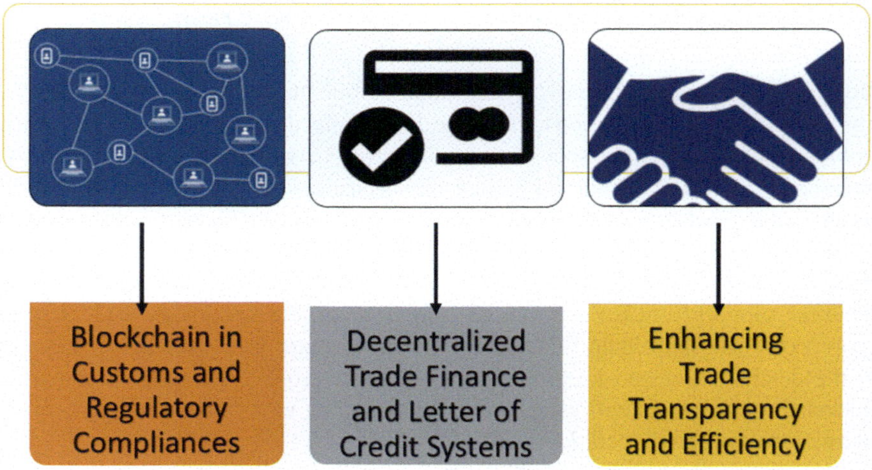

Fig. 1 Blockchain global and trade management

This system decreases search costs and reduces enforcement expenses. Businesses must accept Trade Lens widely for maximum benefit but must work through application issues, according to Louw-Reimer et al. [81]. This new technology provides a substantial opportunity to improve worldwide trade operations through joint data processing. Blockchain enhances supply chain management activities by improving relationship-building with all dependent parties. The technology lets everyone create verified transaction records via a protected system [45]. Proofs from different industries show that blockchain makes supply chains better at tracking products while working faster and building connections between parties [90]. Lenovo has shown how blockchain increases supply chain collaboration speed during its industry's best practices [140]. The leading blockchain features, including unchanging records and consensus protocols, solve standard supply chain security problems and trust issues. Research shows that blockchain applications mainly occur in farming sectors, yet organisations want to explore blockchain's IT possibilities with international partners. Further research will make blockchain systems revolutionise supply chains worldwide.

7 Blockchain-Driven Financial Transactions and Cross-Border Payments

7.1 *Role of Cryptocurrencies and Stablecoins*

The blockchain system now makes international payments faster and safer while showing all transaction details and cutting the costs of middlemen [16]. Its capacity for digital currency and safety enhancements helps establish better market performance in both banking processes [62]. Across Africa, people use the Stellar blockchain to provide financial opportunities and solve local payment challenges with stablecoins, as described by Au et al. [17]. Although blockchain needs work to succeed, implementation teams must solve different issues, from technology to rules and social acceptance, while all parties participate [112]. As Zheng et al. [144] state, blockchain technology provides an effective solution to Belt and Road transactions, but requires appropriate rules to work well. Eurasian nations can benefit from this technology to solve problems in money transfers and finance deals according to Zheng et al. [144].

7.2 *Cross-Border Payment Solutions: Ripple, Stellar, and CBDCs*

Central Bank Digital Currencies offer solutions to improve cross-border payment technology over the corresponding banking system [70]. Different cross-border CBDC payment models include compatible and interlinked systems alongside one system serving multiple currencies [124]. These system designs seek to enable quick and inexpensive payments between different nations [111]. According to Minaei et al. [85], Universal Payment Channels (UPC) provide a method for instant fund transfers between various CBDC ledgers. Although progress continues with CBDCs, obstacles still block access, such as safeguarding personal data while ensuring that systems work well together and meet legal needs [124]. Many central banks are developing cross-border CBDC payment systems because they face growing challenges that multiple initiatives aim to solve [124].

7.3 *Reducing Costs and Improving Efficiency in Global Financial Transactions*

The high prices of financial transactions create significant problems for markets worldwide, especially in developing nations [35]. Firms can improve their bottom line and service quality by reducing transaction expenses [135]. Tourniquet standards are

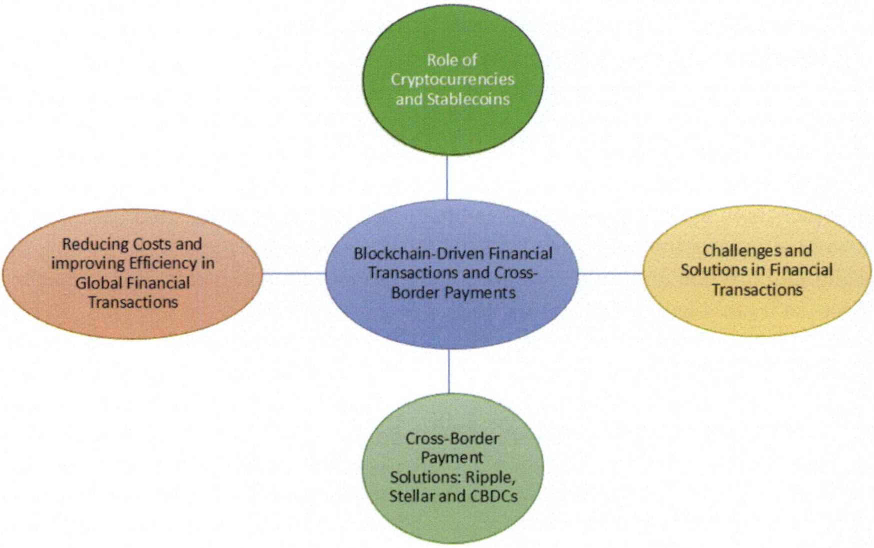

Fig. 2 Role of blockchain transaction in cross-border payments

needed in global financial markets to bring these benefits, yet market failures slow this advancement [104]. Willett et al. [137] Suggests that senior leaders should lead, and research on this topic needs fresh emphasis to help remove market obstacles. Effective industrial growth requires investment in technologies that do not demand quick cash returns. However, these projects become possible in low-cost financial markets that assist investors with their funding needs [127]. Lowering costs for financial market deals shapes technology decisions that improve production quality and economic expansion according to Levine [76]. The effect of decreasing transaction costs on growth depends on the type of capital structure [97] (Fig. 2).

8 Smart Contracts for International Agreements and Legal Frameworks

8.1 Automating Trade Agreements and Contract Execution

New technology allows us to program legal and trade contracts for international use. System-based programs help parties reach and maintain their contracts through automatic program steps [132]. The beagle framework wants to place smart contracts inside legal contracts to help execute parts of the deal automatically and keep digital records [11]. Digital smart contracts help businesses save money and speed up work processes, yet raise problems for legal systems trying to work with new digital methods [24]. The study of Sinha and Roy Chowdhury [116] provide information related to construct a new framework for developing blockchain-ready legal agreements that can be automatically checked on the Ethereum blockchain [4]. This method uses online browsing tools and formal exam checks running on Ethereum's blockchain platform. The development of smart contracts will change how countries work together in terms of both economic relations and trade rules [32].

8.2 Legal Implications and Smart Contract Enforceability

Blockchain technology with smart contracts brings value to international trade and legal agreements, because it makes operations faster and cheaper [72]. Their adoption meets challenges related to laws that vary from one place to another [86]. The official status of smart contracts requires attention to basic contract law standards, including intent to form contracts and determination of who can join such agreements [117]. Some authorities have begun to set up methods to fold smart contracts into their current legal processes. Contracts involving international business should include specific rules that approve the use of smart contracts, especially in cross-border transactions [56]. Smart contracts gain legal backing when they use digital signing systems according to Philip and Saravanaguru [101]. Developments in technology require continuous laws and standards creation through team work that combines multiple sectors to protect digital agreements in worldwide trade [99].

8.3 Case Study: Real-World Applications of Smart Contracts in Global Business

Businesses and governments now use smart contracts to change international operations because technology allows basic tasks to run better and safely while staying open and clear. Blockchains allow digital contracts to run themselves when agreed-upon criteria are fulfilled to achieve savings in both transactions and middlemen use. Multiple working programs demonstrate how smart contracts benefit international business operations. Using TradeLens technology, Maersk and IBM set up smart contracts to run operations that promptly send documents, which saves time and boosts performance in global trading. Quorum from JPMorgan proves that smart contracts deliver value to financial services because their system handles inter-bank deals and loans faster than before while lowering operating fees. Through its blockchain network, Propy helps clients make safe property deals that happen faster and protects against fraudulent actions. Healthcare organizations receive better patient record protection and record sharing through the blockchain technology in MedRec. The insurance company AXA helps customers with flight delay claims through smart contracts that complete processing and payouts instantly with full traceability. The listed situations explain how smart contracts wean businesses away from normal deals through digital frameworks that protect both parties while operating faster. As companies and law-enforcing organizations team up, they need to solve legal barriers to ensure that smart contracts follow worldwide rules and compliance requirements.

9 Regulatory and Compliance Challenges in Blockchain Adoption

9.1 Global Regulatory Landscape for Blockchain and Digital Assets

The use of blockchain technology proves useful for different fields in making digital payments and regulatory systems more effective. Blockchain assists in making sanctions and AML compliance more open to everyone to check and perform tasks automatically [60]. The technology encounters major difficulties in making it available across many areas because it needs to work better and work under strict government requirements [88]. Digital payment systems now work faster than ever before to protect customer data and money, however, society faces problems regarding how personal information is handled [13]. Research teams study Layer-2 solutions and shared technologies to improve crypto system speed [89]. Countries have diverse rules regarding blockchain, which produce unclear legal guidelines that slow down its acceptance [13]. People in finance must work together to make rules that allow blockchain to grow without harming money security and public safety [60].

9.2 Data Privacy and Security in Cross-Border Blockchain Networks

The operation of blockchain activities between multiple countries leads to major problems in keeping the data private and secure. The study by Zhang et al. [143] introduce blockchain architecture with two layers to help international businesses handle various rules using zero-knowledge proofs for privacy and smart contracts for payment-history detection. Garg and Mondal [44] discovered that blockchain's secure core elements enable data protection, but users oppose its use because governments lack rules on how to use it. Further studies are needed to make blockchain technology safer for use. Haugum et al. [53] examined blockchain interoperability by discussing the protection flaws of notary schemes, and hashed time-lock contracts. The authors reviewed the research materials to find ways to reduce risks and highlight the remaining unanswered issues. These research projects show that managing data security in cross-border blockchains depends on multiple elements to stay compliant and operate between systems.

9.3 Aligning Blockchain with International Standards

Multinational blockchain systems encounter strong security and private data protection problems. The study of Peng et al. [100] suggested running blockchain systems with two levels for privacy protection and transaction tracking to help service providers across borders manage different privacy rules. In their research, Garg and Mondal [44] explained why blockchain protects sensitive data, yet organizations resist using it because of privacy issues and uncertain laws. Research needs to focus on reducing blockchain technology risks when employed in finance. According to Rieger et al. [107], the General Data Protection Regulation creates implementation challenges for blockchain systems because of data correction and deletion requirements. The distributed storage features and security controls built into blockchain networks make it possible to enhance data privacy and create protected data transfers [71]. Under GDPR requirements, firms must determine who handles data and the reasons for processing personal information, according to Tikkinen-Piri et al. [126]. Organizations need to quickly create new standards because blockchain technology outpaces established norms and standards [145]. Successful blockchain implementation requires merging its standards with worldwide banking requirements because these standards must work together as explained by Farrell [38]. The process of linking blockchain to worldwide standards demands attention to current laws and research on the development of new standards. Tajti [121] studied blockchain interoperability issues that armoured blockchain designs, such as Notary Schemes, Sidechains, Hashed Time-Lock Contracts, face security, and privacy protection. They reviewed different sources of insight to investigate solutions to problems and new challenges. The analyzed studies demonstrate that securing data privacy across different countries in blockchain systems remains challenging without breaking legal rules and enabling network connections.

9.4 Addressing Regulatory Uncertainties and Governance Models

Companies face two main hurdles when using blockchain technology to improve operations, because governments have not finalized global rules [73] for this technology, and most businesses still follow traditional management approaches. Different national authorities use separate methods to control blockchain deals, creating many rules across the world. Many countries have different legal standards that limit the use of smart contracts when two parties conduct business internationally. Lawmakers in different locations acknowledge smart contracts as enforceable documents even though official policies remain incomplete. These varied legal standards block straightforward international business operations, therefore, standard rules must be created. Organizations must design blockchain network governance methods to create trust and effective ways to handle and settle differences [92]. Several business groups and government testing areas have appeared to connect blockchain technology with present standards from multiple sectors. The International Chamber of Commerce and World Economic Forum use their resources to create global blockchain standards [58]. Regulatory bodies from different nations test smart contract effects in their official programmes before applying new laws. For blockchain adoption between international parties, businesses and lawmakers must work together to make legal frameworks easier to accept and use. When regulatory challenges are solved, smart contracts show their real power in handling global deals better and reducing expenses with no legal issues [66].

10 Integration of AI, IoT, and Blockchain in Cross-Border Collaboration

10.1 The Synergy Between Blockchain and Artificial Intelligence (AI)

These three technologies join forces to modernize multiple industries through better security and business procedures [30]. This integrated technology system helps sectors, including supply chains, healthcare, and smart cities, gather data in real time while protecting transactions and making better decisions from improved data analysis [47]. All of these technologies work together to solve data security issues while providing flexible and decentralized systems to users [8]. Implementing these three key technologies is difficult because of connectivity problems, data protection requirements, and infrastructure requirements [2]. Both technological advancements enable businesses to bring out new ways to serve customers, so companies must stay flexible to use them effectively [109].

10.2 Role of the Internet of Things (IoT) in Smart Logistics and Supply Chains

The Internet of Things (IoT) improves SCM and logistics practices through better operations and lower expenses. By using Internet-connected devices, IoT systems provide immediate access to data from all supply chain operations, such as transport delivery and storage [79]. Research on how IoT links with SCM has gained importance, particularly when studying RFID technology, Industry 4.0, and reverse logistics [106]. IoT operations in supply chain management help automate business procedures while delivering secure wireless data networks to support better operational choices [98]. Most studies today explore how IoT affects supply chain management, however, they lack analytical models and test results [21]. More advanced IoT technology will resolve supply chain issues by handling uncertain situations and making operations simpler and safer for SCM to operate more powerfully.

10.3 Predictive Analytics and Fraud Detection in Blockchain Networks

Many blockchain networks face threats from fraudulent activities and system glitches that weaken their security systems [94]. These documents explain all aspects of how blockchain works, what fraudulent activities can happen, and how to track them down. Multiple detection systems such as statistics, AI models, game theory, digital tracing, reputational evaluation, and risk rating methods operate to combat threats to blockchain networks [94]. Network analytics helps identify fraud cases and find the origins of cryptocurrency transactions while revealing how market prices are set [57]. This study investigates pathways to transform transaction data into entity networks and explains their development alongside practical uses [130]. Research teams have provided practical examples of working blockchain networks in their publications in 2023 and 2024 to guide future studies and production activities [94]. These resources teach experts and scientists how to improve the security and reliability of blockchain networks.

10.4 Case Study: AI and Blockchain for International Trade Optimization

Artificial Intelligence technology helps businesses today by making trade operations run better and improving financial processes and market reach [96]. AI enhances business value by improving supply chain planning and logistics, as well as automating document handling and risk analysis for trade finance services according to Ozturk et al. [96]. Blocking technology combined with artificial intelligence helps organizations build better data records while increasing public confidence, according to SUN research. Current research on global trade shows how Alibaba, DHL, and Maersk put AI into practical use, according to Ozturk et al. [96]. AI shows benefits in developing business tools for smart contracts, data tracking, and supply chain control [115]. The global use of AI provides better online transactions and reduces expenses [15], however, technological obstacles as well as legal and security challenges persist. Research needs to study AI system issues while investigating different trade handling methods using artificial intelligence technology according to Fiorentini [40] and Stone et al. [119].

11 Challenges and Risks of Blockchain Adoption

11.1 Scalability and Energy Consumption Concerns

Global businesses find it difficult to implement blockchain technology because it has both performance and energy use problems. Network expansion creates three main scaling problems: slower performance, reduced handling capacity, and greater storage space requirements [65]. The present limitations obstruct how blockchain technology can help practical operations in digital networks, healthcare services, and money-handling systems [65]. Energy usage is a primary challenge for proof-of-work mechanisms, particularly in Bitcoin operations, as stated by Alshahrani et al. [9]. Research suggests that sharding works within blockchain platforms while lightning networks operate beyond their limits [28]. A full toolkit to solve these issues requires product evaluation along with the examination of technical elements and factors linked to money, society, and nature at work [7]. The use of blockchain technology needs development alongside electricity markets combined with research into user experience to test in temporary regulatory settings as described by Ahl et al. [7]. Research shows that blockchain technology makes distributed electricity swaps more secure and cuts down business expenses while using decentralized platforms.

11.2 Interoperability Issues Between Different Blockchain Networks

Blockchain interoperability research grew vital because separate blockchain networks slow blockchain development and use [59]. Scientists and colleagues have developed hash time-locked contracts for safe money movement between platforms that use the same technology. Current efforts to link blockchain systems face difficulties because they lack common security standards and code structures, especially when dealing with multiple blockchain types [67]. Despite many proposed solutions appearing in published research, only a small number have been successfully put into practical use, with most focused-on cryptocurrency transfers rather than sharing various types of data. Scientists have developed solutions to overcome shared network trust issues and explored ways to make blockchain work better on large networks with weak connections. Future blockchain development will depend on research improvements to standardization standards for all businesses [142].

11.3 *Adoption Barriers: Lack of Standardization and Trust Issues*

Companies need blockchain interoperability to join several networks because this feature solves important scalability and connection problems with independent platforms [103]. Specialists provide different methods for moving assets across multiple blockchain networks. The study of Abebe et al. [3] showed how independent networks can share trusted data through a system based on consensus network protocols. The work of Gosu et al. [46] detailed how hash time-locked contracts make it safe to move tokens among homogenous blockchains. Existing interoperability solutions employ different methodologies with varying characteristics such as throughput, block confirmation time, and consensus mechanisms [1]. Despite progress, blockchain interoperability needs better standards and users do not readily trust other network platforms [103]. Developers now focus on creating blockchain solutions for cross-platform operations even though these projects currently face technical barriers [1].

11.4 *Strategies to Overcome Challenges and Ensure Effective Implementation*

For blockchain to reach mass use and grow further, blockchain systems must be able to communicate safely within their networks [1]. Challenges exist in reaching interoperability because developers are yet to agree on common blockchain standards [19]. Researchers have suggested two ways to solve this problem: designing blockchain layers [139] and setting timed password security levels [34]. These solutions help different blockchains communicate securely while preserving their core design concepts, and do not require central authorities to operate. Interoperability restrictions mainly affect specific business associations and technological platforms because research efforts (Fig. 3).

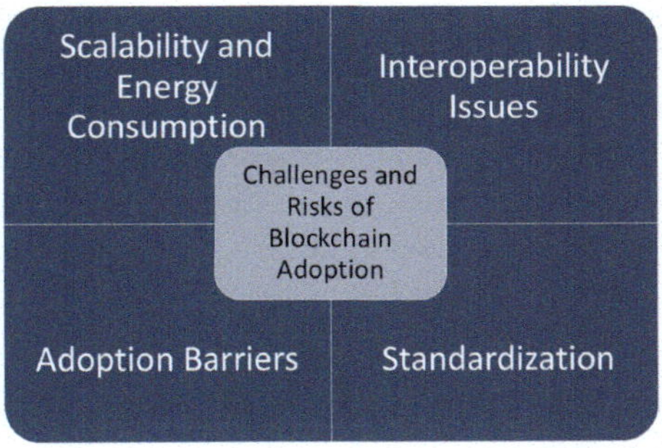

Fig. 3 Risks and challenges of blockchain adoption

12 Future Perspectives and Strategic Roadmap

12.1 The Future of Cross-Border Collaboration with Blockchain

Through blockchain transactions, blockchain technology creates new ways for businesses worldwide to work together by strengthening the trust between participants [68]. This makes trade finance easier while also improving the processing of trade transactions and handling intellectual property, as described by Maltsev and Maltseva [83]. Blockchain technology partnered with IoT devices enables parties to monitor operations without trust in central systems [54]. Research centers around present obstacles with blockchain systems related to their ease of use and weak security measures [54]. International banking processes benefit from blockchain, which simplifies transaction procedures between different nations, especially when dealing with remittance trade financing and global payments [64]. The study of Kashyap and Saxena [64] proposed central bank digital currencies as another payment option, alongside other payment systems. Many countries need to work together to adopt blockchain in trade processes with solutions for legal and policy problems before it can be completely used globally [31].

12.2 Potential Innovations: Decentralized Autonomous Organizations (DAOs) in Trade

According to Hannemann and Buchman [51], decentralized autonomous organizations (DAOs) represent the latest innovations that use blockchain to build new ways of running and managing organizations. DAOs run without leadership authority because their decision-making power comes solely from group approval [114]. Business operations will change, and corporate management will become more transparent and fair when DAOs work with blockchain technology [23]. Through DAOs, communities gain control and independent coordination, which may create more balanced and equal social systems [113]. DAOs experience difficulties because industry authorities and laws recognize them as distinct entities. Experts find that the combination of DAOs and artificial intelligence is worth studying as of 2023, according to Hannemann. DAO progress will reshape many sectors and administrative settings, while creating new possibilities for distributed collective management systems.

12.3 Policy Recommendations for Governments and International Organizations

The translation of business activities across international borders moves toward seamless operation through blockchain technology thanks to its ability to increase openness while maintaining safe and effective processes. G20 nations should implement blockchain to create an accessible digital economy that solves tax and sustainability problems while helping more people benefit from finance. Blockchain improves document handling and connects trading partners more securely so that they can send goods faster without needing middlemen. Achieving the complete benefits of blockchain requires both strong phase-moved decisions and operational support. Canada stands out as an important blockchain technology investor who can lead global policy development efforts. Business models that use blockchain require joint action from industry groups and educational institutions along with government agencies to develop new solutions for international commerce and supply networks.

13 Conclusion and Discussion

Digital modernization allows people to work together virtually across international distance. Virtual collaboration among different sectors, including the music-making process and agricultural operations. The growing number of cyber threats requires the US and Canada to combine their efforts against these threats. New technological innovations support better international payments through Bluetooth system updates, DLT, and CBDCs. Both blockchain technology and artificial intelligence support consumers in completing payments in real time and expediting customs processes effectively. Digital access and financial inclusion become possible through Bluetooth and IoT device connections, which help businesses grow worldwide. Such technologies will lead to stronger connections between international businesses and markets.

Small economies gain a competitive advantage through border cooperation as international trade systems have developed over many years. Regular border trade involves property protection and transaction security. New technologies allow secure and expedited transactions, but companies must build complete IP defense plans to use them effectively. The new industrial technology revolution changes how businesses operate and distribute goods, but lacks enough workers and demands heavy upfront spending. Networks in international business are developing through connections built using Bluetooth technology, and researchers are seeking direct tests between international networks and real-world company links. Its three basic components are transparent records, fixed-state changes, and distributed access. Every blockchain system falls into the public–private-consortium category with unique advantages for users. Users can exchange information in open and decentralized public blockchains or find rapid secure processing in private blockchains. The technology streamlines day-long transactions, raises efficiency, and delivers cost savings for global partnerships. Companies use blockchain technology to defend themselves against currency value shifts to improve their international team performance.

The crypto technology platform is changing how global business and product delivery systems are run through better trade visibility and faster operations. Distributed technology systems and digital agreements through smart contracts simplify the trading processes and make products easier to follow. Blockchain technology works better with AI systems, making trade procedures more effective. The technology strengthens trade confidence and security for decentralized finance systems and the letter-of-credit processes. By managing the correct data and adhering to rules, blockchain makes business approval easier. Business cases at IBM Trade Lens and Maersk with Ve Chain show how blockchain helps shipper activities and supply chain operations improve. Future research can improve supply chain operations across all business networks. Blockchain powers better international money exchange and cross-border payments, which now finish faster and with higher security. Due to its payment solutions, Stellar provides financial services throughout Africa that eliminate local money issues by using crypto and stablecoin systems. CBDCs create platforms and linked systems with multiple currency options to manage cross-border payments effectively. The main barrier is protecting private information while complying with legal rules. Economic growth in underdeveloped countries depends on better ways to manage money transfers in the international market. Implementing tourniquet standards will deliver these benefits, while issues in the market are slowing the development process. Senior executives should initiate research to remove market limitations and put money into systems that deliver results without immediate demand for cash revenue. The Beagle framework enables parties to build and maintain smart contracts for international deals by running automatic program actions according to new technological processes. Businesses use these contracts to streamline work activities while providing legal systems with technical problems to solve. Certain authorities now support smart contracts by creating official guidelines for their implementation, particularly in international business deals. Several real-world examples demonstrate that smart contracts enhance international businesses by handling basic work more efficiently and securely and giving full transparency to operations. Today, companies and government agencies deploy smart contracts to enhance their worldwide management methods and to confirm adherence to global standards. Organizations need to unite law enforcement to solve problems that prevent smart contracts from meeting global business standards and legal requirements.

Widespread adoption of blockchain technology in digital payments and regulations is difficult to achieve because governments enforce strict rules, and different nations follow distinct professional standards. When blockchain links multiple international systems, it creates problems while protecting personal data and keeping them secure. The two-layer blockchain system proposed by researchers helps international businesses meet rules; however, users avoid it because they lack official guidance. Countries need to establish blockchain standards, including GDPR, FATF, and ISO 20022, to protect information in international blockchain systems. Organizations find it difficult to protect data privacy when transferring it between nations while abiding by all legal requirements and maintaining online links.

Organizations need to solve the current legal chaos and management requirements before blockchain technology can be widely employed in business operations. Blockchain deals face different control methods in each national authority, leading to worldwide rule disparities. Every region now permits smart contracts as official documents, but the national policy is still being completed. Both organizations and governments need to work together to strengthen the acceptance of blockchain legal frameworks for enterprise progress. Companies should solve these problems by combining blockchain with AI and IoT during international teamwork. Connecting AI with IoT and blockchain technology helps business owners and rule makers build an effective global operational platform that protects sensitive data.

Both technologies merge to mend modern industries by improving business processes, protecting data safety, and analyzing results. These technologies allow different industry sectors to use real-time data to gain transaction security and improve their decision-making power. The implementation process faces difficulties owing to link problems, data security standards, and technology infrastructure requirements. The Internet of Things (IoT) helps supply chain management to perform real-time monitoring of its entire operation. Advanced IoT systems help businesses handle unexpected problems by creating smoother and safer operations. Blockchain networks face hacks from rogue users while normal operations break down; therefore, statisticians apply several defensive methods to keep the systems safe. Network analysis shows that fraud occurs and finds the market value for products. Businesses across different countries use AI and blockchain technology to enhance their trade by designing better supply chains and helping with paperwork while analyzing trading risks. Business operations receive elevated value from AI by improving supply chain activities and enhancing both network logistics and data utilization. Despite the progress in digital trade research, more testing is necessary to solve the present technology barriers and security problems.

Global businesses find it difficult to use blockchain because of system weaknesses, demand for energy plus connectivity problems, and the need to integrate new systems. The large-scale deployment and power usage of blockchain make it impractical to use in digital networks, healthcare operations, and money-handling technologies. Differently built blockchain systems cannot easily communicate because the security standards and code patterns do not match. Research teams have found ways to address trust problems between linked networks and create better system connections. The development of blockchain interoperability research is essential for solving the connection and processing problems experienced by independent platforms. Research has shown that networks work better when blockchain engineers create multiple security layers and set automatic password limit changes for user safety. Researchers study potential future uses, such as blockchain cross-border cooperation, and suggest new technologies, such as DAOs, and propose government and international organization policy changes. The G20 nations need to put blockchain to work for setting up a global digital commerce system and fixing tax and ecological deficiencies, while making document exchange and safe transactions simpler. All parties in business group education and the government must work together to build new ways to run international trade and supply chains successfully.

Acknowledgements No funding received. All authors structured the entire chapter with brainstorming discussion.

References

1. Abbas, S., Talib, M.A., Ahmed, A., Khan, F., Ahmad, S., Kim, D.-H.: Blockchain-based authentication in internet of vehicles: a survey. Sensors **21**(23) (2021)
2. Abdel Hakeem, S.A., Hussein, H.H., Kim, H.: Security requirements and challenges of 6G technologies and applications. Sensors **22**(5) (2022)
3. Abebe, R., Aruleba, K., Birhane, A., Kingsley, S., Obaido, G., Remy, S.L., Sadagopan, S.: Narratives and counternarratives on data sharing in Africa. In: Proceedings of the 2021 ACM Conference on Fairness, Accountability, and Transparency (2021)
4. Abichandani, P., Lobo, D., Kabrawala, S., McIntyre, W.: Secure communication for multiquadrotor networks using ethereum blockchain. IEEE Internet Things J. **8**(3), 1783–1796 (2021). https://doi.org/10.1109/JIOT.2020.3015716
5. Abor, J. Y., Subramanian, L., Iddrisu, K., Nsor-Ambala, R.: Multilateral development banks: contributions and challenges. In: Abor, J.Y., Ofori-Sasu, D. (eds.) Perspectives on Development Banks in Africa: Case Studies and Emerging Practices at the National and Regional Level, pp. 277–298. Springer Nature Switzerland (2024). https://doi.org/10.1007/978-3-031-59511-0_12
6. Aggarwal, S., Kumar, N.: Chapter Sixteen—hyperledger working model. In Aggarwal, S., Kumar, N., Raj, P. (eds.) Advances in Computers, vol. 121, pp. 323–343. Elsevier (2021)
7. Ahl, A., Goto, M., Yarime, M., Tanaka, K., Sagawa, D.: Challenges and opportunities of blockchain energy applications: interrelatedness among technological, economic, social, environmental, and institutional dimensions. Renew Sustain Energy Rev 166:112623. https://doi.org/10.1016/j.rser.2022.112623
8. Alsamhi, S.H., Myrzashova, R., Hawbani, A., Kumar, S., Srivastava, S., Zhao, L., Wei, X., Guizan, M., Curry, E.: Federated learning meets blockchain in decentralized data sharing: healthcare use case. IEEE Internet Things J. **11**(11), 19602–19615 (2024). https://doi.org/10.1109/JIOT.2024.3367249
9. Alshahrani, H., Islam, N., Syed, D., Sulaiman, A., Al Reshan, M.S., Rajab, K., Shaikh, A., Shuja-Uddin, J., Soomro, A.: Sustainability in blockchain: a systematic literature review on scalability and power consumption issues. Energies **16**(3) (2023)
10. Alsheyab, M.S.A.: Legal analysis of the merits of electronic transferable records: toward cross-border trade digitalization. Int. J. Law Manage. **67**(1), 145–163 (2025). https://doi.org/10.1108/IJLMA-09-2023-0209
11. Amato, F., Cozzolino, G., Moscato, F., Moscato, V., Xhafa, F.: A model for verification and validation of law compliance of smart contracts in IoT environment. IEEE Trans. Indus. Inf. **17**(11), 7752–7759 (2021). https://doi.org/10.1109/TII.2021.3057595
12. Aoun, A., Ilinca, A., Ghandour, M., Ibrahim, H.: A review of Industry 4.0 characteristics and challenges, with potential improvements using blockchain technology. Comput. Indus. Eng. **162**, 107746 (2021). https://doi.org/10.1016/j.cie.2021.107746
13. Arif, Z., Zulfitria, Supyadillah, A., Irfan, A., Taufik, A.: The revolution of blockchain in digital payment systems: legal implications and regulatory challenges. J. Ecohumanism **3**(8), 12269–12284 (2024). https://doi.org/10.62754/joe.v3i8.5833

14. Aránguiz, M., Margheri, A., Xu, D., Tran, B.: International trade revolution with smart contracts. In: The Digital Transformation of Logistics, pp. 169–184 (2021). https://doi.org/10.1002/9781119646495.ch12
15. Ashta, A., Herrmann, H.: Artificial intelligence and fintech: an overview of opportunities and risks for banking, investments, and microfinance. Strateg. Change **30**(3), 211–222 (2021). https://doi.org/10.1002/jsc.2404
16. Attarde, K., Jaiswal, C., Khatwani, R., Pinto, G., Kumar, V.: A novel central bank digital currency framework design for offline and foreign transactions based on blockchain. Digit. Policy, Regul. Gov. **27**(2), 201–220 (2025). https://doi.org/10.1108/DPRG-10-2023-0146
17. Au, C.H., Li, G., Hsu, W.-S., Shieh, P.-H., Law, K.M.Y.: Characteristics of proliferating cryptocurrencies: a comparative study between stable and non-stable cryptocurrencies. Enterp. Inf. Syst. **18**(7), 2356771 (2024). https://doi.org/10.1080/17517575.2024.2356771
18. Avinadav, T., Levy, P.: Improving efficiency in supply chains with a capital-constrained app developer under the agency contract. Eur. J. Oper. Res. **315**(3), 991–1005 (2024). https://doi.org/10.1016/j.ejor.2024.01.015
19. Azeem, M., Khan, S.U.R., Ibrahim, N., Yousafzai, A., Wang, W.L., Kareem, A.: A blockchain-based requirement management framework for distributed software development environment. In: 2023 IEEE 23rd International Conference on Software Quality, Reliability, and Security Companion (QRS-C) (2023)
20. Bag, S., Telukdarie, A., Pretorius, J.H.C., Gupta, S.: Industry 4.0 and supply chain sustainability: framework and future research directions. Benchmarking: Int. J. **28**(5), 1410–1450 (2021). https://doi.org/10.1108/BIJ-03-2018-0056
21. Ben Daya, B., Audy, J.-F.: Port access fluidity management during a major extension project: a simulation-based case study. Sustainability **16**(7) (2024)
22. Bhat, J.R., AlQahtani, S.A., Nekovee, M.: FinTech enablers, use cases, and role of future internet of things. J. King Saud Univ. Comput. Inf. Sci. **35**(1), 87–101 (2023). https://doi.org/10.1016/j.jksuci.2022.08.033
23. Bonnet, S., Teuteberg, F.: Impact of blockchain and distributed ledger technology for the management of the intellectual property life cycle: a multiple case study analysis. Comput. Indus. **144**, 103789 (2023). https://doi.org/10.1016/j.compind.2022.103789
24. Böhmecke-Schwafert, M.: The role of blockchain for trade in global value chains: a systematic literature review and guidance for future research. Telecommun. Policy **48**(9), 102835 (2024). https://doi.org/10.1016/j.telpol.2024.102835
25. Chander, B., Pal, S., De, D., Buyya, R.: Artificial intelligence-based internet of things for Industry 5.0. In: Pal, S., De, D., Buyya, R. (eds.) Artificial Intelligence-based Internet of Things Systems, pp. 3–45. Springer International Publishing (2022). https://doi.org/10.1007/978-3-030-87059-1_1
26. Chang, Y., Iakovou, E., Shi, W.: Blockchain in global supply chains and cross border trade: a critical synthesis of the state-of-the-art, challenges and opportunities. Int. J. Prod. Res. **58**(7), 2082–2099 (2019). https://doi.org/10.1080/00207543.2019.1651946
27. Cordieri, S.A., Bordin, C., Mishra, S.: A bottom-up optimization model for solar organic Rankine cycle in the context of transactive energy trading. Energy Syst. (2025). https://doi.org/10.1007/s12667-025-00723-w
28. Dasaklis, T.K., Malamas, V.: A review of the lightning network's evolution: unraveling its present state and the emergence of disruptive digital business models. J. Theor. Appl. Electron. Commer. Res. **18**(3), 1338–1364 (2023)
29. Deepa, N., Pham, Q.-V., Nguyen, D.C., Bhattacharya, S., Prabadevi, B., Gadekallu, T.R., Maddikunta, P.K.R., Fang, F., Pathirana, P.N.: A survey on blockchain for big data: approaches, opportunities, and future directions. Future Gener. Comput. Syst. **131**, 209–226 (2022). https://doi.org/10.1016/j.future.2022.01.017
30. Destro, F., Barolo, M.: A review on the modernization of pharmaceutical development and manufacturing—trends, perspectives, and the role of mathematical modeling. Int. J. Pharm. **620**, 121715 (2022). https://doi.org/10.1016/j.ijpharm.2022.121715

31. Divyashree, K.S., Mishra, A.: Blockchain technology in financial sector and its legal implications. In: Chaurasia, M.A., Juang, C.-F. (eds.) Emerging IT/ICT and AI Technologies Affecting Society, pp. 219–231. Springer Nature Singapore (2023). https://doi.org/10.1007/978-981-19-2940-3_15
32. Doepke, M., Hannusch, A., Kindermann, F., Tertilt, M.: Handbook of the Economics of the Family, vol. 1, pp. 151–254. North-Holland (2023). https://doi.org/10.1016/bs.hefam.2023.01.003
33. Du, Y., Li, X., Liang, Y., Tan, Z.: Two-stage multi-objective distributionally robust optimization of the electricity-hydrogen coupling system under multiple markets. Energy **303**, 131960 (2024). https://doi.org/10.1016/j.energy.2024.131960
34. D'Cunha, N.M., Isbel, S.T., Frost, J., Fearon, A., McKune, A.J., Naumovski, N., Kellett, J.: Effects of a virtual group cycling experience on people living with dementia: a mixed method pilot study. Dementia **20**(5), 1518–1535 (2021)
35. Ediagbonya, V., Tioluwani, C.: The role of fintech in driving financial inclusion in developing and emerging markets: issues, challenges and prospects. Technol. Sustain. **2**(1), 100–119 (2023). https://doi.org/10.1108/TECHS-10-2021-0017
36. Errico, F., Corallo, A., Spennato, A., Berlingerio, G.E.: Spatial proximity versus social distance: partnership development in the cross-border cooperation. J. Knowl. Econ. **15**(1), 461–486 (2024). https://doi.org/10.1007/s13132-022-01077-9
37. Farah, M.B., Ahmed, Y., Mahmoud, H., Shah, S.A., Al-kadri, M.O., Taramonli, S., Bellekens, X., Abozariba, R., Idrissi, M., Aneiba, A.: A survey on blockchain technology in the maritime industry: challenges and future perspectives. Future Gener. Comput. Syst. **157**, 618–637 (2024). https://doi.org/10.1016/j.future.2024.03.046
38. Farrell, S.: Blockchain standards in international banking: understanding standards deviation. J. ICT Standard. **7**(3), 209–224 (2019). https://doi.org/10.13052/jicts2245-800X.732
39. Finke, C., Schumann, M.: Is DLT the solution for current supply chain problems? Bridging the gap between EDI and DLT. Electron. Mark. **35**(1), 15 (2025). https://doi.org/10.1007/s12525-024-00750-y
40. Fiorentini, F.: Legal transplants in the law of secured transactions. Current problems and comparative perspectives. In: Fiorentini, F., Infantino, M. (eds.) Mentoring Comparative Lawyers: Methods, Times, and Places : Liber Discipulorum Mauro Bussani, pp. 3–23. Springer International Publishing (2020). https://doi.org/10.1007/978-3-030-34754-3_1
41. Fleith de Medeiros, J., Bisognin Garlet, T., Duarte Ribeiro, J.L., Nogueira Cortimiglia, M.: Success factors for environmentally sustainable product innovation: an updated review. J. Cleaner Prod. **345**, 131039 (2022). https://doi.org/10.1016/j.jclepro.2022.131039
42. Fleming, M.M.: The imperative of building and fostering a strong compliance culture in the evolving cross-border payments space. J. Payments Strategy Syst. **18**(1), 73–79 (2024)
43. Foster, C., Thelen, K.: Brandeis in Brussels? Bureaucratic discretion, social learning, and the development of regulated competition in the European Union. Regu. Gov. **18**(4), 1083–1103 (2024). https://doi.org/10.1111/rego.12570
44. Garg, S., Mondal, T.: Review on data privacy, protection, and security challenges in blockchain adoption across diverse domains. Int. J. Manage. Human. (IJMH) **10**(7) (2024)
45. Ghazali, R., Ali, F.H.M., Bakar, H.A., Ahmad, M.N., Haron, N.S., Omar, A.H., Ahmadian, A.: Blockchain for record-keeping and data verifying: proof of concept. Multimedia Tools Appl. **81**(25), 36587–36605 (2022). https://doi.org/10.1007/s11042-021-11336-7
46. Gosu, P., Tanguturu, R., Aenugutala, S.P., Cunha, T. B.D., Manjappa, K.: Decentralised authentication protocol for devices and users to access private network services using blockchain. In: 2023 33rd Conference of Open Innovations Association (FRUCT) (2023)
47. Gupta, I., Singh, A.K., Lee, C.N., Buyya, R.: Secure data storage and sharing techniques for data protection in cloud environments: a systematic review, analysis, and future directions. IEEE Access **10**, 71247–71277 (2022). https://doi.org/10.1109/ACCESS.2022.3188110
48. Gupta, S., Modgil, S., Choi, T.-M., Kumar, A., Antony, J.: Influences of artificial intelligence and blockchain technology on financial resilience of supply chains. Int. J. Prod. Econ. **261**, 108868 (2023). https://doi.org/10.1016/j.ijpe.2023.108868

49. Habib, G., Sharma, S., Ibrahim, S., Ahmad, I., Qureshi, S., Ishfaq, M.: Blockchain technology: benefits, challenges, applications, and integration of blockchain technology with cloud computing. Future Internet **14**(11) (2022)
50. Han, D., Zhu, Y., Li, D., Liang, W., Souri, A., Li, K.C.: A blockchain-based auditable access control system for private data in service-centric IoT environments. IEEE Trans. Indus. Inf. **18**(5), 3530–3540 (2022). https://doi.org/10.1109/TII.2021.3114621
51. Hannemann, A., Buchmann, E.: Is homomorphic encryption feasible for smart mobility? In: 2023 18th Conference on Computer Science and Intelligence Systems (FedCSIS) (2023)
52. Haricha, K., Khiat, A., Issaoui, Y., Bahnasse, A., Ouajji, H.: Recent technological progress to empower smart manufacturing: review and potential guidelines. IEEE Access **11**, 77929–77951 (2023). https://doi.org/10.1109/ACCESS.2023.3246029
53. Haugum, T., Hoff, B., Alsadi, M., Li, J.: Security and privacy challenges in blockchain interoperability-a multivocal literature review. In: Proceedings of the 26th International Conference on Evaluation and Assessment in Software Engineering (2022)
54. Henry, T., Laga, N., Hatin, J., Gaaloul, W., Boughzala, I.: Cross-collaboration processes based on blockchain and IoT: a survey. In: HICSS 2021: 54th Hawaii International Conference on System Sciences (2021)
55. Herbe, A., Estermann, Z., Holzwarth, V., vom Brocke, J.: How to effectively use distributed ledger technology in supply chain management? Int. J. Prod. Res. **62**(7), 2522–2547 (2024). https://doi.org/10.1080/00207543.2023.2218947
56. Hongmei, Z.: A cross-border E-commerce approach based on blockchain technology. Mobile Inf. Syst. **2021**(1), 2006082 (2021). https://doi.org/10.1155/2021/2006082
57. Hossain, M.R., Nirob, F.A., Islam, A., Rakin, T.M., Al-Amin, M.: A comprehensive analysis of blockchain technology and consensus protocols across multilayered framework. IEEE Access **12**, 63087–63129 (2024). https://doi.org/10.1109/ACCESS.2024.3395536
58. Ibrahim, I.A., Truby, J.: Governance in the era of blockchain technology in Qatar: a roadmap and a manual for trade finance. J. Bank. Regul. **23**(4), 419–438 (2022). https://doi.org/10.1057/s41261-021-00165-1
59. Imteaj, A., Shahid, A.R., Zaman, S.: Leveraging blockchain interoperability for interdependent networks. IEEE Consum. Electron. Mag. **13**(2), 81–86 (2024). https://doi.org/10.1109/MCE.2023.3245283
60. Jain, V., Wadhwani, K., Eastman, J.K.: Artificial intelligence consumer behavior: a hybrid review and research agenda. J. Consum. Behav. **23**(2), 676–697 (2024). https://doi.org/10.1002/cb.2233
61. Jenkins, J., Graves, L.: Do more with less: minimizing competitive tensions in collaborative local journalism. Digit. J. **12**(2), 101–120 (2024). https://doi.org/10.1080/21670811.2022.2026237
62. Jiao, J., Yan, A.: Management of the governance and social aspects of ESG in the financial industry. Rev. Financ. Econ. **43**(1), 111–131 (2025). https://doi.org/10.1002/rfe.1222
63. Kamran, R., Khan, N., Sundarakani, B.: Blockchain technology development and implementation for global logistics operations: a reference model perspective. J Glob. Oper. Strateg. Sourc. **14**(2), 360–382 (2021). https://doi.org/10.1108/JGOSS-08-2020-0047
64. Kashyap, R., Saxena, D.B.: Revolutionizing cross-border banking: the imperative and hurdles of blockchain technology. Available at SSRN 4698172 (2024)
65. Khan, S.N., Abdullah, M.A., Nadeem, A., Arshad, N.: Sustainable energy transition optimization through decentralized hybrid energy systems with various energy storage technologies under multi-criterion indices: A mid-career repowering scenarios. J. Energy Storage **94**, 112490 (2024). https://doi.org/10.1016/j.est.2024.112490
66. Kirli, D., Couraud, B., Robu, V., Salgado-Bravo, M., Norbu, S., Andoni, M., Antonopoulos, I., Negrete-Pincetic, M., Flynn, D., Kiprakis, A.: Smart contracts in energy systems: a systematic review of fundamental approaches and implementations. Renew. Sustain. Energy Rev. **158**, 112013 (2022). https://doi.org/10.1016/j.rser.2021.112013

67. Kotey, S.D., Tchao, E.T., Ahmed, A.-R., Agbemenu, A.S., Nunoo-Mensah, H., Sikora, A., Welte, D., Keelson, E.: Blockchain interoperability: the state of heterogenous blockchain-to-blockchain communication. IET Commun. **17**(8), 891–914 (2023). https://doi.org/10.1049/cmu2.12594
68. Kouskoura, A., Kalliontzi, E., Skalkos, D., Bakouros, I.: Evaluating experts' perceptions on regional competitiveness based on the ten key factors of assessment. Sustainability **16**(14), 5944 (2024). https://doi.org/10.3390/su16145944
69. Kowalski, M., Lee, Z.W.Y., Chan, T.K.H.: Blockchain technology and trust relationships in trade finance. Technol. Forecast. Soc. Change **166**, 120641 (2021). https://doi.org/10.1016/j.techfore.2021.120641
70. Krivosheya, E.: Network effects in retail payments market: evidence from individuals. In Karminsky, A.M., Mistrulli, P.E., Stolbov, M.I., Shi, Y. (eds.) Risk Assessment and Financial Regulation in Emerging Markets' Banking: Trends and Prospects, pp. 369–394. Springer International Publishing (2021). https://doi.org/10.1007/978-3-030-69748-8_18
71. Kshetri, N.: Blockchain's roles in strengthening cybersecurity and protecting privacy. Telecommun. Policy **41**(10), 1027–1038 (2017). https://doi.org/10.1016/j.telpol.2017.09.003
72. Kuehnlenz, S., Orsi, B., Kaltenbrunner, A.: Central bank digital currencies and the international payment system: the demise of the US dollar? Res. Int. Bus. Finance **64**, 101834 (2023). https://doi.org/10.1016/j.ribaf.2022.101834
73. Kumar, K., Kumar, V., Seema, Sharma, M.K., Khan, A.A., Idrisi, M.J.: A systematic review of blockchain technology assisted with artificial intelligence technology for networks and communication systems. J. Comput. Netw. Commun. **2024**(1), 9979371. https://doi.org/10.1155/2024/9979371
74. Lee, D.K.C., Yan, L., Wang, Y.: A global perspective on central bank digital currency. China Econ. J. **14**(1), 52–66 (2021). https://doi.org/10.1080/17538963.2020.1870279
75. Lee, H., Yeon, C.: Blockchain-based traceability for anti-counterfeit in cross-border E-commerce transactions. Sustainability **13**(19)
76. Levine, R.: International financial liberalization and economic growth. Rev. Int. Econ. **9**(4), 688–702 (2001). https://doi.org/10.1111/1467-9396.00307
77. Li, Y., Zheng, Z., Zhao, M., Liu, Z.: How does digital trade impact urban carbon emissions efficiency? Evidence from China's cross-border e-commerce pilot zones. J. Cleaner Prod. **456**, 142363. https://doi.org/10.1016/j.jclepro.2024.142363
78. Lin, W.L., Yang, Y., Leow, N.X., Lim, W.M., Khee, P.C., Yip, N.: Trends in the dynamic evolution of blockchain and circular economy: a literature review and bibliometric analysis. Bus. Strategy Environ. **34**(1), 1062–1084 (2025). https://doi.org/10.1002/bse.4027
79. Liu, A., Zhu, Q., Xu, L., Lu, Q., Fan, Y.: Sustainable supply chain management for perishable products in emerging markets: an integrated location-inventory-routing model. Transp. Res. Part E: Logistics and Transp. Rev. **150**, 102319. https://doi.org/10.1016/j.tre.2021.102319
80. Lobo, V.B., Rao, M.: A conceptual framework for building robust blockchain technology-enabled business processes in commodity markets: a systems approach. Smart Trends in Computing and Communications, Singapore (2024)
81. Louw-Reimer, J., Nielsen, J.L.M., Bjørn-Andersen, N., Kouwenhoven, N.: Boosting the effectiveness of containerised supply chains: a case study of TradeLens. In: Lind, M., Michaelides, M., Ward, R., Watson, R.T. (eds.) Maritime Informatics: Additional Perspectives and Applications, pp. 95–115. Springer International Publishing (2021). https://doi.org/10.1007/978-3-030-72785-7_6
82. Malhotra, G., Kharub, M.: Elevating logistics performance: harnessing the power of artificial intelligence in e-commerce. Int. J. Logistics Manage. **36**(1), 290–321 (2025). https://doi.org/10.1108/IJLM-01-2024-0046
83. Maltsev, Y., Maltseva, K.: Fatty acids of microalgae: diversity and applications. Rev. Environ. Sci. Bio/Technol. **20**(2), 515–547 (2021). https://doi.org/10.1007/s11157-021-09571-3
84. Merlec, M.M., Lee, Y.K., Hong, S.-P., In, H.P.: A smart contract-based dynamic consent management system for personal data usage under GDPR. Sensors **21**(23) (2021)

85. Minaei, M., Chatzigiannis, P., Jin, S., Raghuraman, S., Kumaresan, R., Zamani, M., Moreno-Sanchez, P.: Unlinkability and interoperability in account-based universal payment channels. In: Financial Cryptography and Data Security. FC 2023 International Workshops, Cham (2024)
86. Misra, S.K., Sharma, S.K., Gupta, S., Das, S.: A framework to overcome challenges to the adoption of artificial intelligence in Indian Government Organizations. Technol. Forecast. Soc. Change **194**, 122721 (2023). https://doi.org/10.1016/j.techfore.2023.122721
87. Moghavvemi, S., Mei, T.X., Phoong, S.W., Phoong, S.Y.: Drivers and barriers of mobile payment adoption: Malaysian merchants' perspective. J. Retail. Consum. Serv. **59**, 102364 (2021). https://doi.org/10.1016/j.jretconser.2020.102364
88. Mohamed, A.-M.O., Mohamed, D., Fayad, A., Al Nahyan, M.T.: Enhancing decision making and decarbonation in environmental management: a review on the role of digital technologies. Sustainability **16**(16) (2024)
89. Mohammed Abdul, S.S.: Navigating blockchain's twin challenges: scalability and regulatory compliance. Blockchains **2**(3), 265–298 (2024)
90. Mohsen, B.M.: Developments of digital technologies related to supply chain management. Procedia Comput. Sci. **220**, 788–795 (2023). https://doi.org/10.1016/j.procs.2023.03.105
91. Mustafa, G., Rafiq, W., Jhamat, N., Arshad, Z., Rana, F.A.: Blockchain-based governance models in e-government: a comprehensive framework for legal, technical, ethical and security considerations. Int. J. Law Manage. **67**(1), 37–55 (2025). https://doi.org/10.1108/IJLMA-08-2023-0172
92. Naef, S., Wagner, S.M., Saur, C.: Blockchain and network governance: learning from applications in the supply chain sector. Prod. Planning Control **35**(9), 932–946 (2024). https://doi.org/10.1080/09537287.2022.2044072
93. Nguyen, C.T., Hoang, D.T., Nguyen, D.N., Niyato, D., Nguyen, H.T., Dutkiewicz, E.: Proof-of-stake consensus mechanisms for future blockchain networks: fundamentals, applications and opportunities. IEEE Access **7**, 85727–85745 (2019). https://doi.org/10.1109/ACCESS.2019.2925010
94. Osterrieder, J., Chan, S., Chu, J., Zhang, Y., Misheva, B.H., Mare, C.: Enhancing security in blockchain networks: anomalies, frauds, and advanced detection techniques (2024). arXiv preprint arXiv:2402.11231
95. Othman, N.H., Alias, N.H., Fuzil, N.S., Marpani, F., Shahruddin, M.Z., Chew, C.M., David Ng, K.M., Lau, W.J., Ismail, A.F.: A review on the use of membrane technology systems in developing countries. Membranes **12**(1) (2022)
96. Ozturk, O., Akturk, E., Gultekin, S.: An empirical examination of trade policy and food security in MENA countries: evidence from quantile regressions. Glob. Bus. Econ. Rev. **31**(2), 259–272 (2024). https://doi.org/10.1504/GBER.2024.140236
97. O'Brien, J.P., David, P., Yoshikawa, T., Delios, A.: How capital structure influences diversification performance: a transaction cost perspective. Strateg. Manage. J. **35**(7), 1013–1031 (2014). https://doi.org/10.1002/smj.2144
98. Pal, K., Yasar, A.-U.-H.: Internet of things impact on supply chain management. Procedia Comput. Sci. **220**, 478–485 (2023). https://doi.org/10.1016/j.procs.2023.03.061
99. Patrickson, B.: What do blockchain technologies imply for digital creative industries? Creativity Innov. Manage. **30**(3), 585–595 (2021). https://doi.org/10.1111/caim.12456
100. Peng, S., Zhu, L., Hu, S., Cai, Z., Liu, W.: Enhancing global blockchain privacy via a digital mutual trust mechanism. Mathematics **12**(10) (2024)
101. Philip, A.O., Saravanaguru, R.K.: Smart contract based digital evidence management framework over blockchain for vehicle accident investigation in IoV era. J. King Saud Univ. Comput. Inf. Sci. **34**(7), 4031–4046 (2022). https://doi.org/10.1016/j.jksuci.2022.06.001
102. Prabowo, S., Putrada, A.G., Oktaviani, I.D., Abdurohman, M., Janssen, M., Nuha, H.H., Sutikno, S.: Privacy-preserving tools and technologies: government adoption and challenges. IEEE Access **13**, 33904–33934 (2025). https://doi.org/10.1109/ACCESS.2025.3540878
103. Qassem, L.M.A., Stouraitis, T., Damiani, E., Elfadel, I.A.M.: OSµS: An open-source microservice prototyping platform. In: 2023 IEEE International Conference on Cloud Computing Technology and Science (CloudCom) (2023)

104. Rahman, A., Khan, M.A.: Role of consistent regime-specific policies in recovering the negative relationship between financial development and economic growth. Econ. Chang. Restruct. **57**(4), 142 (2024). https://doi.org/10.1007/s10644-024-09722-w
105. Rani, D., Gill, N.S., Gulia, P., Yahya, M., Ahanger, T.A., Hassan, M.M., Abdallah, F.B., Shukla, P.K.: A secure digital evidence preservation system for an IoT-enabled smart environment using IPFS, blockchain, and smart contracts. Peer-to-Peer Netw. Appl. **18**(2), 5 (2024). https://doi.org/10.1007/s12083-024-01855-z
106. Rejeb, A., Rejeb, K., Zailani, S., Treiblmaier, H., Hand, K.J.: Integrating the Internet of Things in the halal food supply chain: a systematic literature review and research agenda. Internet of Things **13**, 100361 (2021). https://doi.org/10.1016/j.iot.2021.100361
107. Rieger, A., Stohr, A., Wenninger, A., Fridgen, G.: Reconciling blockchain with the GDPR: insights from the german asylum procedure. In: Reddick, C.G., Rodríguez-Bolívar, M.P., Scholl, H.J. (eds.) Blockchain and the Public Sector: Theories, Reforms, and Case Studies, pp. 73–95. Springer International Publishing (2021). https://doi.org/10.1007/978-3-030-55746-1_4
108. Saleem, B., Ahmed, M., Zahra, M., Hassan, F., Iqbal, M.A., Muhammad, Z.: A survey of cybersecurity laws, regulations, and policies in technologically advanced nations: a case study of Pakistan to bridge the gap. Int. Cybersecur. Law Rev. **5**(4), 533–561 (2024). https://doi.org/10.1365/s43439-024-00128-y
109. Saritha, K., Sunitha, T.: I want to quit! exploring determinants of turnover intention among employees of microfinance institutions. Contemp. Manag. Res. **18**(2), 165–197 (2023). https://doi.org/10.7903/cmr.22157
110. Schimmelfennig, F.: Rebordering Europe: external boundaries and integration in the European Union. J. Eur. Publ. Policy **28**(3), 311–330 (2021). https://doi.org/10.1080/13501763.2021.1881589
111. Shah, M.A., Raj, N.: Examining the role of blockchain and public-private partnerships in design and deployment of blockchain-enabled CBDC. Digit. Bus. **5**(1), 100111 (2025). https://doi.org/10.1016/j.digbus.2025.100111
112. Shahzad, M.F., Xu, S., Baheer, R.: Assessing the factors influencing the intention to use information and communication technology implementation and acceptance in China's education sector. Human. Soc. Sci. Commun. **11**(1), 283 (2024). https://doi.org/10.1057/s41599-024-02777-0
113. Sharma, K., Anand, H., Nandanwar, H., Chauhan, A.: Taxonomy of routing protocols. In: 2022 International Conference for Advancement in Technology (ICONAT) (2022)
114. Sheikh, S., Sifat, I.: Built to last, not to scale: the long run of decentralised autonomous organisations. J. Innov. Knowl. **9**(3), 100513 (2024). https://doi.org/10.1016/j.jik.2024.100513
115. Shen, L., Zang, Z.: Enterprise supply chain network optimization algorithm based on blockchain-distributed technology under the background of digital economy. Intell. Decis. Technol. **18**(3), 1827–1839 (2024)
116. Sinha, D., Roy Chowdhury, S.: Blockchain-based smart contract for international business—a framework. J. Glob. Oper. Strateg. Sourc. **14**(1), 224–260 (2021). https://doi.org/10.1108/JGOSS-06-2020-0031
117. Soavi, M., Zeni, N., Mylopoulos, J., Mich, L.: From legal contracts to formal specifications: a systematic literature review. SN Comput. Sci. **3**(5), 345 (2022). https://doi.org/10.1007/s42979-022-01228-4
118. Sravan, S.S., Mandal, S., Alphonse, P.J.A.: LIO-PAY: Sustainable low-cost offline payment solution. Electron. Commer. Res. Appl. **67**, 101440 (2024). https://doi.org/10.1016/j.elerap.2024.101440
119. Stone, D.L., Lukaszewski, K.M., Johnson, R.D.: Will artificial intelligence radically change human resource management processes? Organ. Dyn. **53**(1), 101034 (2024). https://doi.org/10.1016/j.orgdyn.2024.101034
120. Sun, P., Gu, L.: Optimization of cross-border e-commerce logistics supervision system based on internet of things technology. Complexity **2021**(1), 4582838 (2021). https://doi.org/10.1155/2021/4582838

121. Tajti, T.: The efficient enforcement challenge of secured transactions law reforms in civil law systems—self-help repossession, strict foreclosure, and other methods for the acceleration of enforcement of security interests. Uniform Law Rev. **29**(2), 139–188 (2024)
122. Tan, T.M., Saraniemi, S.: Trust in blockchain-enabled exchanges: future directions in blockchain marketing. J. Acad. Mark. Sci. **51**(4), 914–939 (2023). https://doi.org/10.1007/s11747-022-00889-0
123. Tan, W.K.A., Sundarakani, B.: Assessing blockchain technology application for freight booking business: a case study from technology acceptance model perspective. J. Glob. Oper. Strateg. Sourc. **14**(1), 202–223 (2021). https://doi.org/10.1108/JGOSS-04-2020-0018
124. Themistocleous, M., Rupino da Cunha, P., Tabakis, E., Papadaki, M.: Towards cross-border CBDC interoperability: insights from a multivocal literature review. J. Enterp. Inf. Manag. **36**(5), 1296–1318 (2023). https://doi.org/10.1108/JEIM-11-2022-0411
125. Tian, M.W., Wang, L., Yan, S.R., Tian, X.X., Liu, Z.Q., Rodrigues, J.J.P.C.: Research on financial technology innovation and application based on 5G network. IEEE Access **7**, 138614–138623 (2019). https://doi.org/10.1109/ACCESS.2019.2936860
126. Tikkinen-Piri, C., Rohunen, A., Markkula, J.: EU General data protection regulation: changes and implications for personal data collecting companies. Comput. Law Secur. Rev. **34**(1), 134–153 (2018). https://doi.org/10.1016/j.clsr.2017.05.015
127. Tirumala, R.D., Tiwari, P.: Innovative financing mechanism for blue economy projects. Mar. Policy **139**, 104194 (2022). https://doi.org/10.1016/j.marpol.2020.104194
128. Tomczyk, Ł.: Evaluation of digital piracy by youths. Future Internet **13**(1) (2021)
129. Trienekens, S., Escobar Campos, J.C., Schutte, L., Bremmer, M.: Who has access to a career in Western classical music? Building a tool to evaluate intersectionality in barriers to music education and careers. Res. Stud. Music Educ. **46**(3), 388–403 (2024). https://doi.org/10.1177/1321103X231224229
130. Trimborn, S., Peng, H., Chen, Y.: Influencer detection meets network autoregression—influential regions in the bitcoin blockchain. J. Empirical Financ. **78**, 101529 (2024). https://doi.org/10.1016/j.jempfin.2024.101529
131. Varriale, V., Cammarano, A., Michelino, F., Caputo, M.: Critical analysis of the impact of artificial intelligence integration with cutting-edge technologies for production systems. J. Intell. Manuf. **36**(1), 61–93 (2025). https://doi.org/10.1007/s10845-023-02244-8
132. Vionis, P., Kotsilieris, T.: The potential of blockchain technology and smart contracts in the energy sector: a review. Appl. Sci. **14**(1) (2024)
133. Vivoda, V., Matthews, R., McGregor, N.: A critical minerals perspective on the emergence of geopolitical trade blocs. Resourc. Policy **89**, 104587 (2024). https://doi.org/10.1016/j.resourpol.2023.104587
134. Walsh, K., Wallace, A., Pavis, M., Olszowy, N., Griffin, J., Hawkins, N.: Intellectual property rights and access in crisis. IIC Int. Rev. Intellect. Prop. Competition Law **52**(4), 379–416 (2021). https://doi.org/10.1007/s40319-021-01041-1
135. Wang, B., Wang, P., Tu, Y.: Customer satisfaction service match and service quality-based blockchain cloud manufacturing. Int. J. Prod. Econ. **240**, 108220 (2021). https://doi.org/10.1016/j.ijpe.2021.108220
136. Wielgos, D.M., Homburg, C., Kuehnl, C.: Digital business capability: its impact on firm and customer performance. J. Acad. Mark. Sci. **49**(4), 762–789 (2021). https://doi.org/10.1007/s11747-021-00771-5
137. Willett, J.F., LaGree, D., Shin, H., Houston, J.B., Duffy, M.: The role of leader communication in fostering respectful workplace culture and increasing employee engagement and well-being. Int. J. Bus. Commun. (2023). https://doi.org/10.1177/23294884231195614
138. Wirtz, J., Kunz, W.H., Hartley, N., Tarbit, J.: Corporate digital responsibility in service firms and their ecosystems. J. Serv. Res. **26**(2), 173–190 (2023)
139. Wu, J., Cui, X., Hu, W., Gai, K., Liu, X., Zhang, K., Xu, K.: A New Sustainable Interchain Design on Transport Layer for Blockchain. Smart Blockchain, Cham (2018)
140. Wu, X.-Y., Chi, X., Pu, X.: Blockchain adoption and collaborative emission reduction in vertical-shareholding supply chain. Manag. Decis. Econ. **46**(2), 1012–1028 (2025). https://doi.org/10.1002/mde.4417

141. Yalli, J.S., Hasan, M.H., Badawi, A.A.: Internet of Things (IoT): origins, embedded technologies, smart applications, and its growth in the last decade. IEEE Access **12**, 91357–91382 (2024). https://doi.org/10.1109/ACCESS.2024.3418995
142. Yang, Y., Chen, N., Chen, H.: The digital platform, enterprise digital transformation, and enterprise performance of cross-border E-commerce—from the perspective of digital transformation and data elements. J. Theor. Appl. Electron. Commer. Res. **18**(2), 777–794 (2023)
143. Zhang, B., Xu, J., Wang, X., Zhao, Z., Chen, S., Zhang, X.: Research on the construction of grain food multi-chain blockchain based on zero-knowledge proof. Foods **12**(8) (2023)
144. Zheng, S., Ahmed, D., Xie, Y., Majeed, M.T., Hafeez, M.: Green growth and carbon neutrality targets in China: do financial integration and ICT matter? J. Cleaner Prod. **405**, 136923 (2023). https://doi.org/10.1016/j.jclepro.2023.136923
145. Zwitter, A.: Cybernetic governance: implications of technology convergence on governance convergence. Ethics Inf. Technol. **26**(2), 24 (2024). https://doi.org/10.1007/s10676-024-09763-9

Blockchain for Digital Tourism in Northeast India: Understanding Tourist Priorities Using Best–Worst Scaling

Rajat Bhattacharjee and Rimakhi Borah

Abstract The Northeast region of India has immense tourism potential despite the challenges of underdevelopment, lack of infrastructure and security issues persisting over the years. There has been an alarming boost in reliance on digital platforms, starting from planning, to booking and payment related to travel, which has unveiled a number of issues concerning security, trust and transparency in transactions of online tourism which can likely be solved with the help of Blockchain technology, owing to its unique and dispersed architecture. The present work makes use of the latest method of multi-criteria decision-making (MCDM) known as the Best Worst Method (BWM) to highlight the critical essence of different aspects or criteria integrating blockchain that are considered by potential tourists to evaluate the official websites operated by Indian state government across North-eastern states, while elucidating their effectiveness on the relevant website traits. Results indicate that Arunachal Pradesh and Meghalaya occupy a relatively superior position to the other states; but not far behind with existing lacunas when it comes to quantify the inflows through the tourism websites. Tourism stakeholders and website managers often spend significant resources collecting data and analyzing the strengths and weaknesses of business websites. Even though the adoption of blockchain-led solutions is deterred by the lack of awareness, acute technical expertise and limited relevant policies, the adoption of the Best–Worst Method (BWM) which is an approach based on criteria, such as the one illustrated in this study may provide fruitful utility. The current work is confined to the official websites owned by the governments only, irrespective of the private-run digital platforms and focuses solely on the perspectives of the tourists rather than the views from industry professionals. Future studies should concentrate on supply chain management, cost–benefit analyses and integration of artificial intelligence (AI) and Internet-of-Things (IoT) to supplement blockchain-enabled solutions in digital tourism.

R. Bhattacharjee (✉)
Department of Finance, Nalbari Commerce College, Nalbari, Assam, India
e-mail: rbhattacharjee@nccnalbari.in

R. Borah
Department of Accountancy, Nalbari Commerce College, Nalbari, Assam, India

A. Kumar et al. (eds.), *Blockchain in the Tourism Industry: A New Era of Secure and Transparent Travel Solutions*, Information Systems Engineering and Management 57,
https://doi.org/10.1007/978-3-031-95341-5_13

Keywords Digital tourism · Blockchain · Northeast India · Best–Worst scaling · Tourists · Perspective

1 Introduction

The role of tourism in driving the growth of an economy is indispensable and more so for geographical locations with abundance of natural heritage and rich culture. Northeast India is one such region unique for its unparalledled natural endowments and supreme cultural heritage. Despite the turmoil of insurgency and underdevelopment existing in the terrain, the Northeast contains immense potential for economic advancement by means of tourism. The rich cultural tapestry, scenic landscapes and treasure of biodiversity add to the diversified potential of this region. The Northeast is a storehouse of indigenous communities contributing to such unique culture owing to a variety of languages, traditions and festivals which serve as paramount attraction to exclusive cultural experiences including traditional music, instruments, and dances to culinary forms highlighting the region's diversity. The scope for cultivating cultural exchange by dint of its mix of cultures is huge. Northeast is recognized as the world's hotspot for its biodiversity which draws the attention of wildlife enthusiasts and conservationists. Adventure tourism thrives in the Northeast, offering a plethora of activities such as trekking, river rafting, and wildlife safaris. Starting from the vibrant terrain of Arunachal Pradesh to the scenic gorges of Nagaland, it serves as the primal zone for adventurous experiences amidst nature. Further, the archeological sites and historical landmarks contribute to its allure besides its natural splendor.

In order to boost the tourism sector, Government participation through policy formulation, infrastructure development, destination marketing, and tourism planning is quintessential [23]. Governments ought to promote tourism based on its potential to encourage economic development and hence, bridge the difference in income gap by creating job opportunities [8]. Governments can potentially safeguard and improve the quality of tourism services by placing proper standards and regulatory measures while promoting the sustainable growth of destinations [33]. Developing the logistics and infrastructural facilities support accessibility which leads to better tourist experiences. Moreover, marketing initiatives and support provided by government to small businesses and local communities enrich the tourism sector. Government involvement can foster the growth of the tourism sector which is reflected in the Indian Prime Minister's addressal at the North Eastern Council in 2016 which stressed on the natural scenic beauty and unique cultural, historical and ethnic traditions and customs of the Northeast states. Tourism in this region can contribute immensely to the opportunities of income and employment, thereby promoting well-being of the economy.

The Northeast can become a premier tourist destination by highlighting its holistic treasury of natural, historical, cultural treasures thus fostering economic prosperity and promoting cross-cultural understanding. The existing constraints of inadequate infrastructure, issues of connectivity and security hinder the optimal use of the tourism potential of this area. However, with the internet revolution, digital technology has borne fruitful results for all sectors marking the dawn of Artificial Intelligence (AI) in tourism as well. The strategic positioning of tourism industry in Northeast region can fetch greater revenue and drive economic activities, thereby empowering local people. India has one of the largest tourism industry in the world which contributes around 9% of the national GDP [7]. The state governments and union territories have a significant role in offering diverse tour and travel information through their official websites meant for tourism. These websites profoundly impact the delivery of vital information to prospective travelers and provide a competitive advantage in the tourism sector. Hence, examining the website efficacy holds a significant role in meeting the increasing customer expectations as prospective tourists depend upon the government websites for reliable and accurate data with real-time information. These websites are highly responsible for influencing their decisions and shaping their overall experience. As a result, these websites have a greater obligation to ensure quality and relevant information with evolution of modern tourism and harnessing competitive advantage. The existing literature reveals that while researchers have worked on some parts of India including the North, nothing much relevant was observed in the context of North eastern states. Fortunately, the concept of digital tourism has been imperative for destination marketing, thereby making it convenient for the state governments to attract tourists through strategizing on their official websites and elevating their respective offerings. Blockchain technology has a lot to contribute to the tourism industry in this transition process while harnessing digital intiatives. There has been an alarming boost in reliance on digital platforms, starting from planning, to booking and payment related to travel [14]. This has unveiled a number of issues concerning security, trust and transparency in transactions of online tourism which can be likely solved with the help of Blockchain technology, owing to its unique and dispersed architecture. Blockchain offers security in transactions, automated smart contract and verifiability to tourism information [26], thus magnifying the efficacy as well as credibility of the tourism websites and making them reliable. However, there is little evidence of blockchain-enabled strategies in the Indian tourism sector, and especially in Northeast India, which makes it suitable for exploration.

2 Review of Literature

Digitalisation has transformed tourists' plans for travel and also redesigned the importance of destinations in marketing themselves for the tourism industry. The integration of Information and Communication Technology (ICT) in tourism industry has seamlessly facilitated real-time access to information related to destination, making online bookings and unified travel experience [6]. The role of digital platforms, particularly websites have acquired instrumental role in promotion of tourism, shaping the perception of tourists and decision making [1]. Northeast India suffers from the constraints of physical accessibility due to lack of reliable information and enriching digital engagement with potential visitors. Digital tourism undoubtedly has benefits, but it faces several limitations in developing regions. The digital platform is faced with issues of usability, security, credibility along with inefficiency of the marketing strategies. The official tourism websites of many states in India operated by the states face the severities of trust, irregular updates and low engagement of the users [12]. Such limitations call for modern technologies to safeguard security, openness and better experience of the digital user. The decentralised and unalterable features of Blockchain provide security, enhance trust and promote efficiency in digital transaction which has added to its popularity in the tourism industry [31].

Blockchain technology has been coined as a revolutionary force destined to reshape financial transactions in the future, with profound implications for the tourism industry [10]. Further, blockchain has been explained as a digital, decentralized and distributed ledger in which transactions are logged and added in chronological order with goal of creating permanent and tamper-proof records [31]. The applications of Blockchain in tourism industry encompass payment security, automated smart contracts, prevention of fraud and decentralised data management [15]. These enhance the credibility of tourism websites by removing intermediaries from the picture and facilitating secured record keeping, thereby improving the overall experience of the digital user. Blockchain plays an immense role in refining different attributes of tourism. The application of cryptographic security mechanisms in payment systems based on blockchain have controlled fraudulent practices in online transactions leading to increase in trust of the touristers [28]. Smart contracts simplify automation in bookings, refunds and service agreements which finally enhances the satisfaction of the consumer [24]. Blockchain helps in controlling document fraud and making online bookings secured owing to the in-built verified traveller identities [30]. Blockchain can assist the hotels and tour operators for validating the reviews made by guests, prevention of fake bookings and certifying ethical service sourcing [5]. Blockchain have opened unique opportunities for the tourism agencies to line up with the customer requirements and preferences, encouraging more customized and P2P interactions leading to increased customer loyalty [32]. Another advantage of blockchain technology is its decentralized nature that eliminates the involvement from third parties or other organizations. Blockchain solutions in tourism industry can reduce the need for intermediaries leading to lower-cost distribution channels and more competitive rates.

Numerous researchers have utilised Multi Criteria Decision Making (MCDM) approach to study the efficiency of tourism services and digital platforms. Best Worst Method is a popular MCDM approach which locates the best and worst attributes for prioritizing decision criteria [27]. Researchers have made use of BWM in the context of effectiveness of tourism website, e-commerce platforms and strategies used in digital marketing [25], however its application to assess blockchain enabled tourism strategies remains an underexplored area, especially in Northeast India. A review of existing literature reveals the studies made on significance of digital tourism, usage of blockchain technology and application of MCDM methods, however, integration of these dimensions is scanty. There seems to be a lack of empirical research in evaluating the effectiveness of official tourism websites with blockchain-enhanced solutions.

The present study is an attempt to evaluate the efficacy of Northeast India's official tourism websites by integrating blockchain-based decision criteria. It uses the multi-criteria decision making (MCDM) approach of Best Worst Method (BWM) to prioritize key traits influencing the effectiveness of the websites inclusive of transparency, security, integration of smart contract and user experience. The research work tries to locate the key factors in this regard so that innovative strategies can be formulated with the aid of blockchain in order to improve the tourism scenario in Northeast India, thereby offering practical recommendation for all the stakeholders. This research contributes to the current body of knowledge by interconnecting blockchain technology with optimization of tourism websites while proffering the utility of such innovative technologies to the governments and tourism boards for improving engagement digitally and restructure the transactions, thereby securing the confidence of tourists in the digital platforms.

3 Methodology

The present study utilizes the Best–Worst Method (BWM) developed [27] to prioritize the evaluation criteria for examining the digital strategies in tourism. The process begins with identifying the best and the worst criterion representing the most important and least important criteria respectively out of the identified factors for determining effectiveness of tourism websites. The seven alternatives are the official websites of the state government for Arunachal Pradesh, Assam, Manipur, Meghalaya, Mizoram, Nagaland, and Tripura. The criteria or factors used by the researcher to evaluate the effectiveness of the selected websites comprise of trust and security, response time and website performance, information fit-to-task, innovativeness in design, smart contract integration, decentralized identity verification and user engagement and accessibility. Following the past bodywork from literature, the implementation of a number of steps were required to be initiated, namely, (1) Articulation of the criteria of the alternatives; (2) Filtering the best criterion as well as the worst criterion; (3) Analyzing the choice of the best criterion in comparison to all other criteria; (4) Examining the preference of all other criteria in comparison to the worst criterion; (5) Estimating optimal weights; and (6) Calculating aggregate scores of alternatives. Previous research works regarding BWM analytical technique validate the quality of data over the quantity. Hence, the present research has adopted purposive sampling in line with earlier studies; selecting 12 pertinent and capable participants to draw out reliable findings [20]. Participants with frequent travel history evidenced from social platforms like LinkedIn and others made into the study. The participants navigated through the website URLs specified to them within the next 15 days only after obtaining their consent. Out of the 12 respondents, 3 had to be dropped due to straight lining. All the 9 participants comprising of three men and six women ultimately considered for the study were employed people and digitally equipped, mostly within the age of 28–45 years. Subsequently, pairwise comparisons are made by rating the preference of best criterion over all other criteria and then comparing the worst criterion against the others using a 1–9 scale, 1 being 'not at all effective' and 9 being 'exceptionally effective'. Ultimately, the BWM model is used to calculate optimal weights for assessing the relative importance of each criterion. This method warrants a structured and quantitative evaluation of the attributes of tourism website, thus facilitating the identification of key areas requiring careful consideration and improvement. The results of BWM are examined to ascertain the most influential factors in digital strategies of tourism by blockchain integration. This will provide insights into the key criteria for designing effective websites of tourism. Moreover, the research undergoes a comparative evaluation of the official tourism websites across states of Northeast India in order to identify strengths and weaknesses so that improvements can be made.

4 Results and Discussion

Determining the criteria of evaluation for aiding the decision-making process about the alternatives, namely, the tourism websites form the first step. The seven alternatives are the official websites of the state government for Arunachal Pradesh, Assam, Manipur, Meghalaya, Mizoram, Nagaland and Tripura. The criteria used by the researcher to evaluate the effectiveness of the selected websites comprise of trust and security, response time and website performance, information fit-to-task, innovativeness in design, smart contract integration, decentralized identity verification, and user engagement and accessibility.

The groundwork for selecting the criteria for the current study is based on the WebQual framework, renowned for its diverse dimensions and widespread adoption by past researchers in assessing website quality. Additionally, blockchain-powered features have also been incorporated to identify and construct a holistic framework for evaluating blockchain-integrated strategies for digital tourism in Fig. 1 based on review of existing literature.

Trust and Security: The application of blockchain improves data security, minimizes fraud and makes digital transactions transparent which makes it trustworthy [19]. Due to its decentralised nature, it is resilient to cyber-attacks which increases credibility to the users [22]. By rendering travel transactions more transparent and secure, blockchain has the potential to mitigate fraud and simplify the tracking of tourism related environmental impacts [24].

Response Time and Website Performance: Websites entailing a longer loading time negatively impact user retention and the overall performance of the website [21]. Consumer trust in digital platforms is greatly influenced by speed and reliability [29]. Blockchain's real time transaction tracking enhances efficiency, transparency and prevents fraud in travel operations.

Fig. 1 Framework for evaluation of tourism website

Information Fit-to-Task: A website is deemed effective when its content meets the informational needs of the users [17]. The utility of the website is enhanced by the accuracy and quality of the information and retrieval accessibility which in turn influences consumer decision-making [13].

Innovativeness in Design: Websites with innovative designs tend to engage more users due to better aesthetics and usability. The blending of personalised AI and interactive elements provide a competitive edge to the digital platforms [17].

Smart Contract Integration: Automation in transactions is facilitated in tourism platforms by integration of smart contracts in blockchain as these are self-executed. This leads to diminished risk of frauds and enhancement of operational efficiency [2], thereby bringing transparency in payments and bookings [3].

Decentralized Identity Verification: The use of block chain for identity verification eradicates the authentication by third-party as users are provided with secure digital identities which further ensure user privacy and seamless verification [9]. As blockchain operates without central authority, rendering it highly resistant to fraudulent activities. The absence of a single point of control makes it less vulnerable to malicious exploitation [24].

User Engagement and Accessibility: User engagement is a metric that measures how much the users interact with a product, service or content. User engagement rate can help inform product development and marketing strategies. The engagement of users is significantly impacted by user-friendly navigation [4], personalised and adaptive features since these enrich experience of the users and dish out better accessibility [18].

Relative advantage: This refers to the benefits a website has over its competitors. This concept, widely studied in the e-commerce sector [16], is also crucial for tourism websites. When users recognize a clear relative advantage of one site over others, it enhances the site's effectiveness and competitive edge.

In the second step, two extreme points from the seven criteria so identified were filtered out by the participants. These points refer to the most important criterion and the least important criterion representing the best and the worst respectively. Subsequently in the third step, in order to obtain the best to others criteria vector, participants were required to rank the importance of the best criterion in comparison to all other criteria using a 1–9 scale as presented in → Table 1. The best to others vector can be expressed as here under:

$$A_B = (a_{b1}, a_{b2}, \ldots, a_{bn}) \tag{1}$$

Here, a_{bj} denotes the preference of the best criterion B over the criterion j.

Again, in the fourth step the participants were asked to rank the importance of all other criteria over the worst criterion on a 1–9 scale, which is expressed below in → Table 2.

$$A_W = (a_{1w}, a_{2w}, \ldots, a_{nw}) \tag{2}$$

Here, a_{jw} denotes the preference of the criterion j over the worst criterion w.

Further, in the fifth step the maximum absolute differences $\left(\left|w_b - a_{bj}w_j\right|, \left|w_j - a_{jw}w_{wt}\right|\right)$ for all j are minimized to ascertain the optimal

Table 1 Assessment of best with vectors

Participants	Best	TS	RWP	IFT	INV	SCI	DIV	UEA	RA
1	INV	4	4	5	1	5	5	7	8
2	DIV	8	8	6	8	7	7	1	6
3	TS	1	6	7	8	4	6	8	5
4	INV	9	9	8	1	5	9	7	7
5	TS	1	5	7	9	8	7	7	9
6	INV	9	9	9	1	8	9	9	9
7	TS	1	8	8	8	9	9	9	8
8	TS	1	6	8	7	5	7	8	7
9	INV	8	8	8	1	6	9	5	5

Table 2 Assessment of others with worst vector

Participants	Worst	TS	RWP	IFT	INV	SCI	DIV	UEA	RA
1	RA	8	7	3	6	7	2	3	1
2	IFT	8	7	1	8	7	7	8	6
3	SCI	7	7	6	8	1	7	9	5
4	SCI	9	9	8	9	1	9	7	7
5	RWP	9	1	7	8	6	6	7	8
6	UEA	7	3	3	8	3	8	1	2
7	UEA	9	8	8	8	8	8	1	7
8	UEA	9	6	6	7	5	8	1	7
9	RA	9	8	8	9	7	8	6	1

weights of a criterion. The minimization problem can be expressed [27] as:

$$min\left[max\left(\left|w_b - a_{bj}w_j\right|, \left|w_j - a_{jw}w_{wj}\right|\right)\right]$$

$$s.t. \sum\nolimits_f wf = 1 \tag{3}$$

$w_f \geq 0$, for all j

In order to solve Eq. (3), a linear optimization model as directed by [27] is shown below in (4).

min δ^L s.t

$$\left|w_{b-}a_{bjW_j}\right| \leq \delta^L, \text{ for all} j$$

$$\left|w_{1-}a_{jw}w_{wj}\right| \leq \delta^L, \text{ for all } j \tag{4}$$

$$\sum_f w_{f=}1.$$

wj ≥ 0, for all j

A solution to (4) gives the optimal weights (w_1, w_2, ….w_j) along with the value of δ^L which represents the consistency ratio of the pair-wise comparison procedure in BWM. → Table 3 reports the criteria-wise optimal weights for individual participant. Trust and Security scored the highest making it the most important criterion with Innovativeness in Design following closely behind; succeeded by UEA, Information fit-to-task, Smart Contract Integration and DIV which seem to be the least important ones. A mean consistency ratio of 0.105 is obtained which is well within the acceptable limit of a δ^L upto 0.459 with eight criteria.

Table 3 Optimal weights calculation

Respondents (full sample)	TS	RWP	IFT	INV	SCI	DIV	UEA	RA
1	0.175	0.131	0.105	**0.311**	0.088	0.049	0.075	0.066
2	0.080	0.080	0.037	0.080	0.092	0.092	**0.431**	0.107
3	**0.384**	0.104	0.089	0.078	0.036	0.104	0.078	0.125
4	0.078	0.078	0.087	**0.443**	0.037	0.078	0.100	0.100
5	**0.436**	0.039	0.097	0.075	0.084	0.097	0.097	0.075
6	0.077	0.077	0.077	**0.490**	0.087	0.077	0.035	0.077
7	**0.483**	0.084	0.084	0.084	0.075	0.075	0.033	0.084
8	**0.423**	0.099	0.074	0.085	0.118	0.085	0.032	0.085
9	0.081	0.081	0.081	**0.412**	0.108	0.072	0.130	0.035
Mean	**0.246**	0.086	0.081	0.229	0.081	0.081	0.112	0.084

The highest scores provided by each participant with respect to the criteria have been highlighted accordingly

Lastly, the effectiveness scores for all seven alternatives are calculated to ascertain the final priority of the tourism websites. Participants rated the effectiveness of the seven alternatives for all seven criteria with a scale of 1–9; 1 being 'not at all effective' and 9 being 'exceptionally effective'. Taking participant 1 as an example, the entire procedure has been demonstrated in → Table 4. Linear normalized approach to normalize the values obtained (divide individual values by the maximum values column-wise). Each normalized value is multiplied by their respective weights to compute the priority of the seven alternatives for each participant. Finally, the final priorities of the seven alternative websites are ascertained by including the row-wise total which is explained below:

$$z_t - \sum_{f=1}^{n} w_f x_{tf}^{norm}$$

Here, Z_i is the final priority value of the alternative i and xij^{norm} which denotes the normalized values of the criterion j under the alternative i. → Table 5 presents priorities values for the seven alternatives (websites) for all participants by repeating the process explained earlier in Table 4; the highest scores for the various state websites by each participant are highlighted accordingly. The priority values for all the alternatives for the full sample are presented in → Table 6 along with criteria wise highest scores for all the alternatives.

Table 4 Analysis for first participant (steps 1–6)

Stepwise quantitative depiction								
Best	TS	RWP	IFT	INV	SCI	DIV	UEA	RA
IFT	4	4	5	1	5	5	7	8
Worst	TS	RWP	IFT	INV	SCI	DIV	UEA	RA
RA	8	7	3	6	7	2	3	1
Weights								
	0.175	0.131	0.105	0.311	0.088	0.049	0.075	0.066
Rating of alternatives								
AP	7	8	7	8	7	8	7	8
AS	8	7	7	8	7	8	9	6
MZ	9	8	7	9	8	7	8	8
MN	6	7	6	8	7	9	8	8
NL	8	8	9	8	8	9	8	9
TP	7	8	7	9	8	7	8	7
MG	8	7	9	7	8	8	8	9
Normalized values								
AP	0.778	1.000	0.778	0.889	0.875	0.889	0.778	0.889
AS	0.889	0.875	0.778	0.889	0.875	0.889	1.000	0.667
MZ	1.000	1.000	0.778	1.000	1.000	0.778	0.889	0.889
MN	0.667	0.875	0.667	0.889	0.875	1.000	0.889	0.889
NL	0.889	1.000	1.000	0.889	1.000	1.000	0.889	1.000
TP	0.778	1.000	0.778	1.000	1.000	0.778	0.889	0.778
MG	0.889	0.875	1.000	0.778	1.000	0.889	0.889	1.000
Priority calculation								
AP	0.136	0.131	0.082	0.277	0.077	0.043	0.058	0.058
AS	0.156	0.115	0.082	0.277	0.077	0.043	0.075	0.044
MZ	0.175	0.131	0.082	0.311	0.088	0.038	0.067	0.058
MN	0.117	0.115	0.070	0.277	0.077	0.049	0.067	0.058
NL	0.156	0.131	0.105	0.277	0.088	0.049	0.067	0.066
TP	0.136	0.131	0.082	0.311	0.088	0.038	0.067	0.051
MG	0.156	0.115	0.105	0.242	0.088	0.043	0.067	0.066

TS = Trust and Security, RWP = Response Time and Website Performance, IFT = Information Fit-to-Task, INV = Innovativeness in Design, SCI=Smart Contract Integration, DIV = Decentralized Identity Verification, UEA = User Engagement and Accessibility, RA = Relative Advantage
AP = Arunachal Pradesh, AS = Assam, MZ = Mizoram; MN = Manipur; NL = Nagaland; TP = Tripura; MG = Meghalaya;

Table 5 Calculation of priorities of all participants

Participants	Alternatives						
	AP	AS	MZ	MN	NL	TP	MG
1	0.863	0.868	**0.950**	0.829	0.938	0.904	0.881
2	0.865	0.918	**0.981**	0.843	0.860	0.859	0.966
3	**1.000**	0.942	0.900	0.806	0.849	0.929	0.990
4	**1.000**	0.889	0.898	0.778	0.863	0.898	0.898
5	0.816	**0.955**	0.903	0.661	0.825	0.810	0.901
6	0.961	0.920	0.969	0.889	**0.983**	0.906	0.906
7	0.970	0.899	0.861	0.899	0.841	0.770	**1.000**
8	0.970	**0.989**	0.892	0.892	0.903	0.939	0.955
9	**0.948**	0.893	0.707	0.847	0.701	0.935	0.896
Mean	**0.933**	0.919	0.896	0.827	0.863	0.883	**0.933**

The highest scores provided by each participant with respect to the criteria have been highlighted accordingly

Table 6 Computation of priority of alternatives for all criterion

Alternatives/criteria (full sample)	TS	RWP	IFT	INV	SCI	DIV	UEA	RA	Sum
AP	0.227	**0.085**	0.068	**0.222**	**0.078**	**0.075**	0.100	0.078	**0.933**
AS	0.236	0.077	0.070	0.204	0.075	0.074	**0.109**	0.075	0.919
MZ	0.226	0.072	0.068	0.198	0.077	0.072	0.107	0.075	0.896
MN	0.191	0.071	0.063	0.189	0.074	0.069	0.098	0.073	0.827
NL	0.212	0.075	0.066	0.196	0.070	0.067	0.104	0.073	0.863
TP	0.216	0.076	0.067	0.212	**0.078**	0.065	0.098	0.072	0.883
MG	**0.237**	0.079	**0.077**	0.203	0.077	0.073	0.108	**0.079**	**0.933**

The highest scores provided by each participant with respect to the criteria have been highlighted accordingly

5 Discussion and Implications

Amidst the herald of digital era, transparency and trust which play a pivotal role in engaging the users, it is further enhanced by the transformative potential offered by blockchain technology in the platforms of digital tourism. Since tourism websites are the initial window of access to information for potential tourists, hence their traits significantly influence the decision of the users. Blockchain runs on decentralization which fosters authenticity of data and security, which in turn builds confidence of the tourists. Based on the observations of the study, the following recommendations have been put forward.

Trust and Security along with Decentralized Identity Verification were identified as the major attributes for blockchain-led digital tourism websites in Northeast India particularly in the context of Meghalaya and Assam which reflect prioritizing on trust and security the most. This would refrain digital transaction records from being tampered, thereby securing the digital identities leading to increased confidence of users and protection of data. Visitor engagement as well as retention can be improved by making the websites fast loading, and navigation user friendly. Blockchain enables use of smart contracts for efficient processing of transactions which result in reduction of website latency to a great extent thereby enhancing the overall user experience as reflected by the outcome of Arunachal Pradesh and Meghalaya.

Decision making by the tourists is greatly facilitated by the presence of well-structured presentation of tourism information as reflected by the scores of Meghalaya, thus indicating demand for verifiable data by using blockchain in tourism websites. This can also be leveraged to verify travel reviews and tour guide credentials which builds confidence of the tourists in reliable content being provided. Visitor interaction can be enhanced by use of blockchain led configurations in the form of decentralized applications, tokenized incentives and community driven contents on tourism which is exercised by Assam and Meghalaya. Innovation in website design is imminent as depicted by the scores of Arunachal Pradesh and Tripura calling for integration of blockchain in the form of virtual experiences and recommendation powered by AI. Smart contracts enabled by blockchain lead to streamlining of operations while providing clarity to financial transactions by means of digital payment and tourism insurance. Blockchain enabled management of tourism data makes the process efficient by providing access to real time updates on hotspots, accommodations and related data.

Government-led tourism websites serve as digital gateways to Northeast India's rich heritage and natural beauty. By incorporating blockchain technology, these platforms can enhance trust, improve efficiency, and offer secure, user-friendly experiences. The adoption of blockchain for identity verification, fraud prevention, and smart contract implementation will revolutionize the digital tourism ecosystem in the region, fostering greater tourist engagement and sustainable growth.

6 Conclusion

The application of blockchain technology has the potential to transform the entire gambit of tourism by increasing the efficiency and security of digital platforms used for tourism in Northeast India. As observed, Arunanchal Pradesh and Meghalaya outperformed the other states in blockchain infused areas leading to greater trust, decentralized identity verification and smart contract integration while Manipur lagged behind. Despite the potential of tourism in advancing the Northeast economy it is constrained by the gaps existing in technology. Blockchain can easily address the constrictions by bringing transparency, data security and seamless transactions, thereby building confidence and interaction of tourists. The integration of smart contracts can improve user experience substantially in terms of automated bookings payments and visitor authentication. However, the adoption of blockchain is limited greatly on account of low awareness, limited expertise and lack of policy frameworks. A multi-stakeholder approach involving the government, private enterprises, and technology firms is essential for successful implementation. Future research should explore Artificial Intelligence, Internet of Things, and Supply Chain Applications to strengthen blockchain-driven tourism, positioning Northeast India as a secure and competitive digital travel destination.

References

1. Afren, S.: The role of digital marketing promoting tourism business: a study of use of the social media in prompting travel. World J. Adv. Res. Rev. **21**(1), 272–287 (2024)
2. Alshahrani, N.M., Kiah, M.L., Zaidan, B.B., Alamoodi, A.H., Saif, A.: A review of smart contract blockchain based on multi-criteria analysis: challenges and motivations (2023). arXiv preprint arXiv:2302.08496
3. Banerji, D., Rashideh, W., Arora, B., Pratihari, A.R.: Application potential of blockchain technologies in the travel and tourism industry. In: Blockchain Applications in IoT Ecosystem, pp 289–299. Springer International Publishing, Cham (2021)
4. Bastos, B.C.: Impact of website navigation on user experience. Master's thesis, Universidade NOVA de Lisboa (Portugal) (2023)
5. Baydeniz, E.: Blockchain technology in tourism: pioneering sustainable and collaborative travel experiences. J. Tourismol. **10**(1), 1–12 (2024)
6. Buhalis, D., O'Connor, P.: Information communication technology revolutionizing tourism. Tour. Recreat. Res. **30**(3), 7–16 (2005)
7. Chatterjee, J., Dsilva, N.R.: A study on the role of social media in promoting sustainable tourism in the states of Assam and Odisha. Tour. Critiques: Pract. Theory **2**(1), 74–90 (2021)
8. Coccossis, H., Parpairis, A.: Assessing the interaction between heritage, environment, and tourism: Mykonos. In: Coccossis, H., Nijkamp, P. (eds.) Sustainable Tourism Development, pp. 127–140. Avebury, Hong Kong (1995)
9. Cresitello-Dittmar, B.: Application of the blockchain for authentication and verification of identity. Independent Paper (2016)
10. Gelter, H.: Digital tourism-an analysis of digital trends in tourism and customer digital mobile behaviour (2017)
11. Visit Artic Europe, Finland, May: Report.
12. Gnanasambandam, C., Madgavkar, A., Kaka, N., Manyika, J., Chui, M., Bughin, J., Gomes, M.: Online and upcoming: the internet's impact on India. McKinsey & Company, Bangalore (2012)
13. Griffith, D.A., Krampf, R.F., Palmer, J.W.: The role of interface in electronic commerce: consumer involvement with print versus on-line catalogs. Int. J. Electron. Commer. **5**(4), 135–153 (2001)
14. Lallukka, T.: The impact of user-generated online reviews on pre-travel planning and accommodation selection processes of Finnish millennial consumers (2021)
15. Laroiya, C., Saxena, D., Komalavalli, C.: Applications of blockchain technology. In: Handbook of Research on Blockchain Technology, pp. 213–243. Academic press (2020)
16. Liang, F., Brunelli, M., Rezaei, J.: Consistency issues in the best worst method: measurements and thresholds. Omega **96**, 102175 (2019)
17. Loiacono, E.T., Watson, R.T., Goodhue, D.L.: WebQual: an instrument for consumer evaluation of website. Int. J. Electron. Commer. **11**(3), 51–87 (2007)

18. Longstreet, P., Brooks, S., Featherman, M., Loiacono, E.: Evaluating website quality: which decision criteria do consumers use to evaluate website quality? Inf. Technol. People (2021)
19. Mougayar, W.: The Business Blockchain: Promise, Practice, and Application of the Next Internet Technology. Wiley (2016)
20. Munim, Z.H., Saha, R., Schøyen, H., Notteboom, T.: Autonomous ships for container shipping in the Arctic routes. J. Mar. Sci. Technol. 1–15 (2021)
21. Muylle, S., Moenaert, R., Despontin, M.: The conceptualization and empirical validation of website user satisfaction. Inf. Manage. **41**(5), 543–560 (2004)
22. Nofer, M., Gomber, P., Hinz, O., Schiereck, D.: Blockchain. Bus. Inf. Syst. Eng. **59**(3), 183–187. https://doi.org/10.1007/s12599-017-0467-3
23. Pearce, P.L.: The Ulysses Factor, Evaluating Visitors in Tourist Settings. Springer Verlag, New York (1988)
24. Pradhan, B., Roy, M., Bhatia, S.: Enhancing tourists' satisfaction through blockchain technology. PUSA J. Hosp. Appl. Sci. **9**(2), 37–47 (2023)
25. Praneeth, B.B., Nadeem, S.P., Vimal, K.E.K., Kandasamy, J.: Performance measurement of e-commerce supply chains using BWM and fuzzy TOPSIS. Int. J. Qual. Reliab. Manage. **40**(5), 1259–1291 (2023)
26. Rashideh, W.: Blockchain technology framework: current and future perspectives for the tourism industry. Tour. Manage. **80**, 104125 (2020)
27. Rezaei, J.: Best-worst multi-criteria decision-making method: some properties and a linear model. Omega **64**, 126–130 (2016)
28. Rodríguez Bolívar, M.P., Sepe, F., Nanu, L., Roberto, F.: Perceptions and challenges of blockchain adoption in tourism industry: a study on trust, privacy and security. J. Hosp. Tour. Technol. (2024)
29. Rose, G.M., Straub, D.W.: The effect of download time on consumer attitude toward the e-service retailer. e-Service **1**(1), 55–76 (2001)
30. Sharma, M., Sehrawat, R., Daim, T., Shaygan, A.: Technology assessment: Enabling Blockchain in hospitality and tourism sectors. Technol Forecast Soc Chang **169**, 120810 (2021)
31. Treiblmaier, H.: The impact of the blockchain on the supply chain: a theory-based research framework and a call for action. Supply Chain Manag. Int. J. **23**(6), 545–559 (2018)
32. Treiblmaier, H., Önder, I.: The impact of blockchain on the tourism industry: a theory-based research framework. In: Treiblmaier, H., Beck, R. (eds.) Business Transformation Through Blockchain, vol. II, pp. 3–21. Palgrave Macmillan, Cham (2019)
33. Wahab, S., Pigrim, J.J. (eds.): Tourism, Development and Growth: The Challenge of sustainability. Routledge, London (1997)

Emerging Trends and Innovation of Blockchain in Tourism

Srikanta Kumar Mohapatra and Jayashree Mohanty

Abstract The travel industry is transforming quickly, harnessing the most recent technologies to facilitate efficiency, safety, and transparency. Blockchain technology has emerged as a game changer with decentralized solutions to enhance operations, trust, and customer experience. With its capability to support immutable, distributed, and verifiable records, blockchain can potentially eradicate fraud, simplify transactions, and enhance supply chain management. This chapter addresses blockchain's primary trends and breakthroughs in tourism, such as smart contracts, safe payment, decentralized identity, and supply chain transparency. It also looks into the challenges and potential futures of blockchain applications within the industry. Further integration with artificial intelligence (AI) and the Internet of Things (IoT) increases blockchain's applications within tourism further by making processes efficient, secure, and more accessible. Based on case studies and field applications, the chapter hopes to present a holistic image of the revolutionary role in forming the future of tourism. User manual provides instructions on how to use the Springer Nature Word template effectively. It is intended as a guide for Springer Nature book and chapter authors in understanding how the template functions. This user guide has been prepared in the form of a sample chapter which at the same time explains the different features offered by this tool.

Keywords Blockchain · Tourism · Smart contracts · Loyalty · Cross border

S. K. Mohapatra (✉)
Chitkara University Institute of Engineering Technology, Chitkara University, Rajpura, Punjab, India
e-mail: srikanta.2k7@gmail.com

J. Mohanty
Department of Computer Science and Engineering, Chandigarh University, Chandigarh, Punjab, India
e-mail: jayashree.simi2k5@gmail.com

A. Kumar et al. (eds.), *Blockchain in the Tourism Industry: A New Era of Secure and Transparent Travel Solutions*, Information Systems Engineering and Management 57,
https://doi.org/10.1007/978-3-031-95341-5_14

1 Introduction

Tourism is the world's biggest industry and plays a significant role in the global economy. However, it is based on issues such as fraud, high transaction costs, inefficiencies in booking processes, and stakeholder trust. Travelers tend to book through third-party intermediaries, which is costly and hinders the resolution of conflicts. The traditional banking system also affects the transaction system, slows the transaction process, and increases fraudulent activity. Blockchain technology addresses these problems and is an emerging solution due to its decentralized nature. It is also immutable and transparent throughout the system. Blockchain technology generally operates in a peer-to-peer transaction process and optimizes the entire process. It also proved a strong security to keep all the records safe and not allow them to be manipulated and kept away from malicious activities. In the tourism sector, the main objective of blockchain technology is the ability to track all transactions in real time. This results in updated insurance claims, baggage tracking and transparent tour operation. It also keeps tourists safe from security threats. In some instances, blockchain technology is integrated with AI and IoT, resulting in operations like automated check-in hotels, applicable loyalty schemes, travel recommendations, and expense activities. The overall process of blockchain technology in tourism is described in Fig. 1. The entire chapter from this section onward signifies the literature survey, key innovations of blockchain in tourism, applications, challenges, case study and conclusion.

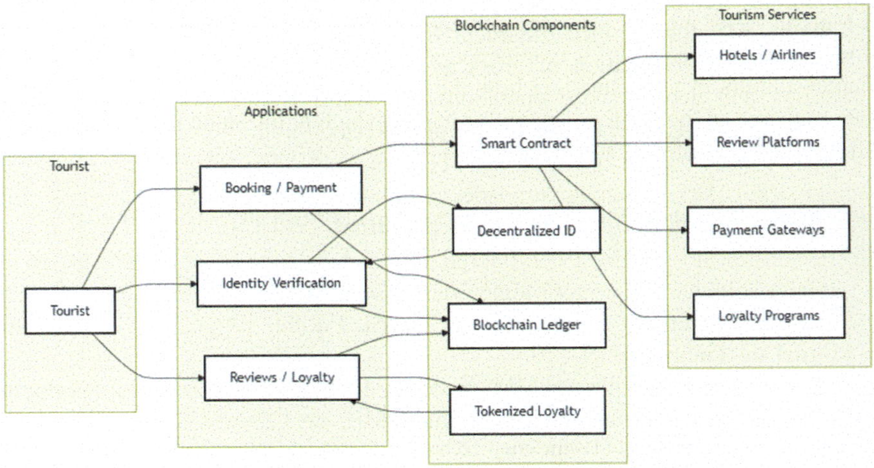

Fig. 1 Blockchain Scenario in Tourism

2 Literature Survey

Several studies have discussed the use of blockchain technology in tourism. Research systematically reviewed blockchain's issues and prospects, stressing its capacity for secure and transparent transactions [1]. The article conducted a systematic literature review of blockchain-based applications, grouping them according to industry and mapping open problems [2]. The article [3] discussed the application of blockchain to sustainable supply chain management, noting its application in tourism logistics and ethical sourcing. Later research examined particular applications in the tourism industry. The author examined how blockchain could affect digital identity verification, where it was able to provide more security and simplify check-in procedures and studied blockchain-based loyalty schemes, highlighting how tokenized incentives increase customer retention and engagement [4, 5]. The study described how blockchain can minimize travel insurance and ticketing system fraud [6]. A number of other studies have broadened the scope of blockchain in tourism. The research also explained the contribution of the blockchain in the field of tourism through carbon footprint. Meanwhile it is also introducing the benefits of blockchain in the field of tourism with the trust and travel benefits in an online system [7, 8]. The research also discussed to the accounting and auditing. They also described the importance of maintaining the ledger in this field. Further enhanced the study by explaining the security aspects of tourism and also explained how these can be managed by the blockchain technology [9, 10]. The authors explained the decentralized aspects of the tourism system by the help of blockchain technology. Smart tourism aspects are explained. Other research has investigated the technology convergence of blockchain with AI and IoT in the tourism sector. They studied how predictive analytic integrated with blockchain can be used to improve individualized travel experiences [11–13]. The viability of blockchain for dynamic pricing models in the airline and hotel sectors. These studies form a strong basis for grasping the revolutionary role of blockchain in tourism, determining some significant advantages and possible implementation hindrances [14]. This chapter relies on these findings by elaborating on upcoming trends, innovations, and future prospects of blockchain-based tourism. The detailed study is given in Table 1.

3 Key Innovations of Blockchain in Tourism

Blockchain technology is introducing a chain of innovations in the tourism sector to enhance efficiency, customer experience, and operational cost savings. Some of the significant innovations include:

Table 1 Detailed study—blockchain based tourism

Study	Key focus	Findings
Zheng et al. [1]	Blockchain challenges and opportunities	Highlights blockchain's security and transparency potential
Casino et al. [2]	Systematic review of blockchain applications	Categorized use cases across industries, including tourism
Saberi et al. [3]	Sustainable supply chain management	Explores blockchain's role in ethical sourcing and logistics
Xu et al. [4]	Digital identity verification	Demonstrates blockchain's ability to enhance security
Kim and Woo [5]	Loyalty programs	Tokenized rewards improve customer retention
Chen et al. [6]	Fraud reduction in insurance and ticketing	Blockchain minimizes fraud risks in travel-related transactions
Gössling et al. [7]	Sustainable tourism	Blockchain aids in carbon footprint tracking
Sun et al. [8]	Online travel agency applications	Enhances trust between service providers and customers
Dai and Vasarhelyi [9]	Blockchain in accounting and auditing	Immutable ledgers improve financial transparency
Wang et al. [10]	Tourist data privacy	Blockchain ensures GDPR-compliant identity verification
Alzahrani and Bulusu [11]	Decentralize applications (DApps)	Reduces intermediary costs in travel bookings
Yang et al. [12]	Blockchain in smart tourism	Automated tourism services enabled by IoT and blockchain
Sharma and Singh [13]	AI and blockchain integration	Predictive analytics enhances personalized travel experiences
Lee et al. [14]	Blockchain in dynamic pricing models	Feasibility of blockchain-driven airline and hospitality pricing

3.1 Decentralized Systems

Blockchain allows decentralized tourist platforms that reduce reliance on intermediaries like travel firms, online reservations, and bank systems. Redistributions of control to a network of peers allow tourists and providers to interact directly with each other, cutting down operating expenses and maximizing levels of trust. The innovation further enhances data security by having it distributed on different nodes instead of vulnerable central servers susceptible to cyber-attacks. Additionally, decentralized systems enhance transparency since tourists can confirm bookings, transactions, and service validity in real time without depending on third-party validations.

3.2 Smart Contracts

Smart contracts are self-executing contracts coded on blockchain networks that automatically implement pre-coded conditions. Smart contracts make travel and tourism transactions easier by automating hotel bookings, flight bookings, and travel insurance claims. For example, a smart contract can offer a quick refund without any manual claim procedure in case of flight cancellation or delay. Not only does this make things more efficient, but it also creates trust between stakeholders. Also, smart contracts minimize conflicts as terms of the agreement are stored immutably and enforced transparently. The researchers introduce that hybrid software reliability also important for choosing the integration [15].

3.3 Digital Identity Verification

Now a days it emphasizes on the digital system, along with the blockchain technology, verifies and enhances the security. Traditional identity verification processes like passports and ID cards are prone to forgery and loss. It also describes blockchain enables self-sovereign identity (SSI) systems, where tourists fully control their personal information. Such systems use encrypted, tamper-proof credentials on a decentralized ledger, allowing tourists to easily verify themselves at airports, hotels, and border control points [16, 17]. This reduces reliance on duplicate verification checks but assists in remaining compliant with regulatory systems like GDPR and the Know Your Customer (KYC) policy.

3.4 Immutable Review Systems

The main objective of the tourism sector is to take the consumer's confidence on a priority basis by eliminating false reviews and providing helpful advice to complete the operations. So, these activities are managed by innovative contract-based blockchain technologies to eliminate customer feedback without tampering with reviews [18]. This preserves the integrity of the reviews and prevents fraudulent activities such as fake ratings and paid endorsements. Blockchain-based review platforms benefit consumers and businesses by promoting transparency, trust, and fair competition in tourism [19].

4 Key Innovations of Blockchain in Tourism

The rapid development of Blockchain in the overall progress of actual uses in the tourism sector is transforming processes, enhancing security, enhancing efficiency, and introducing creative solutions to classic industry challenges. The key uses of Blockchain in tourism are described below.

4.1 Travel Payments and Booking

Blockchain-based platforms like Travala and Winding Tree facilitate direct connections between travelers and service providers without intermediaries like conventional booking sites. These platforms use cryptocurrencies to utilize the travelling costs and execution time [20, 21]. They also use blockchain-based tokens for the transaction. This results in the unauthorized use of foreign currencies and the rapid release of the expense charge-back [22]. The fact that blockchain allows cross-border payments without intermediaries makes it the perfect solution for international travelers who want smooth financial transactions.

4.2 Secure Traveler Identity Management

The identity solution also plays a significant role in tourism. Blockchain technologies also handle this with innovative, decentralized digital identity contract-based solutions. By applying blockchain-driven identity verification, traveler can retain their details securely on a distributed ledger while granting access only to authorized institutions such as airlines, hotels, and immigration services [23]. It reduces the fraudulent activities related to the right person in the tourism sector. It mitigates several documentation versifications, thus enhancing the ease of check-in and boarding [24]. Blockchain improves security by reducing identification theft risks while guaranteeing adherence to international travel regulations.

4.3 Supply Chain Transparency

The tourism sector is based on a sophisticated supply chain incorporating air lines, accommodations, tour companies, and regional businesses. Blockchain makes the supply chain more transparent by allowing the tracking of assets in real-time, guaranteeing responsible sourcing, and authenticating travel-related services [25].

4.4 Automated Check-Ins and Loyalty Programs

Blockchain automates and streamlines check-in procedures at airports and hotels through smart contracts and digital identity platforms. Rather than depending on human verification, the blockchain enables users to utilize secure digital IDs to facilitate smooth check-ins [26]. Blockchain-based loyalty programs also increase customer interaction through tokenized rewards that can be used across various service providers. Unlike conventional reward schemes limited to a single brand, blockchain-based loyalty points are transferable and interchangeable, making them more valuable to travelers[27].

5 Execution Process

The blockchain system in tourism is based upon three categories. The blockchain process in transaction system, in surety and in enhancement the trust of the tourists. A simple blockchain-based booking system is shown in Fig. 2.

Algorithm 1 describes the overall process in the tourism system and represented by the Fig. 3.

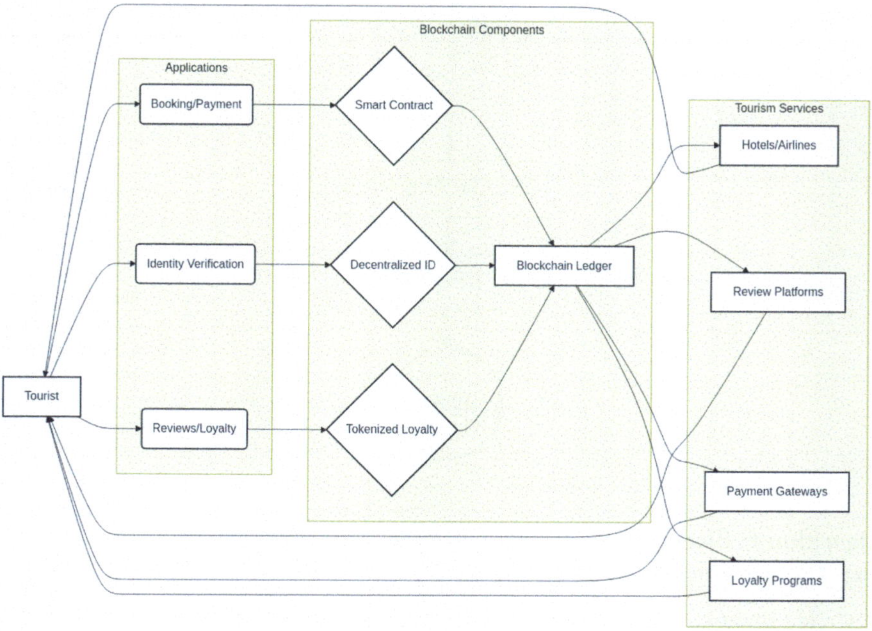

Fig. 2 Blockchain-based booking system for Tourist

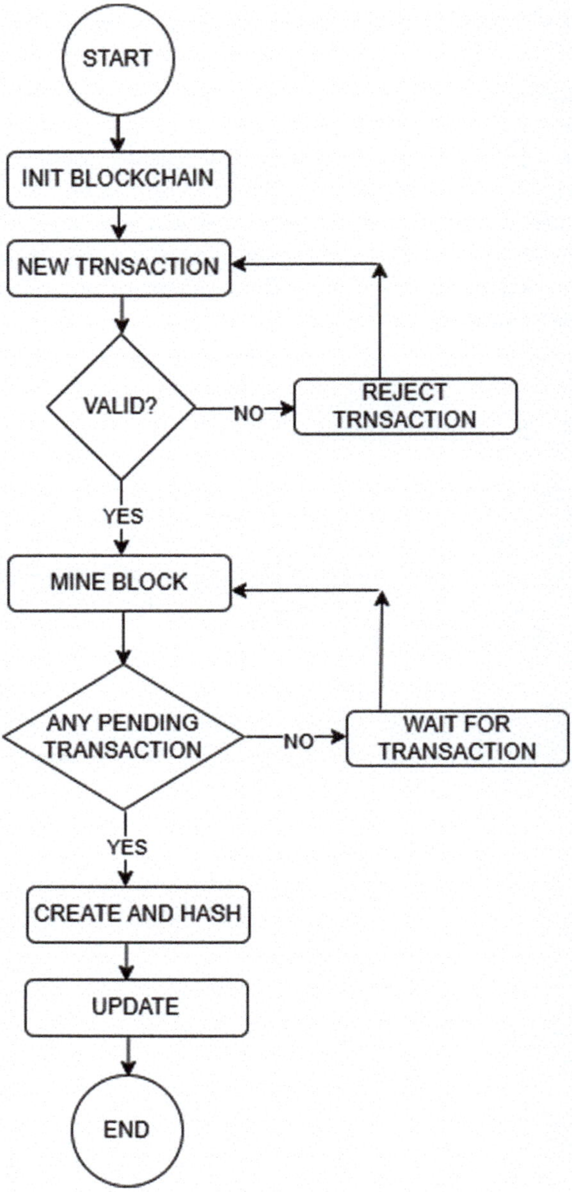

Fig. 3 Blockchain in booking system—tourism

Algorithm 1: Blockchain in Booking System—Tourism
Step 1: Start Blockchain Initialization Process

- Create Empty Chain and mark Pending.Transaction
- Create Genesis block

Step 2: Initiate New transaction

- Booking Details[Cust_inf, Provider,Drtails, Transaction Amount]
- Create: Transaction_Object
- Append: Transaction_object and Pending.Transaction

Step 3: New block Insertion

- Check for the new Pending.Transaction
- Wait:New
- If transaction exist: Create new block and retrive hash of last block

Step 4: Block Creation

- Assigned: Unique Index
- Include: Pending.Transaction
- Retrive:Previous Hash by using SHA-256
- Reset: Pending.Transaction after completion

Step 5: Hash Generation

- Concatenate:Previous.hash and Timestamp
- Generation: New Hash with SHA-256

Step 6: Repeat Process for all transactions

Algorithm 2 describes the overall process in the tourism system and represented by the Fig. 4.

Algorithm 2: Blockchain Application in Security—Tourism
Step 1: Initialize Blockchain:

- Create an empty blockchain and a list for pending transactions.
- Generate the **Genesis Block** (the first block).

Step 2: Customer Booking & Payment:

- Collect **booking details**: Customer info, service provider, service, amount.
- **If** the transaction is valid (verified customer, verified payment), **add** to pending transactions.
- **Else**, **reject** the transaction.

Step 3: Mine New Block:

- **If** there are pending transactions:
  - Retrieve the last block's hash.
  - Create a new block with pending transactions and hash it.
  - **Add** the new block to the blockchain.
  - Reset **pending transactions**.
- **Else**, **wait** for new transactions.

Fig. 4 Blockchain application in security—tourism

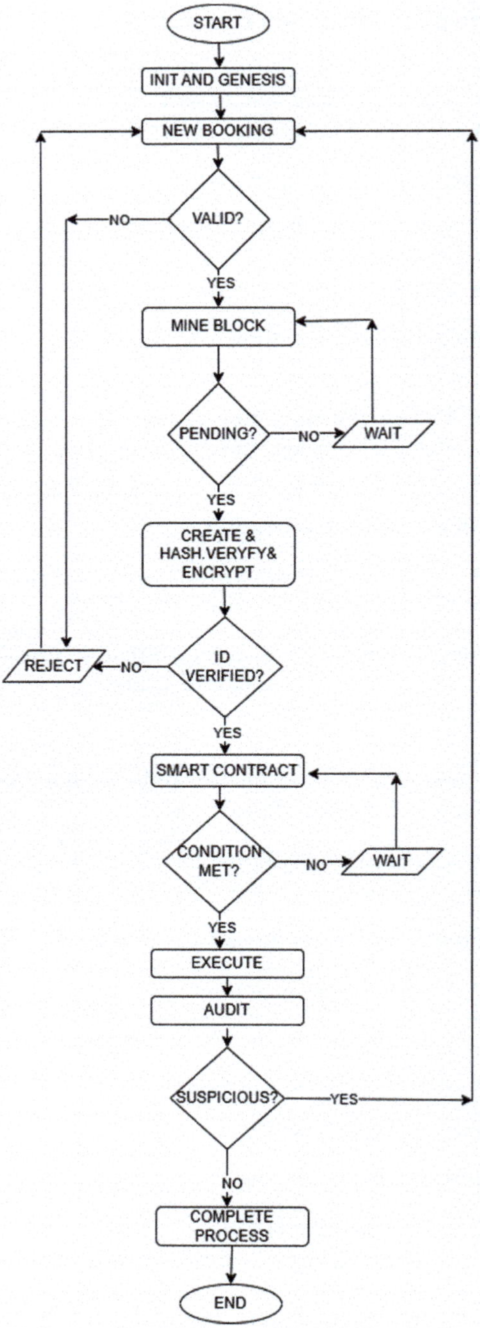

Step 4: Verify Payment & Transaction:

- **If** the payment is valid and processed, mark the transaction as **completed**.
- **Else**, **refund** or handle issues.

Step 5: Encrypt Data:

- Encrypt sensitive customer data and store it in the blockchain securely.

Step 6: Identity Verification:

- **If** the customer's identity is verified, proceed with booking.
- **Else**, **reject** the booking.

Step 7: Smart Contract Execution:

- **If** contract conditions are met (e.g., booking cancellation), **automatically execute** the smart contract.
- **Else**, **wait** for condition fulfillment.

Step 8: Audit & Monitor:

- Continuously **monitor blockchain** for any suspicious activity.
- **If** issues are detected, **audit** and **address** them.

Step 9: Repeat:

- Wait for new transactions, and **repeat** the process.

6 Challenges and Innovations

Blockchain technology can transform the tourism sector, but its implementation is hindered by a number of challenges that need to be overcome for its complete adoption. Concurrently, innovations are on the horizon that can help overcome these challenges and realize the maximum potential of blockchain. The following are the significant challenges and innovations in using blockchain in tourism:

6.1 Blockchain for Sustainable Tourism

Application: Blockchain can be used to track and verify sustainable tourism activities, such as carbon footprint minimization or responsible procurement of goods. Therefore, tourists can confirm that the hotels, tours, and activities they book align with sustainability goals.

Innovation: This application emphasizes a sustainable, environment-friendly booking system. It is transparent and reliable for the green movement.

6.2 Improving Cross-Border Payments

Application: Cross border payment involves transactions within the country with the help of cryptocurrency or currency exchange programmes. However, cryptocurrency implementation reduces the cost of exchange rate or transaction fees [28, 29].

Innovation: The above application leads to cost-effective transactions with a significant travel experience.

6.3 Peer-To-Peer Travel Networks

Application: Per-to-peer network, along with blockchain technology, provides a transparent, decentralized system for the traveler to exchange their services like accommodation, booking or transportation [30, 31].

Innovation: This model is a unique experience for travelers to control themselves without interfering with travel agencies or intermediate persons.

6.4 Tokenization of Travel Assets

Application: It is the conversion of available assets into digital tokens. These tokens can be bought, sold or exchanged in a blockchain environment. This tokenization may involve hotel accommodation or travel expenses.

Innovation: In this scenario, travelers must buy tokens related to the travel experience, hotel booking, or travel expenses. Then, these tokens can be exchanged within a blockchain-based platform to get the required services.

7 Case Study

Travala—A Blockchain Travel Booking Platform Travala is a travel booking website based on cryptocurrency that uses decentralized technology to offer a safe, transparent, and more affordable booking experience. It integrates cryptocurrency payments and smart contracts to establish trust and efficiency in travel transactions. Unlike traditional booking platforms that incur high fees and rely on centralized infrastructure, Travala allows users to pay using cryptocurrencies, reducing transaction fees and enhancing payment security [32]. Travala's decentralized system removes middlemen. Thus, transactions are carried out directly between travelers and service providers. The system also employs blockchain-based rewards in the form of loyalty points, where users get AVA tokens that can be redeemed for subsequent bookings. It also employs unalterable user reviews, providing users with open and verified opinions from real travelers. Travala's success illustrates the potential of blockchain in disrupting the tourism industry by improving efficiency, security, and user experience. (Reference: Travala. (2023). "Travala: Blockchain-Powered Travel Booking." Retrieved from https://www.travala.com) Travala is a travel booking platform powered by blockchain, utilizing decentralized technology to offer a secure, transparent, and economic booking process. It integrates cryptocurrency payments and smart contracts to promote trust and effectiveness in travel transactions. Unlike traditional booking platforms that charge high commissions and employ centralized systems, Travala allows payment in cryptocurrencies, reducing transaction charges and enhancing payment security [20, 33]. Travala's decentralized model eliminates intermediaries, providing direct transactions between travellers and service providers. The website also employs blockchain-based loyalty points, where users are rewarded with AVA tokens that can be redeemed for future bookings. It also employs immutable user reviews, providing open and verified reviews from real travellers. Travala's success demonstrates the potential of blockchain to transform the tourism industry by increasing efficiency, security, and user experience.

8 Future Scope

The future of blockchain technology in the field of tourism is vast and significant instead of achieving the goal by following the following areas.

8.1 Smart Contracts for Seamless Transactions

Smart contracts can automatically release payments to service providers when a traveler completes check-out at a hotel or confirms a booking [34]. The Automating transaction process in the field of tourism can only be achieved by implementing blockchain technology and smart contracts implementation. This can automate the process of booking, payments, searching for the best deal and optimizing customer service with chat-bots instead of real human beings.

8.2 Enhanced Data Security and Privacy

The main disadvantage of the tourism sector is data integrity and security [35]. The privacy and security of data is a challenge in this field. But this can be well managed by implementing blockchain technology it can offer secure and decentralized travel data experience by implementing Self-Sovereign-Identity (SSI) [36].

8.3 Blockchain-Based Loyalty Programs

The loyalty program can be promoted with the help of blockchain technology by adding better transparency and adaptability factors. It can be digitalized for better experience to avail the loyalty tokens [37].

8.4 Decentralized Autonomous Organizations

(DAOs) for Governance DAOs might be employed to develop community-led governance models for tourism decisions. Decentralized systems enable travelers and providers of services to engage in decision-making processes, including establishing sustainable tourism activities or governing travel industry standards [38]. This transformation towards a decentralized government may result in greater democratization and transparency in the tourism industry.

8.5 Streamlined Border Control and Visa Processing

Blockchain can easily simplify border control and visa issuance by safely storing and authenticating traveler information. Blockchain-enabled digital passports would grant authorities immediate, tamper-evident information on travelers' identities, prior visits, and entry/exit history, making it more efficient and secure. This would result in more convenient international travel with fewer forms and waiting times.

8.6 Travel Insurance and Claim Automation

Blockchain technology can also automate the process of claims in travel insurance using smart contracts, so claims get processed immediately and transparently [39]. Travellers will no longer have to wait weeks for approval of their claims since everything will be processed automatically as soon as the contract conditions are fulfilled. Applying blockchain technology to authenticate travel disruptions (e.g., flight delays) makes travel insurance claims faster and more trustworthy [40].

8.7 Increased Trust in Reviews and Rating

Blockchain can resolve problems such as counterfeit reviews and biased feedback [41]. By putting reviews and ratings on the blockchain, travelers can ensure the input is accurate. This fosters trust in online reviews and ratings to the advantage of both consumers and businesses.

8.8 Tokenization of Travel Assets

Tokenization is converting physical properties into digital tokens. This would mean tokenizing hotel rooms, flights, and experiences in tourism. Tourists can purchase tokens to benefit from these services, creating new opportunities for fractional ownership or shared travel experience.

8.9 Integration with Artificial Intelligence (AI)

AI-blockchain integration is a game-changer in the travel sector. AI logic can tailor travel experiences based on preference data stored in blockchain, and blockchain can validate and safeguard AI-recommended travel itineraries [42]. The integration would propel personalized tour packages, real-time pricing, and enhanced customer service experiences with the assurance of data privacy [43].

9 Conclusion

Blockchain technology can ideally address all the issues related to the tourism industry by doing away with certain issues related to security, transparency, and efficiency. It can reduce costs, time, and the use of resources. It can allow the tourism industry to acquire the confidence of tourists. It can improve the decentralized booking platform, apply cryptocurrency for cross-border payment, and apply smart contracts. These transformations make the tourism environment more transparent, secure, and customer-centered. Moreover, the integration of blockchain-based loyalty programs and self-sovereign identity creates new opportunities for individual experiences with guaranteed data privacy. With growing trust and application of blockchain in this area, tourist travel experience will also grow considerably.

This will also enhance data protection and the overall business process. The ability to track sustainability projects through blockchain will also be a key driver in promoting sustainable travel. Lastly, blockchain's potential in the tourism industry extends beyond transactions; it holds the promise of a complete transformation of how travel-related services are provided and consumed, with greater trust, accountability, and innovation. As technology improves and becomes more accepted, we expect blockchain to be a core part of the future of the tourism industry, enhancing operational structures and customer experience.

References

1. Zheng, Z., Xie, S., Dai, H., Chen, X., Wang, H.: An overview of blockchain technology: architecture, consensus, and future trends. In: IEEE International Congress on Big Data (BigData Congress), pp. 557–564 (2018)
2. Casino, F., Dasaklis, T.K., Patsakis, C.: A systematic literature review of blockchain-based applications: current status, classification, and open issues. Telematics Inf. **36**, 55–81 (2019)
3. Saberi, S., Kouhizadeh, M., Sarkis, J., Shen, L.: Blockchain technology and its relationships to sustainable supply chain management. Int. J. Prod. Res. **57**(7), 2117–2135 (2019)
4. Xu, X., Weber, I., Staples, M.: Architecture for Blockchain Applications. Springer International Publishing (2020)
5. Kim, H., Woo, E.: Blockchain in the hospitality industry: current trends and future implications. Int. J. Hosp. Manage. **92**, 102701 (2021)
6. Chen, Y., Xu, X., Zhu, Q.: Blockchain for digital rights management in the media industry: a survey. Comput. Electr. Eng. **93**, 107276 (2022)
7. Gössling, S., Hall, C.M., Scott, D.: Global trends in length of stay: implications for destination management and climate change. J. Sustain. Tour. **27**(5), 579–597 (2019)
8. Sun, J., Yan, J., Zhang, K.Z.K.: Blockchain-based sharing services: what blockchain technology can contribute to smart cities. Financ. Innov. **6**(1), 1–19 (2020)
9. Dai, J., Vasarhelyi, M.A.: Toward blockchain-based accounting and assurance. J. Inf. Syst. **35**(2), 1–21 (2021)
10. Wang, Y., Han, J.H., Beynon-Davies, P.: Understanding blockchain technology for future supply chains: a systematic literature review and research agenda. Supply Chain Manage. Int. J. **26**(2), 170–190 (2021)
11. Alzahrani, N., Bulusu, N.: A new product authentication framework using blockchain technology. Ann. Telecommun. **77**, 1–15 (2022)
12. Yang, Y., Liu, Y., Li, H., Yu, B.: Understanding the impact of blockchain on smart tourism: a systematic literature review. Sustainability **14**(3), 1379 (2022)
13. Sharma, A., Singh, G.: Convergence of blockchain, AI, and IoT in smart cities: a review. IEEE Access **11**, 12345–12360 (2023)
14. Lee, J., Cho, H., Kim, S.: Dynamic pricing in the airline industry: a blockchain approach. J. Air Transp. Manag. **105**, 102126 (2023)
15. Rath, S.K., Sahu, M., Das, S.P., Mohapatra, S.K.: Hybrid software reliability prediction model using feature selection and support vector classifier. In: 2022 International Conference on Emerging Smart Computing and Informatics (ESCI), Pune, India, pp 1–4 (2022). https://doi.org/10.1109/ESCI53509.2022.9758339
16. Rathee, T., Singh, P.: A self-sovereign identity management system using blockchain. In: Cyber Security and Digital Forensics, pp. 371–379. Springer (2022)
17. Rathore, A., Choi, J.: Self-sovereign identity: a systematic review, mapping and taxonomy. Sensors **22**(15), 5641 (2022)
18. Rathore, A., Choi, J.: Self-sovereign identity: a systematic review, mapping and taxonomy. Sensors **22**(15), 5641 (2022)
19. Kaur, H., Sharma, S., Guleria, S.: Navigating the digital landscape with the integration of social media and artificial intelligence (AI) in tourism industry. In: Impact of AI and Tech-Driven Solutions in Hospitality and Tourism, pp. 235–256. IGI Global (2024)
20. Kontogianni, A., Alepis, E., Virvou, M., Patsakis, C.: smart tourism embraces blockchain. In: Smart Tourism—The Impact of Artificial Intelligence and Blockchain, pp. 121–157. Springer Nature Switzerland, Cham (2024)
21. Maythu, Y., Kwok, A.O., Teh, P.L.: Blockchain technology diffusion in tourism: evidence from early enterprise adopters and innovators. Heliyon **10**(2) (2024)
22. Faker, I.: Exploring the disintermediation effect of blockchain technology on the tourism industry (2023)
23. Dandotiya, R., Aggarwal, A.: An examination of tourists' national identity, place attachment and loyalty at a dark tourist destination. Kybernetes **52**(12), 6063–6077 (2023)

24. Mohammad Shafiee, M., Foroudi, P., Tabaeeian, R.A.: Memorable experience, tourist-destination identification and destination love. Int. J. Tour. Cities **7**(3), 799–817 (2021)
25. Song, H.: Tourism Supply Chain Management. Routledge (2012)
26. Başer, M.Y., Büyükbeşe, T., Kizildag, M.: What if we could travel without passport? First sight to blockchain-based identity management in tourism. Asia Pac. J. Tour. Res. **28**(4), 341–363 (2023)
27. Vinod, B.: The impact of emerging technologies on intermediaries. In: Mastering the Travel Intermediaries: Origins and Future of Global Distribution Systems, Travel Management Companies, and Online Travel Agencies, pp. 363–394. Springer Nature Switzerland, Cham (2024)
28. Zhang, Y.: Developing cross-border blockchain financial transactions under the belt and road initiative. Chin. J. Compar. Law **8**(1), 143–176 (2020)
29. Uyduran, B.: The crypto effect on cross border transfers and future trends of cryptocurrencies. Financ. Internet Q. **16**(4), 12–23 (2020)
30. Kiczmachowska, E.: Large deluxe or weekly stay? Diversity of peer-to-peer accommodation offers at an urban tourism destination. Eur. J. Tour. Res. **39**, 3910–3910 (2025)
31. Lee, H., Jang, S., Kim, J.: Spatial coopetition and peer-to-peer accommodation price. Ann. Tour. Res. **109**, 103826 (2024)
32. Ibrahim, M.A.M.: Exploring the usage of cryptocurrency adoption in the egyptian tourism sector: opportunities and challenges. J. Faculty Tour. Hotels-Univ. Sadat City **8**(2/3) (2024)
33. Barkel, C., Kurgun, H., Groen, B.: Blockchain in the hospitality and tourism industry. University of South Florida (USF) M3 Publishing, vol. 17(9781732127593), p. 4 (2021)
34. Deng, Q.: Blockchain and smart contracts for enhanced traceability in cultural and tourism industries. In: Proceedings of the 2024 Guangdong-Hong Kong-Macao Greater Bay Area International Conference on Education Digitalization and Computer Science, pp. 255–262 (2024)
35. Gong, Y., Schroeder, A.: A systematic literature review of data privacy and security research on smart tourism. Tour. Manage. Perspect. **44**, 101019 (2022)
36. Tiwari, V., Mishra, A., Tiwari, S.: Role of data safety and perceived privacy for acceptance of IoT-enabled technologies at smart tourism destinations. Curr. Issue Tour. **27**(19), 3079–3094 (2024)
37. Alam, M., Ali, D., Badshah, S.I.: Transforming travel and tourism: the impact of DeFi and blockchain technologies. J. Bus. Manage. Res. **4**(1), 782–794 (2025)
38. Peña Calvín, A., Duenas-Cid, D., Ahmed, J.: Is DAO governance fostering democracy? Reviewing decision-making in decentraland. In: Proceedings of the 58th Hawaii International Conference on System Sciences. HICSS Conference Office (2025)
39. Genceli, D., Yüksek, G., Uca, S.: Travel insurance behaviors of digital nomads. Worldwide Hosp. Tour. Themes **16**(3), 365–374 (2024)
40. Borketey, B., Aboagye, E.F., Danquah, K.: Optimizing travel insurance purchase detection using predictive models. Int. J. Autom. Artif. Intell. Mach. Learn. **4**(2), 173–207 (2024)
41. Poornima, A.S., Subramanian, S.: Fostering global wellness: harnessing social media to enhance cross-cultural trust to propel medical tourism ensuring economic growth. In: European Conference on Social Media (2024)
42. Singh, S., Verma, R., Kumar, A.: Developing framework of tourism 4.0 using artificial intelligence and blockchain. Doctoral dissertation, Jaypee University of Information Technology, Solan, HP (2024)
43. Youssef, A.B., Dutta, P.K., Doshi, R., Sajnani, M., (eds.): AI, Blockchain, and Metaverse in Hospitality and Tourism Industry 4.0: Case Studies and Analysis. CRC Press (2024)

The manufacturer's authorised representative in the EU is Springer Nature Customer Service Centre GmbH, Europaplatz 3, 69115 Heidelberg, Germany. If you have any concerns regarding our products, please contact ProductSafety@springernature.com

Printed and bound by CPI Group (UK) Ltd, Croydon, CR0 4YY

30/09/2026

02225233-0003